A TEXTBO

VECTOR A

CW00498276

(With applications of Geometry and Statics)

[For the Students of B.A. and B.Sc. (Pass & Hons. Courses)]

SHANTI NARAYAN
Former Dean of Colleges,
Principal, Hans Raj College,
University of Delhi,
DELHI

Revised by

P.K. MITTAL
M.Sc., Ph.D.
Head of Mathematics Deptt.
Govt. Post Graduate College
RISHIKESH
(UTTARANCHAL)

REVISED EDITION

S. CHAND
PUBLISHING
empowering minds

S. CHAND & COMPANY PVT. LTD.

(AN ISO 9001 : 2008 COMPANY)

RAM NAGAR, NEW DELHI - 110 055

S. CHAND & COMPANY PVT. LTD.
(An ISO 9001 : 2008 Company)
Head Office: 7361, RAM NAGAR, NEW DELHI - 110 055
Phone: 23672080-81-82, 9899107446, 9911310888 Fax: 91-11-23677446
www.schandpublishing.com; e-mail: helpdesk@schandpublishing.com

S. CHAND PUBLISHING
empowering minds

Branches

Ahmedabad	:	Ph: 27541965, 27542369, ahmedabad@schandpublishing.com
Bengaluru	:	Ph: 22268048, 22354008, bangalore@schandpublishing.com
Bhopal	:	Ph: 4274723, 4209587, bhopal@schandpublishing.com
Chandigarh	:	Ph: 2725443, 2725446, chandigarh@schandpublishing.com
Chennai	:	Ph. 28410027, 28410058, chennai@schandpublishing.com
Coimbatore	:	Ph: 2323620, 4217136, coimbatore@schandpublishing.com (Marketing Office)
Cuttack	:	Ph: 2332580; 2332581, cuttack@schandpublishing.com
Dehradun	:	Ph: 2711101, 2710861, dehradun@schandpublishing.com
Guwahati	:	Ph: 2738811, 2735640, guwahati@schandpublishing.com
Hyderabad	:	Ph: 27550194, 27550195, hyderabad@schandpublishing.com
Jaipur	:	Ph: 2219175, 2219176, jaipur@schandpublishing.com
Jalandhar	:	Ph: 2401630, 5000630, jalandhar@schandpublishing.com
Kochi	:	Ph: 2378740, 2378207-08, cochin@schandpublishing.com
Kolkata	:	Ph: 22367459, 22373914, kolkata@schandpublishing.com
Lucknow	:	Ph: 4026791, 4065646 lucknow@schandpublishing.com
Mumbai	:	Ph: 22690881, 22610885, mumbai@schandpublishing.com
Nagpur	:	Ph: 6451311, 2720523, 2777666, nagpur@schandpublishing.com
Patna	:	Ph: 2300489, 2302100, patna@schandpublishing.com
Pune	:	Ph: 64017298, pune@schandpublishing.com
Raipur	:	Ph: 2443142, raipur@schandpublishing.com (Marketing Office)
Ranchi	:	Ph: 2361178, ranchi@schandpublishing.com
Siliguri	:	Ph: 2520750, siliguri@schandpublishing.com (Marketing Office)
Visakhapatnam	:	Ph: 2782609 visakhapatnam@schandpublishing.com (Marketing Office)

First Edition 1954
Subsequent Edition and Reprints 1968, 69, 73, 75, 78, 80, 82, 83, 85, 86, 87, 93, 95, 98, 99, 2002, 2003, 2004 (Twice), 2005, 2006, 2010 (Twice), 2012, 2015
Reprint 2016

ISBN : 978-81-219-0952-5 **Code :** 1014B 064

This book is for Sale in India & other SAARC Countries only.

PRINTED IN INDIA

By Vikas Publishing House Pvt. Ltd., Plot 20/4, Site-IV, Industrial Area Sahibabad, Ghaziabad-201010 and Published by S.Chand & Company Pvt. Ltd., 7361, Ram Nagar, New Delhi-110 055.

PREFACE TO THE REVISED EDITION

The book originally about 50 years ago, has during the intervening period been revised and re-printed several times. Due to the changing trends of University studies and the demand of the students a thorough revision of the book was overdue. I very humbly took the challenge of revising a perfect work of Dr. Shanti Narayan and tried to meet the demands of those who always loved and liked the books of Dr. Shanti Narayan.

To make the book up to date and useful to the students interested in self-studies and appearing in different competitive examinations, a large number of illustrative examples have been incorporated.

The book, in the present form, is an humble effort to make it more useful to the students and teachers. I owe my special gratitude to Sri Ravindra Kumar Gupta, Managing Director, S. Chand & Company Ltd., for giving me opportunity to revise the books of late Dr. Shanti Narayan, an eminent Indian Mathematician.

I also acknowledge my sincere thanks to Sri Navin Joshi, General Manager (Sales & Marketing) and Shri Dharmendra Jha, editor, M/s. S. Chand & Company Ltd., for providing necessary assistance in this revision work.

As the need for improvement is never ending, I will look forward to receive valuable and useful suggestions from our users for further improvements in the book.

Rishikesh
Uttaranchal

P.K. Mittal

PREFACE TO THE EIGHTH EDITION

The book has been subjected to a pretty thorough revision. Some sections have been rewritten and the matter has also been rearranged so as to make the book more suitable for the students for whom it is meant. Chapter summaries have also been included in the book.

April, 1980

AUTHOR

PREFACE TO THE SIXTH EDITION

The book has been subjected to a thorough revision consisting in a new arrangement of the material. The different aspects of applications of the notions of the scalar and vector products have been given in different chapters in the hope that this arrangement will help the reader to have a better perspective of the subject matter treated in the book.

A systematic use has also been made of the symbols

$$\Rightarrow, \Leftrightarrow$$

which respectively stand for 'Implies', and 'Is equivalent to'.

It is believed that the employment of these symbols will give a better idea of the relationships between successive pairs of statements. The use of these symbols has, of course, very much reduced the employment of the words 'Or', 'That is', 'For', etc.

The author believes that the book in its present form will help strengthening the understanding and use of vector methods.

The author offers his loving thanks to Kumari Nilima, Department of Mathematics, Miranda House, Delhi, for her willing helps in the preparation of this revised Edition.

THE AUTHOR

"This is an attractive well-written book of 190 pages which deals rigorously and systematically with three-dimensional vector algebra and its applications to Euclidean Geometry and Statics. A valuable feature of the book is the large number of worked examples. There is also a large collection of exercises which should prove useful to the student and teacher alike. The printing is clear."

December, 1955 — *MATHEMATICAL GAZETTE*

"Here are two (Referring to 'A Text Book of Vector Algebra' and 'A Text Book of Cartesian Tensors') excellent text books that treat their respective subjects with ingratiating clarity. One feels on reading these books, that their author has an instinct for exposition and that it would be a pleasure to study under him."

Vol. XXIV, September, 1959 — *SCRIPTA MATHEMATICA*

CONTENTS

1

Multiplication of Vectors by Scalars and Addition of Vectors

Introduction. Vectors constitute one of the several *Mathematical systems* which can be usefully employed to provide mathematical handling for certain types of problems in *Geometry, Mechanics* and other branches of Applied Mathematics. Other such systems are those of *Matrices, Tensors, Quaternions* etc. The system of vectors alone, however, will be the subject of study in this book.

Application of mathematics to a problem or a body of problems consists in constructing a system of entities and equipping the same with some structure [order or/and Algebraic or/and Topologic] such that the system of entities and the corresponding structure have a close correspondence with the objects of study in the problem.

Vectors facilitate mathematical study of such physical objects as possess *Direction* in addition to *Magnitude.* Velocity of a particle, for example, is one such object.

It is true that the set of real numbers also provides mathematical tool for the study of various types of physical problems for which vectors are found useful but the use of vectors is more direct and natural. As a result of limitation to the set of real numbers, we associate not one number but a *set* of numbers to a physical entity involving direction, for we have to split up the entity into components and associate a number with each. The use of vectors, however, avoids this splitting up and leads to a direct study of the objects in question.

In this book, the Algebra of Vectors will be dealt with and applied to the study of *Geometry* and *Statics* alongwith an elementary idea of vector calculus. Vectors Calculus involving topological notions of *Limit, Continuity,* and topological algebraic notions of *Derivability* and *Integrability* has been treated in a *second book.

Note. The study of vectors naturally leads to the notion of directed line segments which we shall now introduce.

* Vector Calculus by Shanti Narayan and J.N. Kapur.

1.1. DIRECTED LINE SEGMENTS

Any given portion of a given straight line where for the two end points are distinguished as Initial and Terminal is called a Directed Line Segment.

The directed line segment with *initial* point A and *terminal* point B is denoted by the symbol

$$\vec{AB}.$$

The two end points of a directed line segment are not interchangeable and the directed line segments

$$\vec{AB} \text{ and } \vec{BA}$$

must be thought of as different.

1.1.1. Length, Support and Sense of a Directed Line Segment

Associated with every directed line segment

$$\vec{AB},$$

we have its

Length, Support and Sense

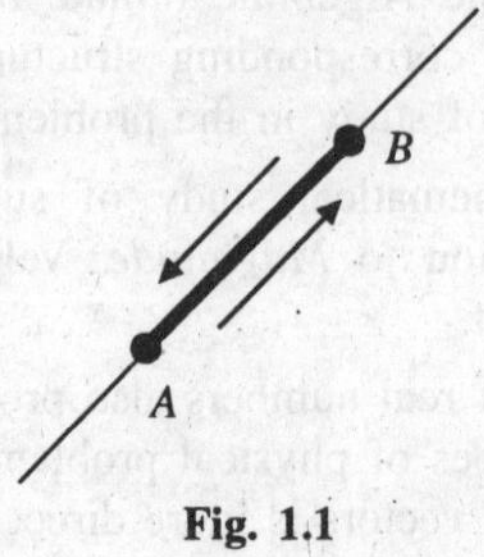

Fig. 1.1

(*i*) **Length.** The length of $\vec{AB}$ will be denoted by the symbol

$$|\vec{AB}|$$

Clearly, we have

$$|\vec{AB}| = |\vec{BA}|.$$

(*ii*) **Support.** The line of unlimited length of which a directed line segment is a part is called its line of support or simply the Support.

(*iii*) **Sense.** The sense of $\vec{AB}$ is from A to B and that of $|\vec{BA}|$ from B to A so that the sense of a directed line segment is from its initial to the terminal point.

The directed line segments

$$\overrightarrow{AB} \text{ and } \overrightarrow{BA}$$

have the same lengths and supports but different senses (Fig. 1.1).

The question of comparison of the senses of two directed line segments arises only when they have the same or parallel support (Fig. 1.2).

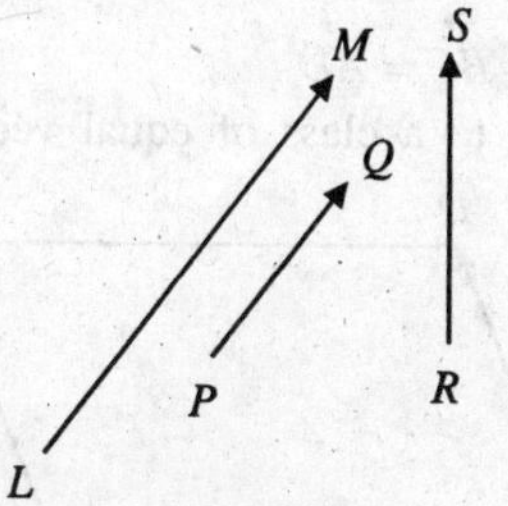

Fig. 1.2

Two directed line segments having the same or parallel supports may have the same or opposite senses.

Ex. How many Directed Line Segments are determined by 2, 3 and 4 given points.

1.2. VECTORS AND SCALARS

1.2.1. Vector

Def. *A directed line segment is called vector.*

1.2.2. Equality of Two Vectors

Def. *Two vectors are said to be equal if they have*

(*i*) *the same length,*

(*ii*) *the same* **or** *parallel supports, and*

(*iii*) *the same sense.*

It may thus be seen that two different Directed Line Segments may correspond to the same vector.

Thus, the vectors

$$\overrightarrow{AB}, \overrightarrow{CD}, \overrightarrow{EF}$$

are equal (Fig. 1.3).

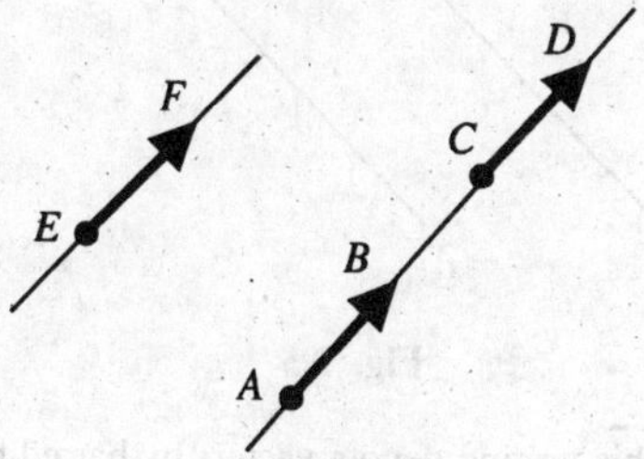

Fig. 1.3

Two vectors will *not* be equal if they have different lengths or inclined supports or again, they will not be equal even if they have the same lengths and parallel supports but different senses (Fig. 1.2).

If *ABCD* is a parallelogram, we have

$$\vec{AB} = \vec{DC}$$

and
$$\vec{BC} = \vec{AD}$$

Every vector belongs to a class of equal vectors.

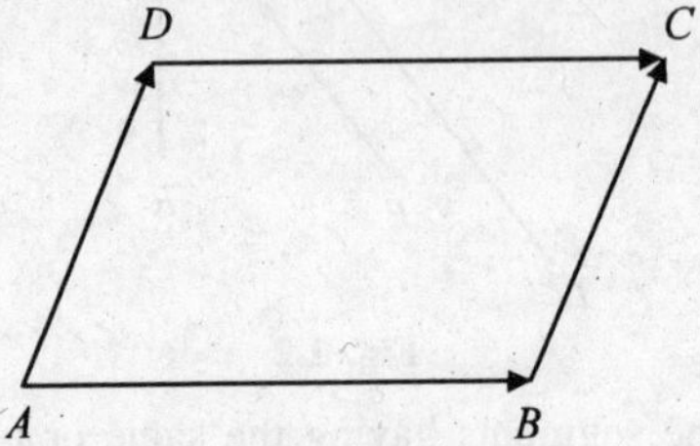

Fig. 1.4

1.2.3. Notation for a Vector

A vector is also denoted by a single letter such that

a, b, c, etc.

by using **bold face*** type so that we may write

$$\mathbf{a} = \vec{AB}.$$

Then the symbol

$$|\,\mathbf{a}\,|$$

denotes the length of the vector, **a**, also called the **Magnitude** of the vector.

1.2.4. Co-initial Vectors

It is possible to replace a given vector by another equal vector having any given point as its initial point.

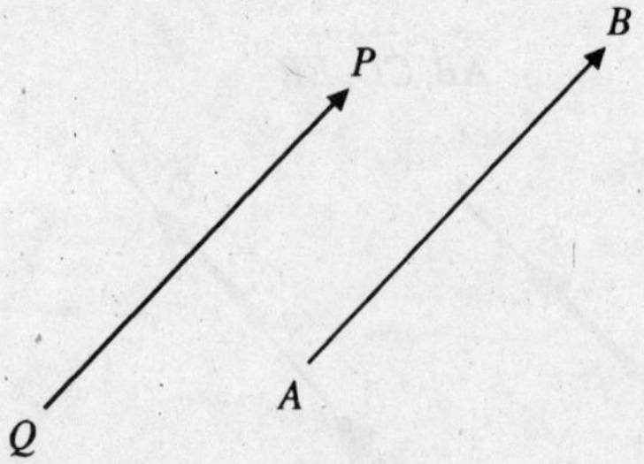

Fig. 1.5

* The reader may, in his writing denote vectors by barred letters such as $\vec{a}$, $\vec{b}$.

Thus, if $\overrightarrow{AB}$ be any given vector and O, any given point, then by drawing through O, a line OP parallel to AB in the same sense as AB of length equal to that of AB, we obtain a vector $\overrightarrow{OP}$ equal to $\overrightarrow{AB}$ and with the given point, O as the initial point (Fig. 1.5).

Vectors with the same initial point may be called Co-initial Vectors.

1.2.5. Zero Vector

A vector whose initial and terminal points are coincident is called the **Zero Vector.**

The length of the zero vector is zero but it can be thought of as having any line as its line of support.

The zero vector is denoted by the bold face type **0**. It will be seen that the zero vector has many properties similar to those of the zero number.

1.2.6. Scalars

In the following, real numbers will be called scalars. The absolute value of a scalar, m, will as usual, be denoted by the symbol $|m|$ so that

$$|m| = m \text{ if } m \geq 0 \quad \text{and} \quad -m \text{ if } m < 0.$$

1.3. ALGEBRA OF VECTORS

It is possible to develop an *Algebra of Vectors* which proves useful in the study of Geometry, Mechanics and other branches of Applied Mathematics.

By the '*Algebra of Vectors*' will be meant a body of work which prescribes various useful manners of combining vectors and scalars satisfying some laws which may be called *Laws of Composition.*

The following manners of composition will be introduced in appropriate places in the book :

I. Multiplication of Vectors by Scalars.

II. Addition of Vectors.

III. Scalar Multiplication of Vectors.

IV. Vector Multiplication of Vectors.

The first two will be introduced in this chapter and the last two in Chapter III and Chapter IV respectively. The first two compositions which will be dealt with in this chapter are also called *Linear Compositions.*

1.3.1. Multiplication of Vectors by Scalars

Let, **a**, be any given vector and, m, be any given scalar. Then the symbol

$$m\,\mathbf{a},$$

called the product of the vector, **a**, by the scalar, *m*, is a vector such that

(*i*) the *length* of $m\,\mathbf{a}$ is given by

$$|\, m\,\mathbf{a} \,| = |\, m \,| \;\; |\, \mathbf{a} \,|$$

i.e., the length of the vector, $m\,\mathbf{a}$ is m times or, $-m$, times that of, **a**, according as m is positive (including zero) or negative.

Thus, $\quad \mathbf{b} = m\,\mathbf{a} \quad \Rightarrow \quad |\, m \,| = |\, \mathbf{b} \,| \,/\, |\, \mathbf{a} \,|.$

(*ii*) the *support* of $m\,\mathbf{a}$, is the same or parallel, to that of **a**.

(*iii*) the *sense* of $m\,\mathbf{a}$, is the same or opposite, to that of **a**, according as *m* is positive or negative.

The symbol, $m\,\mathbf{a}$, is also sometimes written as, $\mathbf{a}\,m$, so that the scalar, *m*, appears on the right of the vector instead of on the left.

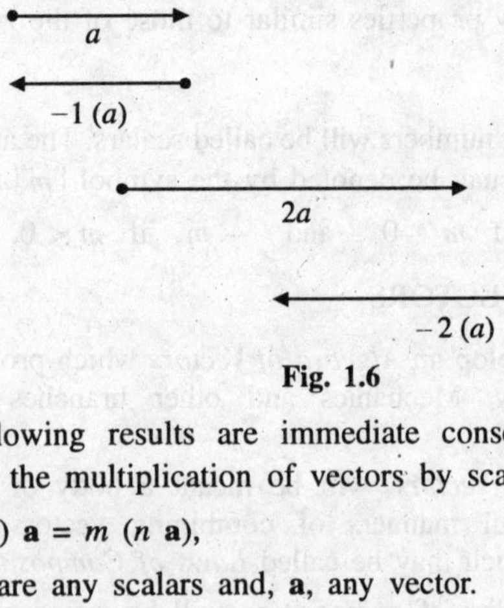

Fig. 1.6

The following results are immediate consequences of the above definition of the multiplication of vectors by scalars.

I. $(m\,n)\,\mathbf{a} = m\,(n\,\mathbf{a})$,

where *m, n* are any scalars and, **a**, any vector.

II. $0\,\mathbf{a} = \mathbf{0}$,

so that the product of a vector, **a**, by the zero scalar is the zero vector.

Here, 0, on the left stands for the zero scalar and **0**, on the right for the zero vector.

III. If two vectors have the same or parallel supports, then each can be thought of as a product of the other by a suitable scalar, the absolute value of the scalar being the ratio of the lengths of the vectors taken in an appropriate order.

Thus, for example, if *C* be the midpoint of a line *AB*, we have

$$\overrightarrow{AB} = 2\,\overrightarrow{AC}, \;\; \overrightarrow{BC} = -\frac{1}{2}\,\overrightarrow{AB}$$

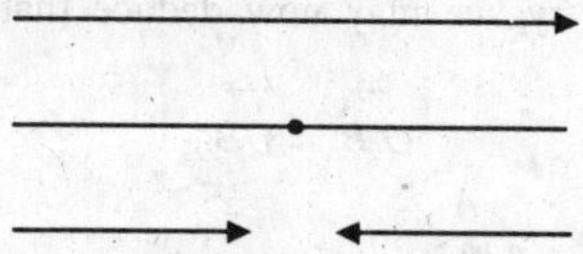

Fig. 1.7

Two parallel vectors may also be described as collinear.

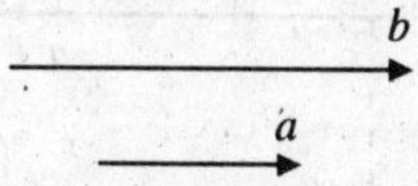

Fig. 1.8

1.3.2. Addition Composition

Let **a, b** be two given vectors. Take a point *O*.

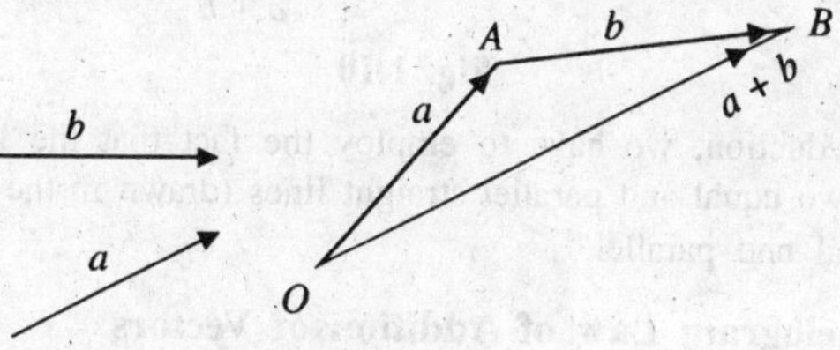

Fig. 1.9

Let

$$\vec{OA} = \mathbf{a},\ \vec{AB} = \mathbf{b},$$

so that the terminal point of the vector **a** is the initial point of the vector **b**.

The vector

$$\vec{OB}$$

is said to be the sum of the vectors, **a** and **b**, and we write

$$\vec{OB} = \vec{OA} + \vec{AB} = \mathrm{a} + \mathrm{b}.$$

We have **a + 0 = a**, for every **a; 0,** being the zero vector.

Note. As a matter of logical necessity, we must show that the sum of two vectors is independent of the choice of the point *O*.

Let *O, O¢* be any two points and let

$$\vec{OA} = \mathbf{a} = \vec{O'A'},$$

$$\vec{AB} = \mathbf{b} = \vec{A'B'}.$$

By elementary geometry, we may now deduce that

$$\overrightarrow{O'B'} = \overrightarrow{OB}.$$

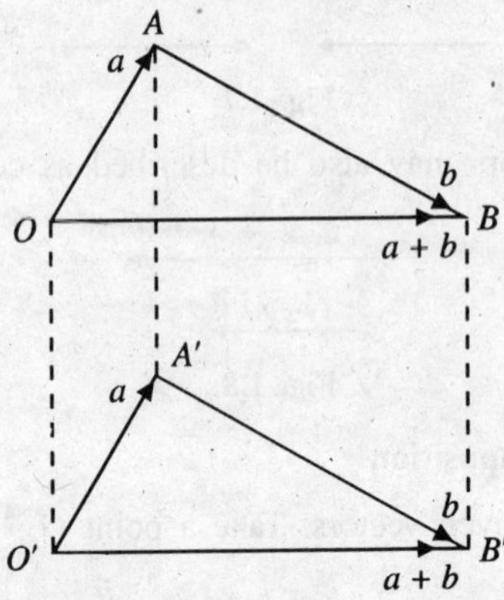

Fig. 1.10

For this deduction, we have to employ the fact that the lines joining the extremities of two equal and parallel straight lines (drawn in the same sense) are themselves equal and parallel.

1.3.3. Parallelogram Law of Addition of Vectors

Consider a parallelogram *OABC*.

The the sum of the vectors

$\overrightarrow{OA}$ and $\overrightarrow{OC}$ is the vector $\overrightarrow{OB}$.

for $\overrightarrow{OB} = \overrightarrow{OA} + \overrightarrow{AB} = \overrightarrow{OA} + \overrightarrow{OC}$.

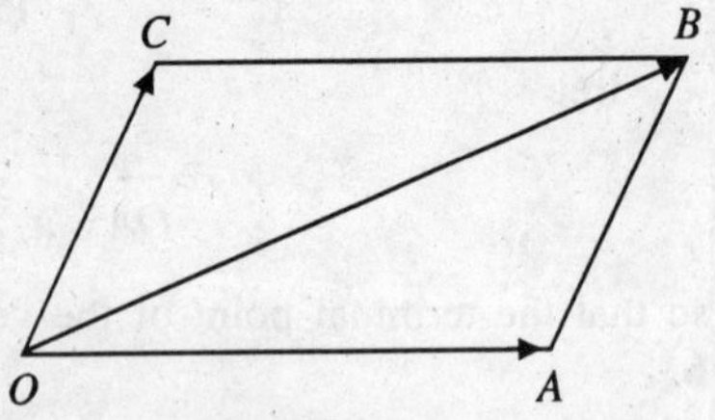

Fig. 1.11

1.4. LAWS OF ADDITION COMPOSITION

1.4.1. *Addition of vectors is Commutative, i.e.,*

$$\mathbf{a} + \mathbf{b} = \mathbf{b} + \mathbf{a}$$

for any pair of vectors, **a, b.**

Let $\overrightarrow{OA} = \mathbf{a}$, $\overrightarrow{AB} = \mathbf{b}$.

We have

$$\mathbf{a} + \mathbf{b} = \overrightarrow{OB}.$$

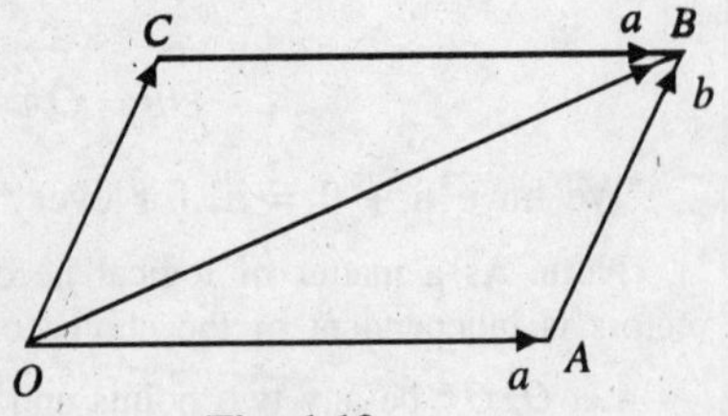

Fig. 1.12

Complete the parallelogram *OABC* having *OA* and *OC* as adjacent sides. Then

$$\overrightarrow{OC} = \overrightarrow{AB} = \mathbf{b},\ \overrightarrow{CB} = \overrightarrow{OA} = \mathbf{a}$$

so that we have

$$\overrightarrow{OB} = \overrightarrow{OC} + \overrightarrow{CB} = \mathbf{b} + \mathbf{a}$$

Hence, $\mathbf{a} + \mathbf{b} = \mathbf{b} + \mathbf{a}$.

1.4.2. *Addition of vectors is Associative, i.e.,*

$$\mathbf{a} + (\mathbf{b} + \mathbf{c}) = (\mathbf{a} + \mathbf{b}) + \mathbf{c}$$

where, **a, b, c** are any three vectors.

Take any point O. Let

$$\overrightarrow{OA} = \mathbf{a},\ \overrightarrow{AB} = \mathbf{b},\ \overrightarrow{BC} = \mathbf{c}.$$

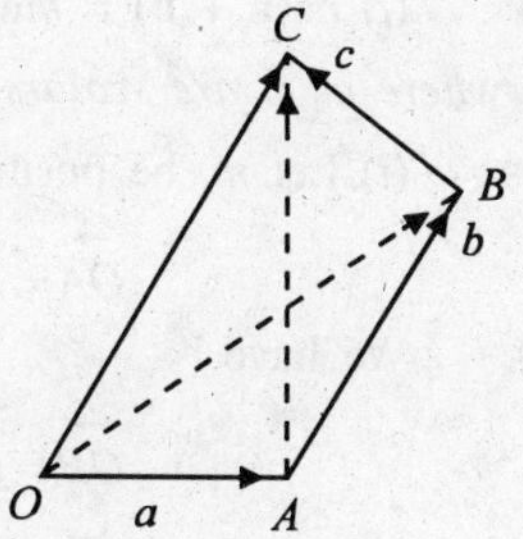

Fig. 1.13

We have

$$\mathbf{b}+\mathbf{c} = \overrightarrow{AB} + \overrightarrow{BC} = \overrightarrow{AC}$$

$$\Rightarrow \quad \mathbf{a}+(\mathbf{b}+\mathbf{c}) = \overrightarrow{OA} + \overrightarrow{AC} = \overrightarrow{OC}.$$

Again

$$\mathbf{a}+\mathbf{b} = \overrightarrow{OA} + \overrightarrow{AB} = \overrightarrow{OB}$$

$$\Rightarrow \quad (\mathbf{a}+\mathbf{b})+\mathbf{c} = \overrightarrow{OB} + \overrightarrow{BC} = \overrightarrow{OC}.$$

Thus

$$(\mathbf{a}+\mathbf{b})+\mathbf{c} = \overrightarrow{OC} = \mathbf{a}+(\mathbf{b}+\mathbf{c}).$$

In view of the equality of the vectors.

$$(\mathbf{a} + \mathbf{b}) + \mathbf{c},\ \ \mathbf{a} + (\mathbf{b} + \mathbf{c})$$

We denote each of these equal vectors by

$$\mathbf{a} + \mathbf{b} + \mathbf{c}.$$

1.4.3. *Negative of a vector*

For every vector **a**,

$$\mathbf{a} + (-1)\,\mathbf{a} = \mathbf{0}.$$

If $\overrightarrow{OA} = \mathbf{a}$, we have, according to the definition of the multiplication of vectors by scalars,

$$\overrightarrow{AO} = (-1)\,\mathbf{a}.$$

Thus,

$$\mathbf{a}+(-1)\,\mathbf{a} = \overrightarrow{OA} + \overrightarrow{AO} = \overrightarrow{OO} = \mathbf{0}.$$

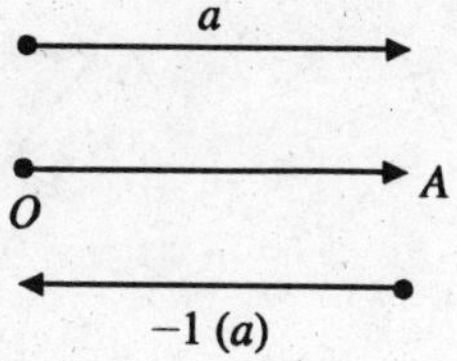

Fig. 1.14

On account of this property, the vector (– 1) **a** is called the negative of the vector **a**, and we write

$$-\mathbf{a} = (-1)\,\mathbf{a}$$

so that the relation $\mathbf{a} + (-1)\,\mathbf{a} = \mathbf{0}$, may also be re-written as

$$\mathbf{a} + (-\mathbf{a}) = \mathbf{0}.$$

1.5. RELATIONS BETWEEN THE TWO COMPOSITIONS

We shall now consider the relations between the two linear compositions considered in § 3.1. and § 3.2.

(*i*) $m(\mathbf{a}+\mathbf{b}) = m\mathbf{a} + m\mathbf{b}$. (*ii*) $(m+n)\,\mathbf{a} = m\mathbf{a} + n\mathbf{a}$,

where m, n are scalars and **a**, **b** *are vectors.*

(*i*) Let *m*, be positive and let

$$\vec{OA} = \mathbf{a},\ \vec{AB} = \mathbf{b},$$

We have

$$\vec{OB} = \mathbf{a}+\mathbf{b}.$$

Let $\vec{OA'} = m\mathbf{a}.$

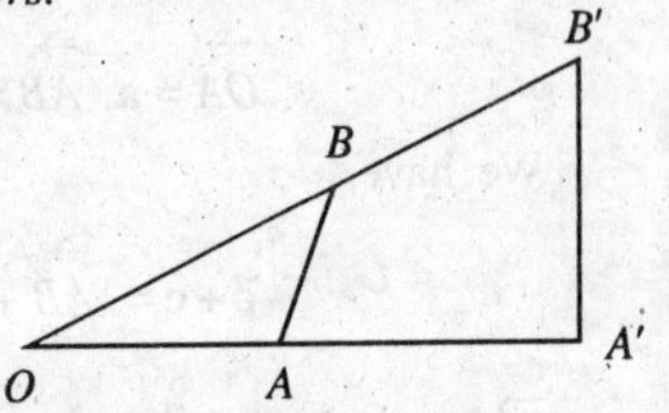

Fig. 1.15

Through A' draw a line parallel to AB and in the same sense as $\vec{AB}$ to meet OB in B'. With the help of Elementary Geometry, we have

$$A'B' = m\,AB \quad \text{and} \quad OB' = m\,OB.$$

Thus, we have

$$\vec{A'B'} = m\vec{AB} = m\mathbf{a} + m\mathbf{b}$$

and

$$\vec{OB'} = m\vec{OB} = m(\mathbf{a}+\mathbf{b}).$$

It follows that

$$m(\mathbf{a}+\mathbf{b}) = \vec{OB'} = \vec{OA'} + \vec{A'B'} = m\mathbf{a} + m\mathbf{b}.$$

The result may similarly be proved with the help of the accompanying figure 16, when *m* is negative.

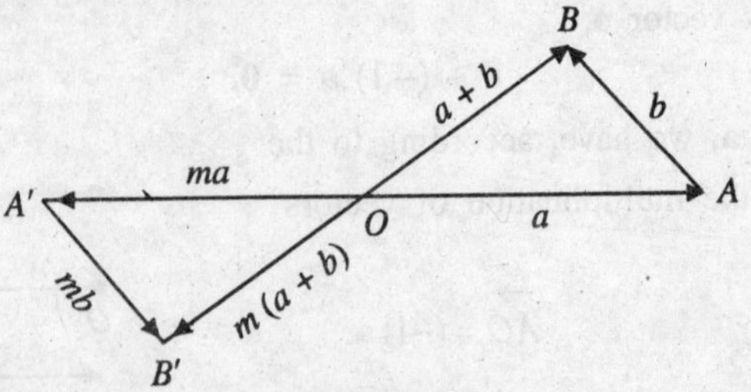

Fig. 1.16

(*ii*) The proof is a simple development of the definitions.

1.6. SUBTRACTION

Difference of Two Vectors. Def. *If* **a**, **b**, *be any two given vectors, then we write*

$$\mathbf{a} + (-\mathbf{b}) = \mathbf{a} - \mathbf{b}.$$

and call the composition Subtraction.

Thus, we have, in particular

$$\mathbf{a} - \mathbf{a} = \mathbf{a} + (-\mathbf{a}) = \mathbf{a} + (-1)\,\mathbf{a} = \mathbf{0}.$$

It is important to see that

$$\vec{OA} = \mathbf{a},\ \vec{OB} = \mathbf{b} \Rightarrow \vec{AB} = \mathbf{b} - \mathbf{a}.$$

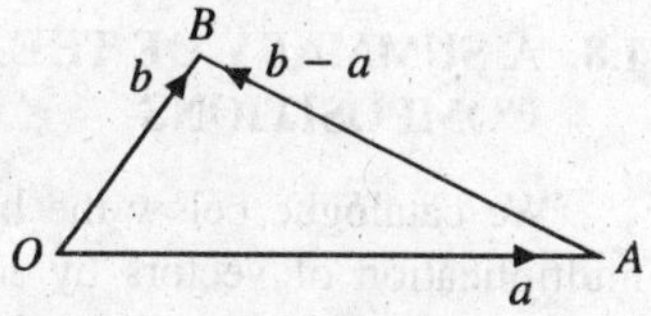

Fig. 1.17

In fact, we have

$$\mathbf{b} - \mathbf{a} = \mathbf{b} + (-\mathbf{a})$$

$$= \vec{OB} + \vec{AO} = \vec{AO} + \vec{OB} = \vec{AB}.$$

1.6.1. – (a + b) = – a – b. We have

$$[(-\mathbf{a}) + (-\mathbf{b})] + [\mathbf{b} + \mathbf{a}]$$
$$= (-\mathbf{a}) + [(-\mathbf{b}) + \mathbf{b}] + \mathbf{a}$$
$$= -\mathbf{a} + \mathbf{0} + \mathbf{a}$$
$$= (-\mathbf{a} + \mathbf{a}) + \mathbf{0} = \mathbf{0}.$$

Thus, $(-\mathbf{a}) + (-\mathbf{b}) = -(\mathbf{a} + \mathbf{b})$

$$\Rightarrow \quad (-\mathbf{a} + \mathbf{b}) = -\mathbf{a} - \mathbf{b}.$$

The proof of the result may be also directly obtained with the help of figure 18 given below.

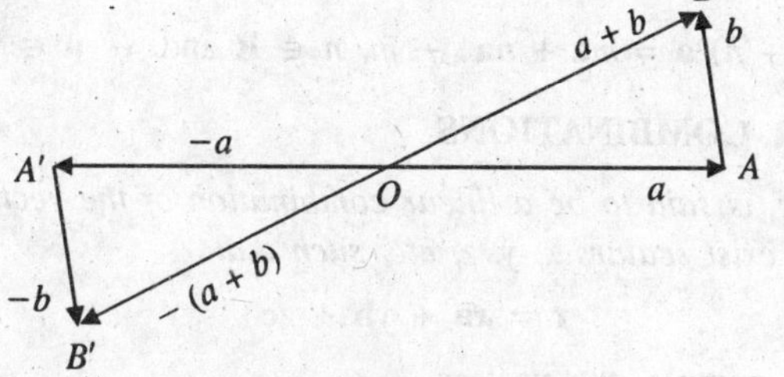

Fig. 1.18

1.7. A VECTOR EQUATION

If **a**, **b** *be two given vectors, then the vector equation*

$$\mathbf{a} + x = \mathbf{b}.$$

is satisfied by one and only one vector, viz.,

$$x = \mathbf{b} - \mathbf{a}.$$

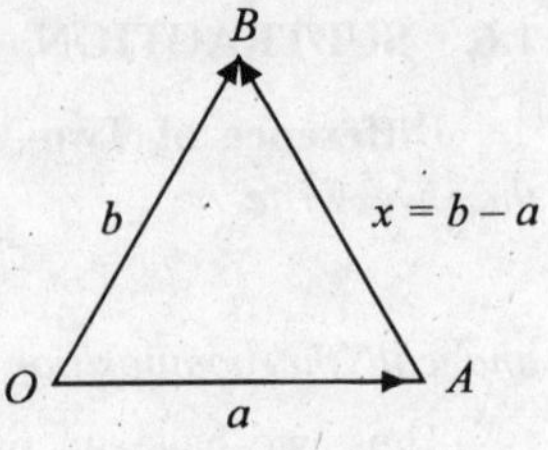

Fig. 1.19

We have

$$\mathbf{a} + x = \mathbf{b}$$

$$\Leftrightarrow \quad (-\mathbf{a}) + (\mathbf{a} + x) = (-\mathbf{a}) + \mathbf{b}$$

$$\Leftrightarrow \quad [(-\mathbf{a}) + \mathbf{a}] + x = \mathbf{b} + (-\mathbf{a})$$

$$\Leftrightarrow \quad \mathbf{0} + x = \mathbf{b} - \mathbf{a}$$

$$\Leftrightarrow \quad x = \mathbf{b} - \mathbf{a}.$$

1.8. A SUMMARY OF THE BASIC PROPERTIES OF THE LINEAR COMPOSITIONS

We catalogue below the basic laws of the Addition of vectors and of multiplication of vectors by scalars, *i.e.*, of the linear compositions. The set of all the vectors will be denoted by **V** and the set of all real numbers by **R**.

I. $\mathbf{a} + \mathbf{b} = \mathbf{b} + \mathbf{a} \ \forall \ \mathbf{a}, \mathbf{b} \in \mathbf{V}$;

II. $\mathbf{a} + (\mathbf{b} + \mathbf{c}) = (\mathbf{a} + \mathbf{b}) + \mathbf{c} \ \forall \ \mathbf{a}, \mathbf{b}, \mathbf{c} \in \mathbf{V}$;

III. $\exists$ **a** vector, *viz.*, **0** such that

$$\mathbf{a} + \mathbf{0} = \mathbf{a} = \mathbf{0} + \mathbf{a} \ \forall \ \mathbf{a} \in \mathbf{V};$$

IV. To each $\mathbf{a} \in \mathbf{V}$ there corresponds $\mathbf{b} \in \mathbf{V}$ such that

$$\mathbf{a} + \mathbf{b} = \mathbf{0} = \mathbf{b} + \mathbf{a};$$

The vector **b** is denoted by − **a**.

V. $(mn)(\mathbf{a}) = m(n\mathbf{a}) \ \forall \ \mathbf{a} \in \mathbf{V}$ and $\forall \ m, n \in \mathbf{R}$;

VI. $1(\mathbf{a}) = \mathbf{a} \ \forall \ \mathbf{a} \in \mathbf{V}$;

VII. $m(\mathbf{a} + \mathbf{b}) = m\mathbf{a} + m\mathbf{b} \ \forall \ m \in \mathbf{R}$ and $\forall \ \mathbf{a}, \mathbf{b} \in \mathbf{V}$;

VIII. $(m + n)\mathbf{a} = m\mathbf{a} + n\mathbf{a} \ \forall \ m, n \in \mathbf{R}$ and $\forall \ \mathbf{a} \in \mathbf{V}$.

1.9. LINEAR COMBINATIONS

A vector, **r**, *is said to be a linear combination of the vectors,* **a, b, c,** *... etc. if there exist scalars x, y, z, etc. such that*

$$\mathbf{r} = x\mathbf{a} + y\mathbf{b} + z\mathbf{c}$$

Thus, for example, the vectors

$$2\mathbf{a} + \mathbf{b} - 4\mathbf{c}, \quad \mathbf{a} + 2\mathbf{b} - 3\mathbf{c}$$

are linear combinations of the vectors **a, b, c.**

A linear combination of vectors involves the two linear compositions of the addition of vectors and the multiplication of vectors by scalars.

In the following linear combinations

$$x\mathbf{a},\ x\mathbf{a} + y\mathbf{b},\quad x\mathbf{a} + y\mathbf{b} + z\mathbf{c}$$

will be of special interest to us.

1.9.1. Parallel Vectors. Collinear Vectors

Of the two vectors having the same or parallel supports, each is a linear combination of the other. Thus, if **a, b** be two parallel vectors, then there exists a scalar x such that $\mathbf{b} = x\mathbf{a}$ or $\mathbf{a} = (1/x)\,\mathbf{b}$.

Parallel vectors are also called collinear vectors, for the support of parallel coinitial vectors is the same line.

1.9.2. Def. Coplanar Vectors.

A set of vectors is said to be coplanar, if their supports are parallel to the same plane, *i.e.*, if there exists a plane parallel to the supports of each of the vectors.

The support of coplanar co-initial vectors are coplanar.

The vector $x\mathbf{a} + y\mathbf{b}$ which is a linear combination of the vectors **a, b** is coplanar with **a** and **b.**

EXAMPLE

ABCDEF is a regular hexagon. Let $\vec{AB} = \mathbf{a}$ *and* $\vec{BC} = \mathbf{b}$. *Find the vectors determined by the other four sides taken in order. Also express the vectors* $\vec{AC}$, $\vec{AD}$, $\vec{AF}$, $\vec{AE}$, $\vec{CE}$ *in terms of* **a** *and* **b.**

Solution. $\vec{AC} = \vec{AB} + \vec{BC} = \mathbf{a} + \mathbf{b}$

$\because$ AD is parallel and double of BC,

$\therefore$ $\vec{AD} = 2\mathbf{b}$.

In ΔACD,

$$\vec{AC} + \vec{CD} = \vec{AD}$$

$$\Rightarrow \quad \vec{CD} = \vec{AD} - \vec{AC}$$

$$= 2\mathbf{b} - (\mathbf{a} + \mathbf{b})$$

$$= \mathbf{b} - \mathbf{a}.$$

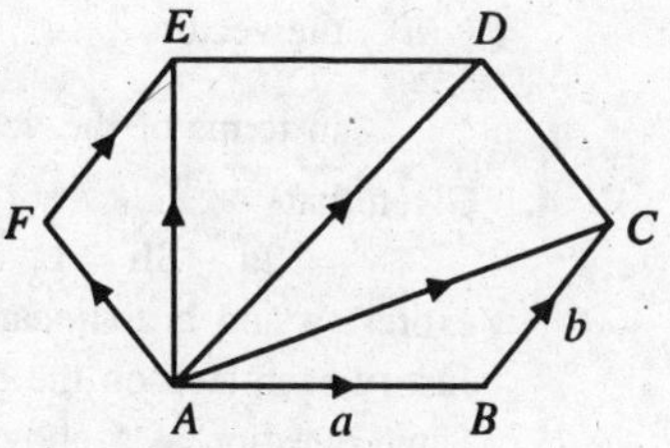

Fig. 1.20

$$\vec{AF} = \vec{CD} = \mathbf{b} - \mathbf{a}.$$

Now, $$\vec{DE} = \vec{BA} = -\mathbf{a}$$

$$\vec{EF} = \vec{CB} = -\mathbf{b}$$

$$\vec{FA} = \vec{DC} = -(\mathbf{b} - \mathbf{a}) = \mathbf{a} - \mathbf{b}$$

Again, $$\vec{AE} = \vec{AD} + \vec{DE} = 2\mathbf{b} + (-\mathbf{a}) = 2\mathbf{b} - \mathbf{a}$$

and $$\vec{CE} = \vec{CD} + \vec{DE} = \mathbf{b} - \mathbf{a} + (-\mathbf{a})$$

$$= \mathbf{b} - 2\mathbf{a}$$

EXERCISES

1. ABC is any triangle and D, E, F are the middle points of its sides BC, CA, AB respectively; express

   (*i*) the vectors $\vec{BC}, \vec{AD}, \vec{BE}$ and $\vec{CF}$ as linear combinations of the vectors $\vec{AB}$ and $\vec{AC}$.

   (*ii*) the vectors $\vec{AB}, \vec{BC}, \vec{CA}, \vec{AD}$ as linear combinations of the vectors $\vec{BE}$ and $\vec{CF}$.

   (*iii*) the vectors $\vec{AC}, \vec{BC}, \vec{AD}$ and $\vec{CF}$ as linear combinations of the vectors $\vec{AB}$ and $\vec{BE}$.

   Show that $\vec{AD} + \vec{BE} + \vec{CF} = 0$.

2. $ABCD$ is a parallelogram and AC, BD are its diagonals. Express

   (*i*) $\vec{AC}$ and $\vec{BD}$ in terms of $\vec{AB}$ and $\vec{AD}$,

   (*ii*) $\vec{AB}$ and $\vec{AD}$ in terms of $\vec{AC}$ and $\vec{BD}$,

   (*iii*) $\vec{AB}$ and $\vec{AC}$ in terms of $\vec{AD}$ and $\vec{BD}$.

   Show that

   $$\vec{AC} + \vec{DB} = 2\,\vec{DC},\quad \vec{AC} - \vec{BD} = 2\,\vec{AB}.$$

3. $OABC$ is a tetrahedron express

   (*i*) the vectors $\vec{BC}, \vec{CA}$, and $\vec{AB}$ in terms of the vectors $\vec{OA}, \vec{OB}$ and $\vec{OC}$;

   (*ii*) the vectors $\vec{OA}, \vec{OB}$ and $\vec{CA}$ in terms of the vectors $\vec{OC}, \vec{AB}$ and $\vec{BC}$

4. Given that

   $$2\mathbf{a} - 3\mathbf{b} = \mathbf{c}; \quad -3\mathbf{a} + 5\mathbf{b} = \mathbf{d},$$

   express **a** and **b** as linear combinations of **c** and **d**.
   Justify each step on the basis of different laws of compositions.

5. Same question as 5 above for the following pairs of equations :

   (*i*) $-\mathbf{a} + 2\mathbf{b} = 2\mathbf{c}; \quad 5\mathbf{a} - 2\mathbf{b} = 3\mathbf{d}$.

   (*ii*) $-\mathbf{a} + \mathbf{b} = -\mathbf{c}; \quad 2\mathbf{a} - \mathbf{b} = \mathbf{c} + \mathbf{d}$.

6. OB and OC are two lines and D is a point on BC such that

   $$\frac{BD}{DC} = \frac{m}{n},$$

show that

$$\vec{OD} = \frac{n\vec{OB} + m\vec{OC}}{m+n}$$

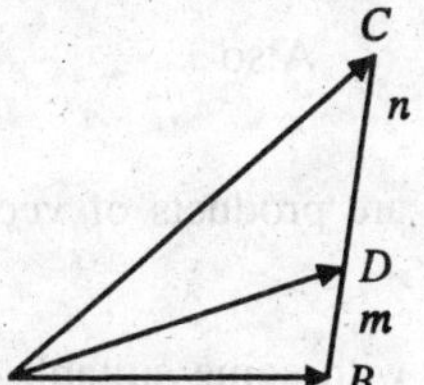

Fig. 1.21

7. Which of the following statements are correct :
If in triangle *OAC*, *B* is the mid-point of *AC* and

$\vec{OA} = \mathbf{a}$ and $\vec{OB} = \mathbf{b}$, then

(*a*) $\vec{OC} = \frac{1}{2}(\mathbf{a}+\mathbf{b})$, (*b*) $\vec{OC} = 2\mathbf{b} - 2\mathbf{a}$,

(*c*) $\vec{OC} = 2\mathbf{b} - \mathbf{a}$, (*d*) $\vec{OC} = 3\mathbf{a} - 2\mathbf{b}$,

(*e*) $\vec{OC} = 3\mathbf{b} - 2\mathbf{a}$.

8. Which of the following statements are correct :
If *M* is the mid-point of *AB* and *O* is any point, then

(*a*) $\vec{OM} = \vec{OA} + \vec{MA}$, (*b*) $OM = \vec{OA} - \vec{MA}$

(*c*) $\vec{OM} = \frac{1}{2}(\vec{OA} - \vec{OB})$, (*d*) $OM = \frac{1}{2}(\vec{OB} + \vec{OA})$,

(*e*) $\vec{OM} = \frac{1}{2}(OB - \vec{OA})$.

1.10. EXPRESSIONS AS LINEAR COMBINATIONS

1.10.1. Coplanar Vectors

If **a**, **b** *be two given non-collinear vectors, then every vector* **r**, *coplanar with* **a** *and* **b** *can be represented as a linear combination*

$$x\mathbf{a} + y\mathbf{b}$$

x, y being some scalars. Also this representation is unique.

Existence. Take any point *O*.

Let $\vec{OA} = \mathbf{a}$, $\vec{OB} = \mathbf{b}$.

Let $\vec{OP} = \mathbf{r}$ be any vector coplanar with $\vec{OA}$ and $\vec{OB}$.

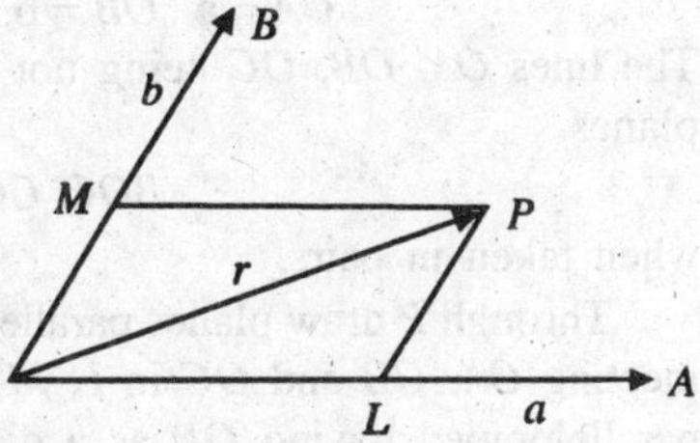

Fig. 1.22

The lines *OA*, *OB* and *OP* are coplanar.

Through *P*, draw lines parallel to *OB* and *OA* to meet *OA* and *OB* respectively in *L* and *M*. We have

$$\mathbf{r} = \vec{OP} = \vec{OL} + \vec{LP} = \vec{OL} + \vec{OM}.$$

Also

$$\vec{OL} \text{ and } \vec{OM}$$

are products of vectors

$$\vec{OA} \text{ and } \vec{OB}$$

with some suitable scalars. Let

$$\vec{OL} = x\vec{OA} = x\mathbf{a}, \ \vec{OM} = y\vec{OB} = y\mathbf{b}.$$

Thus, we have

$$\mathbf{r} = x\mathbf{a} + y\mathbf{b}.$$

Uniqueness. Let, if possible,

$$\mathbf{r} = x'\mathbf{a} + y'\mathbf{b}.$$

We have

$$x\mathbf{a} + y\mathbf{b} = x'\mathbf{a} + y'\mathbf{b}$$

$$\Rightarrow \quad (x - x')\,\mathbf{a} + (y - y')\,\mathbf{b} = \mathbf{0}.$$

If $x - x' \neq 0$, we have

$$\mathbf{a} = \frac{y - y'}{x - x'}\mathbf{b},$$

so that **a** is collinear with **b**. We thus arrive at a contradiction.

Thus, $x = x'$. Similarly $y = y'$.

Hence, the uniqueness of the representation.

1.10.2. Arbitrary System of Vectors

If **a, b, c** *be three given non-coplanar vectors, then any vector,* **r**, *can be represented as a linear combination*

$$x\mathbf{a} + y\mathbf{b} + z\mathbf{c}$$

x, y, z being some scalars. Also this representation is unique.

Take any point O. Let

$$\vec{OA} = \mathbf{a}, \ \vec{OB} = \mathbf{b}, \ \vec{OC} = \mathbf{c}, \ \vec{OP} = \mathbf{r}.$$

The lines *OA, OB, OC* being not coplanar, they determine three different planes

$$BOC, COA, AOB$$

when taken in pairs.

Through P draw planes parallel to the three planes *BOC, COA* and *AOB* meeting *OA, OB* and *OC* in *L, M* and *N* respectively so that we obtain a parallelopiped having *OP* as a diagonal. We have

$$\begin{aligned} \mathbf{r} &= \vec{OP} \\ &= \vec{OL} + \vec{LP} \\ &= \vec{OL} + \vec{LN'} + \vec{N'P} = \vec{OL} + \vec{OM} + \vec{ON}. \end{aligned}$$

There exists scalars x, y and z such that

$$\overrightarrow{OL} = x\overrightarrow{OA} = x\mathbf{a},\ \overrightarrow{OM} = y\overrightarrow{OB} = y\mathbf{b},\ \overrightarrow{ON} = z\overrightarrow{OC} = z\mathbf{c}.$$

Thus, we have

$$\mathbf{r} = x\mathbf{a} + y\mathbf{b} + z\mathbf{c}.$$

Uniqueness. Let, if possible

$$\mathbf{r} = x'\mathbf{a} + y'\mathbf{b} + z'\mathbf{c}.$$

We have

$$x\mathbf{a} + y\mathbf{b} + z\mathbf{c} = x'\mathbf{a} + y'\mathbf{b} + z'\mathbf{c}$$

$$\Rightarrow \quad (x - x')\,\mathbf{a} + (y - y')\,\mathbf{b} + (z - z')\,\mathbf{c} = \mathbf{0}.$$

If $x - x' \neq 0$, we have

$$\mathbf{a} = \frac{y - y'}{x' - x}\mathbf{b} + \frac{z - z'}{x - x'}\mathbf{c}$$

$\Rightarrow$ **a** is coplanar with 1, **b** and **c** and we arrive at a contradiction of the given hypothesis.

Thus, $x - x' = 0 \Leftrightarrow x = x'$. Similarly $y = y'$ and $z = z'$.

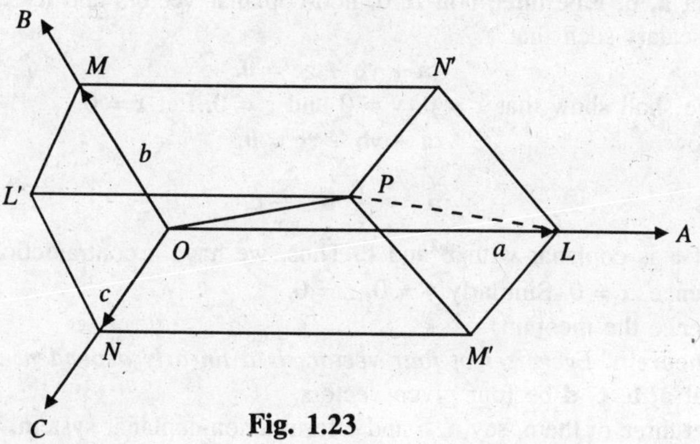

Fig. 1.23

The uniqueness of the representation is thus proved.

Ex. Show that the vector $x\mathbf{a} + y\mathbf{b}$ is coplanar with the vectors **a** and **b**.

1.11. LINEARLY INDEPENDENT AND DEPENDENT SYSTEMS OF VECTORS

Def. *A system of vectors*

$$\mathbf{a}, \mathbf{b}, \mathbf{c}, \ldots$$

is said to be linearly dependent, if there exists a system of scalars

$$x, y, z, \ldots$$

not all zero, such that

$$0 = x\mathbf{a} + y\mathbf{b} + z\mathbf{c} + \ldots, \qquad \ldots(i)$$

A system which is not linearly dependent is said to be linearly independent. For such a system, every relation of the type (*i*) *implies that*

$$x = 0,\ y = 0,\ z = 0, \ldots$$

Note. A pair of collinear vectors is a linearly dependent system. Also a triad of coplanar vectors is a linearly dependent system.

Theorem. *A pair of non-zero, non-collinear vectors is a linearly dependent system.*

Let **a, b** be a pair of non-zero, non-collinear vectors and let x, y be two scalars such that

$$x\mathbf{a} + y\mathbf{b} = \mathbf{0}.$$

We shall show that $x = 0$ and $y = 0$. Let $x \neq 0$. Now

$$x\mathbf{a} + y\mathbf{b} = \mathbf{0} \Rightarrow \mathbf{a} = -\frac{y}{x}\mathbf{b}$$

so that **a** appears as a product of the vector **b** with a scalar. Thus **a** and **b** are collinear and as such we arrive at a contradiction.

Hence, $x = 0$. Similarly $y = 0$.

Hence the theorem.

Theorem. *A triad of non-zero, non-coplanar vectors is a linearly independent system.*

Let **a, b, c** be three non-zero, non-coplanar vectors and let x, y, z be three scalars such that

$$x\mathbf{a} + y\mathbf{b} + z\mathbf{c} = \mathbf{0}.$$

We shall show that $x = 0$, $y = 0$ and $z = 0$. Let $x \neq 0$.

Now $$x\mathbf{a} + y\mathbf{b} + z\mathbf{c} = \mathbf{0}.$$

$$\Rightarrow \quad \mathbf{a} = -\frac{y}{x}\,\mathbf{b} - \frac{z}{x}\,\mathbf{c},$$

so that **a** is coplanar with **b** and **c**. Thus, we have a contradiction.

Hence, $x = 0$. Similarly $y = 0$, $z = 0$.

Hence the theorem.

Theorem. *Every set of four vectors is a linearly dependent system.*

Let **a, b, c, d** be four given vectors.

Let three of them, say **a, b** and **c** form a non-coplanar system. We then have a relation of the form

$$\mathbf{d} = x\mathbf{a} + y\mathbf{b} + z\mathbf{c} \quad \Rightarrow \quad x\mathbf{a} + y\mathbf{b} + z\mathbf{c} + (-\mathbf{d}) = \mathbf{0}$$

so that **a, b, c, d** form a linearly dependent system.

Now suppose that no three of the given vectors forms a non-coplanar system. Consider the triad of vectors **a, b, c** which are coplanar. Supposing that no pair of these three vectors is collinear, we have a relation of the form

$$\mathbf{c} = x\mathbf{a} + y\mathbf{b} \quad \Rightarrow \quad x\mathbf{a} + y\mathbf{b} + (-\mathbf{c}) + (0)\,\mathbf{d} = 0$$

so that **a, b, c, d** form a linearly dependent system.

We may similarly consider the case when a, b, c are all collinear.

1.12. BASES. COMPONENTS AND CO-ORDINATES OF A VECTOR

Let a, b, c be three given non-coplanar vectors so that if r be any given vector, we have the relation

$$\mathbf{r} = x\mathbf{a} + y\mathbf{b} + z\mathbf{c}$$

The vectors

$$x\mathbf{a},\quad y\mathbf{b},\quad z\mathbf{c}$$

are said to be the *components* and the scalars x, y, z the *co-ordinates* of the vector **r**, relatively to the triad of non-coplanar vectors **a, b, c.**

The set of non-coplanar vectors **a, b, c** relatively to which we may decompose any given vector **r** is called a **Base.** Every triad of non-coplanar vectors can thus be thought of as a Base.

Of course, the components and the co-ordinates of a given vector will be different for different Bases.

1.12.1. Co-ordinates of the Sum of Two Vectors and the Product of a Vector with a Scalar

Let **r, r′** be two given vectors with co-ordinates (x, y, z) and (x', y', z') respectively relatively to a given base $\{\mathbf{a}, \mathbf{b}, \mathbf{c}\}$ so that we have

$$\mathbf{r} = x\mathbf{a} + y\mathbf{b} + z\mathbf{c};\qquad \mathbf{r}' = x'\mathbf{a} + y'\mathbf{b} + z'\mathbf{c}.$$

Thus,

$$\mathbf{r} + \mathbf{r}' = (x + x')\,\mathbf{a} + (y + y')\,\mathbf{b} + (z + z')\,\mathbf{c}$$

so that the co-ordinates of $\mathbf{r} + \mathbf{r}'$ are

$$(x + x', y + y', z + z')$$

obtained by co-ordinate-wise addition of the co-ordinates of **r** and **r′**.

Again, if m be any scalar, we have

$$m\mathbf{r} = mx\mathbf{a} + my\mathbf{b} + mz\mathbf{c}$$

so that the co-ordinates of the vector $m\mathbf{r}$ are

$$(mx, my, mz)$$

obtained on co-ordinate-wise multiplication of the co-ordinates of **r** with m.

Note 1. It is easy to see that if (x, y, z) be the co-ordinates of a vector r relatively to a base **(a, b, c)** so that we have a relation

$$\mathbf{r} = x\mathbf{a} + y\mathbf{b} + z\mathbf{c},$$

then $\quad \mathbf{r}$ is coplanar with $\mathbf{a}, \mathbf{b} \quad \Leftrightarrow \quad z = 0,$

$\mathbf{r}$ is coplanar with $\mathbf{a} \quad \Leftrightarrow \quad y = 0, z = 0.$

Note 2. If **a, b** be two non-collinear vectors in a plane, then every vector in that plane is expressible in the form $x\mathbf{a} + y\mathbf{b}$ so that we can think of x and y as the co-ordinates of this vector. We may thus see that a system of coplanar vectors has a base consisting of only two members. Similarly, for a system of collinear vectors, we have a base consisting of only **one** member.

EXAMPLES

Example 1. *Show that if a triad* **a, b, c** *is a base, then the triad*

$$\mathbf{a},\ \mathbf{a} + \mathbf{b},\ \mathbf{a} + \mathbf{b} + \mathbf{c}$$

is also a base.

Given that x, y, z are the co-ordinates of a vector **u** *relatively to the base* **a, b, c,** *what are its co-ordinates relatively to the base* **a, a + b, a + b + c,**

Solution. We write

$$\mathbf{p} = \mathbf{a},\ \mathbf{q} = \mathbf{a} + \mathbf{b},\ \mathbf{r} = \mathbf{a} + \mathbf{b} + \mathbf{c}.$$

and show that **p, q, r** is a linearly independent set.

Now

$$l\mathbf{p} + m\mathbf{q} + n\mathbf{r} = \mathbf{0}$$

$$\Rightarrow \quad l\mathbf{a} + \mathbf{m}\,(\mathbf{a} + \mathbf{b}) + n\,(\mathbf{a} + \mathbf{b} + \mathbf{c}) = \mathbf{0}$$

$$\Rightarrow \quad (l + m + n)\,\mathbf{a} + (m + n)\,\mathbf{b} + n\mathbf{c} = \mathbf{0}$$

$$\Rightarrow \quad l + m + n = 0,\ m + n = 0,\ n = 0$$

[**a, b, c** being a base and therefore a linearly independent set]

$$\Rightarrow \quad n = 0, m = 0 = l = 0$$

$\Rightarrow$ **p, q, r** is a linearly indenendent set

$\Rightarrow$ **p, q, r** is a base.

Again we are given that

$$\mathbf{u} = x\mathbf{a} + y\mathbf{b} + z\mathbf{c}. \qquad ...(i)$$

Let

$$\mathbf{u} = x'\mathbf{a} + y'\,(\mathbf{a} + \mathbf{b}) + z'\,(\mathbf{a} + \mathbf{b} + \mathbf{c})$$

$$= (x' + y' + z')\,\mathbf{a} + (y' + z')\,\mathbf{b} + z'\mathbf{c}. \qquad ...(ii)$$

The expression of **u** as a linear combination of **a, b, c** being unique, we obtain

$$\left.\begin{aligned} x' + y' + z' &= x \\ y' + z' &= y \\ z' &= z \end{aligned}\right\} \Leftrightarrow \left\{\begin{aligned} z' &= z, \\ y' &= y - z, \\ x' &= x - y. \end{aligned}\right.$$

Thus, $x - y,\ y - z,\ z$ are the required co-ordinates relatively to the new base. Thus, we have

$$\mathbf{u} = x\mathbf{a} + y\mathbf{b} + z\mathbf{c}.$$

and $$\mathbf{u} = (x - y)\,\mathbf{a} + (y - z)\,(\mathbf{a} + \mathbf{b}) + z\,(\mathbf{a} + \mathbf{b} + \mathbf{c}).$$

Example 2. *Examine whether the vectors* 5**a** + 6**b** + 7**c**, 7**a** – 8**b** + 9**c** *and* 3**a** + 20**b** + 5**c**, (**a, b, c** *being non-coplanar vectors*) *are linearly independent or dependent.*

Solution. If possible, let txe linearly dependent. Then there exist scalars x_1, x_2, x_3, not all zero, such that

$$x_1\,(5\mathbf{a} + 6\mathbf{b} + 7\mathbf{c}) + x_2\,(7\mathbf{a} - 8\mathbf{b} + 9\mathbf{c}) + x_3\,(3\mathbf{a} + 20\mathbf{b} + 5\mathbf{c}) = \mathbf{0} \qquad ...(i)$$

$$\Rightarrow \quad (5x_1 + 7x_2 + 3x_3)\,\mathbf{a} + (6x_1 - 8x_2 + 20x_3)\,\mathbf{b} + (7x_1 + 9x_2 + 5x_3)\,\mathbf{c} = 0$$

As **a, b, c** are non-coplanar vectors,

$$\Rightarrow \quad \begin{aligned} 5x_1 + 7x_2 + 3x_3 &= 0 \\ 6x_1 - 8x_2 + 20x_3 &= 0 \\ 7x_1 + 9x_2 + 5x_3 &= 0 \end{aligned}$$

From first two equations, we get

$$\frac{x_1}{2} = \frac{x_2}{-1} = \frac{x_3}{-1} = k \qquad \text{(say)}$$

$$\therefore \quad x_1 = 2k,\ x_2 = -k,\ x_3 = -k.$$

These values also satisfy the third equation.

Hence, there exist scalars x_1, x_2, x_3 such that (1) holds. Hence, given vectors are linearly dependent.

Example 3. *If* **a, b, c** *be three non-zero, non-coplanar vectors, find a relation between the vectors* **a** + 3**b** + 4**c**, **a** – 2**b** + 3**c**, **a** + 5**b** – 2**c**, 6**a** + 14**b** + 4**c**.

Solution. Let

$$\mathbf{a} + 3\mathbf{b} + 4\mathbf{c} = x_1 (\mathbf{a} - 2\mathbf{b} + 3\mathbf{c}) + x_2 (\mathbf{a} + 5\mathbf{b} - 2\mathbf{c}) + x_3 (6\mathbf{a} + 14\mathbf{b} + 4\mathbf{c})$$

$$\Rightarrow \quad \mathbf{a} + 3\mathbf{b} + 4\mathbf{c} = (x_1 + x_2 + 6x_3)\,\mathbf{a} + (-2x_1 + 5x_2 + 14x_3)\,\mathbf{b} + (3x_1 - 2x_2 + 4x_3)\,\mathbf{c}$$

Equating the coefficients of **a, b, c** on both sides, we get

$$x_1 + x_2 + 6x_3 = 1, \quad -2x_1 + 5x_2 + 14x_3 = 3, \quad 3x_1 - 2x_2 + 4x_3 = 4$$

$$\Rightarrow \quad x_1 = -2, \quad x_2 = -3, \quad x_3 = 1.$$

$$\therefore \quad \mathbf{a} + 3\mathbf{b} + 4\mathbf{c} = -2(\mathbf{a} - 2\mathbf{b} + 3\mathbf{c}) - 3(\mathbf{a} + 5\mathbf{b} - 2\mathbf{c}) + (6\mathbf{a} + 14\mathbf{b} + 4\mathbf{c}).$$

EXERCISES

1. Given that the vectors **a, b, c** form a base, find the co-ordinates of the vectors.
   (*i*) 3**u** – **v** + **w** if **u** = **a** + **c**, **v** = **b** + **c**, **w** = **a** – **b**;
   (*ii*) 2**u** – 3**v** + 4**w** if **u** = 2**b** + 3**c**, **v** = – 2**a** + **c**, **w** = **a** – **b** + **c**;
   (*iii*) 2**u** – 3**v** + 4**w** if **u** = 2**a** + 3**b** + 4**c**, **v** = 3**a** – 2**b** + 4**c**, **w** = 2**a** – 4**b** + 3**c**.
2. Show that the following set of vector is linearly dependent :
   a – 2**b** + 3**c**, – 2**a** + 3**b** – 4**c**, – **b** + 2**c**.
   Find also the linear relations.
3. **a, b, c** is a system of linearly independent vectors; show that the following sets of vectors are also linearly independent.
   (*i*) **a** – **b** + **c**, **b** + **c** – **a**, 2**a** – 3**b** + 4**c**.
   (*ii*) 2**a** – **b** + 3**c**, **a** + **b** – 2**c**, **a** + **b** – 3**c**.
4. Given that *a, b, c* is a linearly independent set of vectors show that :
   (*i*) **a** + 2**b** + **c**, 2**a** – **b** + **c**, 3**a** + **b** + 2**c**
   is a linearly dependent set.
   (*ii*) 2**a** + **b** + **c**, **a** – **b** + 2**c**, – **a** + **b** – **c**,
   is a linearly independent set.
   (*iii*) *m***a** + **b** + **c**, **a** + *m***b** + **c**, **a** + **b** + *m***c**
   is a linearly independent set if and only if $m \neq -2$.
5. **u, v, w** being a linearly independent set of vectors; examine the set **p, q, r** where

$$\mathbf{p} = \mathbf{u}\cos a + \mathbf{v}\cos b + \mathbf{w}\cos c,$$
$$\mathbf{q} = \mathbf{u}\sin a + \mathbf{v}\sin b + \mathbf{w}\sin c,$$
$$\mathbf{r} = \mathbf{u}\sin(x + a) + \mathbf{v}\sin(x + b) + \mathbf{w}\sin(x + c),$$

   for linear independence.

6. Consider a base **a**, **b**, **c** and a vector

$$-2\mathbf{a} + 3\mathbf{b} - \mathbf{c}.$$

Compute the co-ordinates of this vector relatively to the base **p**, **q**, **r** where

$$\mathbf{p} = 2\mathbf{a} - 3\mathbf{b}, \quad \mathbf{q} = \mathbf{a} - 2\mathbf{b} + \mathbf{c}, \quad \mathbf{r} = -3\mathbf{a} + \mathbf{b} + 2\mathbf{c}.$$

7. With reference to the base, **u**, **v**, **w**, show that the vectors

$$\mathbf{u} + \mathbf{w}, \quad -\mathbf{u} + \mathbf{w}, \quad \mathbf{u} + \mathbf{v} + \mathbf{w}$$

also constitute a base. Give the formulae for the transformation of the co-ordinates of a vector.

8. (Fig. 23). Given that *x*, *y*, *z* are the co-ordinates of a vector in terms of tbe base $\overrightarrow{OL}, \overrightarrow{OM}, \overrightarrow{ON}$, find its co-ordinates in terms of the base $\overrightarrow{OL'}, \overrightarrow{OM'}, \overrightarrow{ON'}$.

9. Find the linear relations between the following system of vectors; **a**, **b**, **c**; being any three non-coplanar vectors;

(*i*) $\mathbf{a} - \mathbf{b} + \mathbf{c}, \quad \mathbf{b} + \mathbf{c} - \mathbf{a}, \quad \mathbf{c} + \mathbf{a} + \mathbf{b}, \quad 2\mathbf{a} - 3\mathbf{b} + 4\mathbf{c}.$

(*ii*) $\mathbf{a} - 2\mathbf{b} + \mathbf{c}, \quad -\mathbf{b} + 2\mathbf{c} + \mathbf{a}, \quad \mathbf{c} - \mathbf{a} + 3\mathbf{b}, \quad \mathbf{a} + \mathbf{b} + \mathbf{c}.$

10. Given that **a**, **b** is a pair of non-collinear vectors such that

(*i*) $(1 + 2h - k)\,\mathbf{a} + (2 - h + 2k)\,\mathbf{b} = \mathbf{0},$

(*ii*) $(3h + k)\,\mathbf{a} + (1 - 2h)\,\mathbf{b} = h\mathbf{a} + 2k\mathbf{b}.$

find *h* and *k* in each case.

11. Given that **a**, **b**, **c** is a triad of three non-coplanar vectors such that

(*i*) $(1 - h - k + l)\,\mathbf{a} + (2h - 3k - l)\,\mathbf{b} + (3 + h + k)\,\mathbf{c} = \mathbf{0},$

(*ii*) $(2h + k)\,\mathbf{a} + (3 - 4h + l)\,\mathbf{b} + (1 + h + k)\,\mathbf{c} = h\mathbf{a} + k\mathbf{b} + l\mathbf{c}.$

find *h*, *k* and *l* in each case.

12. Show that the three vectors with co-ordinates

$$(x_1, y_1, z_1), \quad (x_2, y_2, z_2), \quad (x_3, y_3, z_3),$$

relatively to any given base will be linearly dependent if and only if

$$\begin{vmatrix} x_1 & y_1 & z_1 \\ x_2 & y_2 & z_2 \\ x_3 & y_3 & z_3 \end{vmatrix} = 0.$$

SUMMARY

1. Properties of multiplication of vectors by scalars and of addition of vectors.

I. $\mathbf{a} + \mathbf{b} = \mathbf{b} + \mathbf{a} \;\forall\; \mathbf{a}, \mathbf{b} \in \mathbf{V}.$

II. $\mathbf{a} + (\mathbf{b} + \mathbf{c}) = (\mathbf{a} + \mathbf{b}) + \mathbf{c} \;\forall\; \mathbf{a}, \mathbf{b}, \mathbf{c} \in \mathbf{V}.$

III. $\exists$ a vector, *viz.*, **0** such that

$$\mathbf{a} + \mathbf{0} = \mathbf{a} = \mathbf{0} + \mathbf{a} \;\forall\; \mathbf{a} \in \mathbf{V}.$$

IV. To each $\mathbf{a} \in \mathbf{V}$ there corresponds $\mathbf{b} \in \mathbf{V}$ such that

$$\mathbf{a} + \mathbf{b} = \mathbf{0} = \mathbf{b} + \mathbf{a}.$$

The vector **b** is denoted by – **a**.

V. $(mn)\,\mathbf{a} = m\,(n\mathbf{a})\ \forall\ \mathbf{a} \in \mathbf{V}$ and $\forall\ m, n \in \mathbf{R}$.

VI. $1\,(\mathbf{a}) = \mathbf{a}\ \forall\ \mathbf{a} \in \mathbf{V}$;

VII. $m\,(\mathbf{a} + \mathbf{b}) = m\mathbf{a} + m\mathbf{b}\ \forall\ m \in \mathbf{R}$ and $\forall\ \mathbf{a}, \mathbf{b} \in \mathbf{V}$.

VIII. $(m + n)\,\mathbf{a} = m\mathbf{a} + n\mathbf{a}\ \forall\ m, n\ \mathbf{a} \in \mathbf{R}$ and $\forall\ \mathbf{a} \in \mathbf{V}$.

2. A vector **r** is a linear combination of

   (*i*) two vectors **a**, **b** if ∃ scalars *x*, *y* such that
   $$\mathbf{r} = x\mathbf{a} + y\mathbf{b}.$$
   (*ii*) three vectors **a**, **b**, **c** if ∃ scalars *x*, *y*, *z* such that
   $$\mathbf{r} = x\mathbf{a} + y\mathbf{b} + z\mathbf{c}.$$

3. (*i*) If two vectors **a**, **b** are linearly independent, then
   $$x\mathbf{a} + y\mathbf{b} = 0, \quad \Rightarrow \quad x = 0,\ y = 0.$$
   Two collinear vectors are linearly dependent and two non-collinear vectors are linearly independent.

   (*ii*) If three vectors **a**, **b**, **c** are linearly independent, then
   $$x\mathbf{a} + y\mathbf{b} + z\mathbf{c} = 0, \quad \Rightarrow \quad x = 0,\ y = 0,\ z = 0.$$
   Three coplanar vectors are linearly dependent and three non-coplanar vectors are linearly independent. Every system of four vectors is linearly dependent.

4. If a vector **r** is coplanar with two vectors **a**, **b**, ∃ scalars *x*, *y* such that
   $$\mathbf{r} = x\mathbf{a} + y\mathbf{b}$$
   If **a**, **b**, **c** be three given non-coplanar vectors and **r** is any vector, ∃ scalars *x*, *y*, *z* such that
   $$\mathbf{r} = x\mathbf{a} + y\mathbf{b} + z\mathbf{c}.$$

5. The set of linearly independent vectors such that any given vector is a linear combination of the members of the set is called a Base.

OBJECTIVE QUESTIONS

For each of the following questions, four alternatives are given for the answer. Only one of them is correct. Choose the correct alternative.

1. If **a** and **b** represent the sides $\overrightarrow{AB}$ and $\overrightarrow{BC}$ of a regular hexagon *ABCDEF*, than $\overrightarrow{FA} =$

   (*a*) **b** – **a** (*b*) **a** – **b**

   (*c*) **a** + **b** (*d*) None of these

2. If *c* is the middle point of *AB* and *P* is any point outside *AB*; then

   (*a*) $\overrightarrow{PA} + \overrightarrow{PB} = \overrightarrow{PC}$ (*b*) $\overrightarrow{PA} + \overrightarrow{PB} = 2\overrightarrow{PC}$

   (*c*) $\overrightarrow{PA} + \overrightarrow{PB} + \overrightarrow{PC} = \mathbf{0}$ (*d*) $\overrightarrow{PA} + \overrightarrow{PB} + 2\overrightarrow{PC} = \mathbf{0}$

3. One of the following is not a vector :
(*a*) displacement (*b*) work
(*c*) centrifugal (*d*) gravitational field

4. Which one of the following is not a scalar :
(*a*) temperature (*b*) density
(*c*) mass (*d*) weight

5. Let $\alpha = (x + 4y)\,\mathbf{a} + (2x + y + 1)\,\mathbf{b}$ and

$$\beta = (y - 2x + 2)\,\mathbf{a} + (2x - 3y - 1)\,\mathbf{b},$$

where **a** and **b** are non-zero, non-collinear. If $3\alpha + 2\beta$, then
(*a*) $x = 1, \quad y = 2$ (*b*) $x = 2, \quad y = 1$
(*c*) $x = -1, \quad y = 2$ (*d*) $x = 2, \quad y = -1$

6. Direction of zero vector
(*a*) does not exist (*b*) towards origin
(*c*) indeterminate (*d*) None of these

7. If **a** be an unit vector, then
(*a*) direction of **a** is constant
(*b*) magnitude of **a** is constant
(*c*) direction and magnitude of **a** is constant
(*d*) any one of direction or magnitude is constant.

8. If *a, b, c* are three non-coplanar, non-zero vectors and *r* is any vector, then *r* can be represented as $x\mathbf{a} + y\mathbf{b} + z\mathbf{c}$
(*a*) always (*b*) never
(*c*) is one and only one way (*d*) None of these

9. If vectors $(x - 2)\,\mathbf{a} + \mathbf{b}$ and $(2x + 1)\,\mathbf{a} - \mathbf{b}$ are parallel, then $x =$
(*a*) 1/3 (*b*) 3
(*c*) – 3 (*d*) – 1/3

10. If **a** is a vector and *x* is a non-zero scalar, then
(*a*) $x\mathbf{a}$ is a vector in the direction of **a**
(*b*) $x\mathbf{a}$ is a vector collinear to **a**
(*c*) $x\mathbf{a}$ and **a** have independent directions
(*d*) None of these.

ANSWERS

1. (*b*) **2.** (*b*) **3.** (*b*) **4.** (*d*) **5.** (*d*)
6. (*c*) **7.** (*c*) **8.** (*c*) **9.** (*a*) **10.** (*b*)

2

Geometry with Vectors. Affine Geometry

Introduction. The application to Geometry of the Vector Algebra so far developed will be considered in this chapter. It should be of some importance to notice that geometrical problems of *Affine nature* alone can be treated with the help of the compositions of the addition of vectors and the multiplication of vectors by scalars. These two compositions are inadequate for an unrestricted treatment of Metric relations involving comparisons of lengths, angles, areas and volumes and the compositions of scalars and vector products are needed for the purpose which will be considered in the following chapters.

It is not proposed to consider the detailed meanings and significance of the terms, **Affine** and **Metric** here. It may only suffice to say that the relations which involve comparisons of distances lying along the *same* or *parallel* lines only or deal with the comparison of directions only so far as parallelism is concerned are called *Affine.*

Thus, the scope of Affine geometry in so far as it deals with lengths and angles is very limited.

Some well-known theorems of plane Geometry such as Pappu's theorem, Desargue's theorem on triangles etc., will also be proved with the help of vector methods and included in on appendix to the chapter.

2.1. ORIGIN OF REFERENCE, POSITION VECTOR

The application of vector methods of Geometry depends upon the concept of *Position Vector or* the *Co-ordinate Vector* of a point which will now be introduced.

2.1.1. We take any arbitrary point, O to be called the **Origin of reference.**

The position vector of any point P, with respect to the origin O is the vector $\overrightarrow{OP}$.

Thus, with the choice of any point, O as the origin of reference, we can associate a vector to every point P and, conversely, to every given vector, **r**, there corresponds a point, *viz.*, the point P such that

$$\overrightarrow{OP} = \mathbf{r}.$$

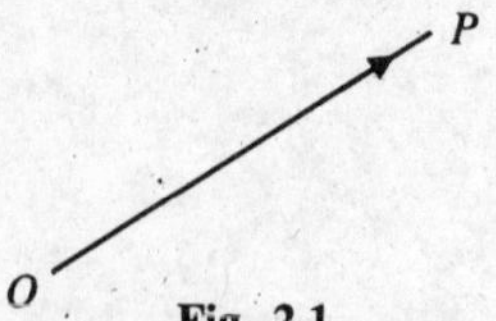

Fig. 2.1

In fact P is the terminal point of the vector **r**, whose initial point is the origin O.

2.1.2. Choice of the Origin of Reference

Origin of reference may be chosen arbitrarily but a suitable choice of the same often facilitates the solution of geometrical problems a good deal.

The origin of reference having been once chosen, a point whose position vector is **r** will usually be referred to as the point **r**.

2.1.3. *A vector expressed in terms of the position vectors of its end points.* The equality

$$\overrightarrow{AB} = \overrightarrow{OB} - \overrightarrow{OA}$$

which expresses any vector

$$\overrightarrow{AB}$$

in terms of the position vectors

$$\overrightarrow{OA} \text{ and } \overrightarrow{OB}$$

of its end points will prove very useful in what follows.

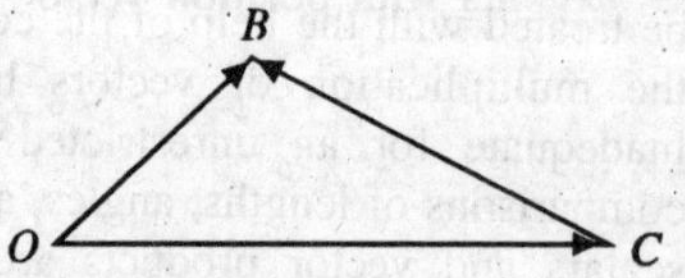

Fig. 2.2

2.2. SECTION FORMULA

To find the position vector of the point which divides the line joining two given points in a given ratio.

Let O, be the origin of reference and let **a, b** be the position vectors of the given points, A, B so that we have

$$\overrightarrow{OA} = \mathbf{a},\ \overrightarrow{OB} = \mathbf{b}.$$

Let P divide AB so that

$$\frac{AP}{PB} = \frac{m}{n}. \quad ...(i)$$

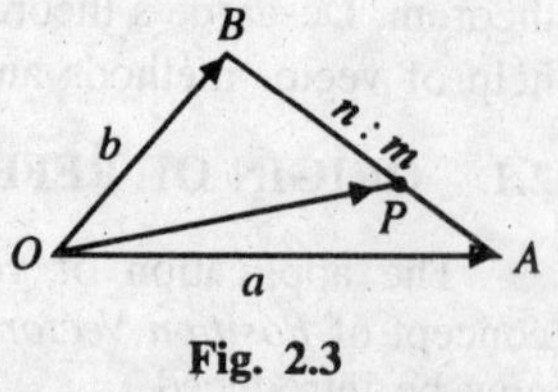

Fig. 2.3

Here m/n is positive or negative according as, P, divides AB internally or externally. We have to express the position vector $\overrightarrow{OP}$ of the point P in terms of the position vectors $\overrightarrow{OA}$ and $\overrightarrow{OB}$ of the points A and B.

We re-write (i) as

$$nAP = mPB$$

and obtain the vector equality

$$n\vec{AP} = m\vec{PB}.$$

Expressing the vector $\vec{AP}$ and $\vec{PB}$ in terms of the position vectors of the end points, we obtain

$$n(\vec{OP} - \vec{OA}) = m(\vec{OB} - \vec{OP})$$

$$\Rightarrow \quad (n+m)\vec{OP} = m\vec{OB} + n\vec{OA}$$

$$\Rightarrow \quad \vec{OP} = \frac{m\vec{OB} + n\vec{OA}}{m+n} = \frac{m\mathbf{b} + n\mathbf{a}}{m+n},$$

This result is analogous to the corresponding one in Cartesian Geometry.

Cor. Middle point. The position vector of the mid-point of the join of two points with position vectors, **a**, and **b**, is

$$\frac{1}{2}(\mathbf{a}+\mathbf{b})$$

obtained on taking $m = n$.

2.3. APPLICATIONS TO GEOMETRY

We shall now consider the applications to geometry of the subject so far developed.

EXAMPLES

Example 1. *Show that the medians of a triangle are concurrent.*

Solution. Let the position vectors of the vertices A, B, C of a triangle ABC with respect to any origin be **a, b, c.**

The position vectors of the mid-points D, E, F of the sides are respectively,

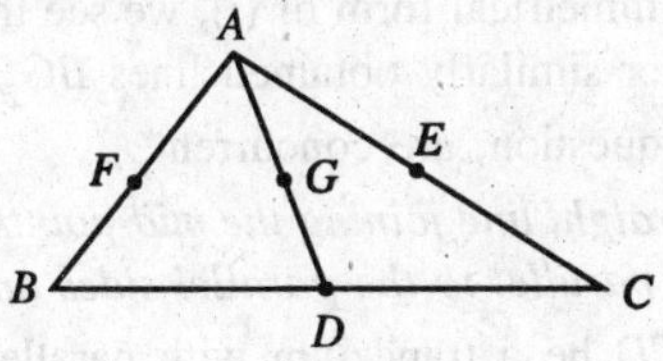

Fig. 2.4

$$\frac{1}{2}(\mathbf{b}+\mathbf{c}), \ \frac{1}{2}(\mathbf{c}+\mathbf{a}), \ \frac{1}{2}(\mathbf{a}+\mathbf{b})$$

Position vector of the point G dividing AD in the ratio 2 : 1 is

$$\frac{2\cdot\frac{1}{2}(\mathbf{b}+\mathbf{c})+1.\mathbf{a}}{2+1}=\frac{1}{3}(\mathbf{a}+\mathbf{b}+\mathbf{c}) \quad ...(i)$$

By symmetry, we see that this point also lies on the other two medians.

Thus, the medians of a triangle are concurrent. Also the position vector of the point of concurrence, is $\frac{1}{3}(\mathbf{a}+\mathbf{b}+\mathbf{c})$; **a, b, c;** being the position vectors of the vertices of the triangle.

The point of concurrence of the medians of a triangle is called its *Centroid.*

Example 2. *Show that the lines joining the vertices of a tetrahedron to the centroids of the opposite faces meet in a point.*

Solution. Let the position vectors of the vertices of a tetrahedron *ABCD*, with respect to any origin of reference *O* be

a, b, c, d.

The position vectors of the centroids

G_1, G_2, G_3, G_4

of the triangular faces

BCD, CDA, DAB, ABC are

$$\frac{1}{3}(\mathbf{b}+\mathbf{c}+\mathbf{d}),\quad \frac{1}{3}(\mathbf{c}+\mathbf{d}+\mathbf{a}),$$

$$\frac{1}{3}(\mathbf{d}+\mathbf{a}+\mathbf{b}),\quad \frac{1}{3}(\mathbf{a}+\mathbf{b}+\mathbf{c}),$$

respectively.

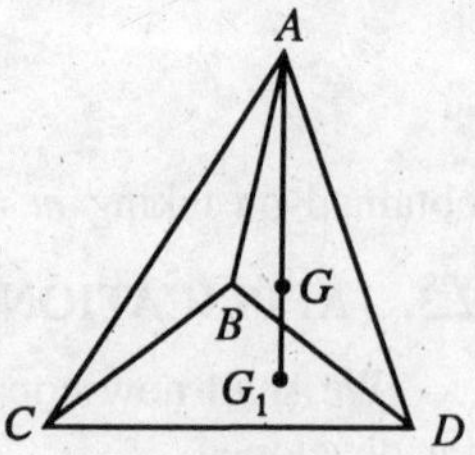

Fig. 2.5

The position vector of the point dividing AG_1 in the ratio 3 : 1 is

$$\frac{3\cdot\frac{1}{3}(\mathbf{b}+\mathbf{c}+\mathbf{d})+1\cdot\mathbf{a}}{3+1}=\frac{\mathbf{a}+\mathbf{b}+\mathbf{c}+\mathbf{d}}{4}. \quad ...(i)$$

By virtue of the symmetrical form of (*i*), we see that this point as well lies on each of the other similarly obtained lines BG_2, CG_3, DG_4.

Thus the lines, in question, are concurrent.

Example 3. *The straight line joining the mid-points of two non-parallel sides of a trapezium is parallel to the parallel sides and half of their sum.*

Solution. Let *ABCD* be a trapezium with parallel sides *AB, CD.*

A as the origin of reference. Let

$$\overrightarrow{AB}=\mathbf{b},\quad \overrightarrow{AD}=\mathbf{d},$$

so that **b, d** are the position vectors of the points *B* and *D* respectively with the point *A* as the origin of reference.

As DC is parallel to AB, the vector $\overrightarrow{DC}$ must be a product of the vector $\overrightarrow{AB}$ by some scalar t. Let

$$\overrightarrow{DC} = t\,\overrightarrow{AB} = t\mathbf{b} \quad ...(i)$$

Now the position vector of C is

$$\overrightarrow{AC} = \overrightarrow{AD} + \overrightarrow{DC} = \mathbf{d} + t\mathbf{b}.$$

The position vectors of the mid-points E and F of BC and AD are $\frac{1}{2}(\mathbf{b}+\mathbf{d}+t\mathbf{b})$ and $\frac{1}{2}\mathbf{d}$ respectively.

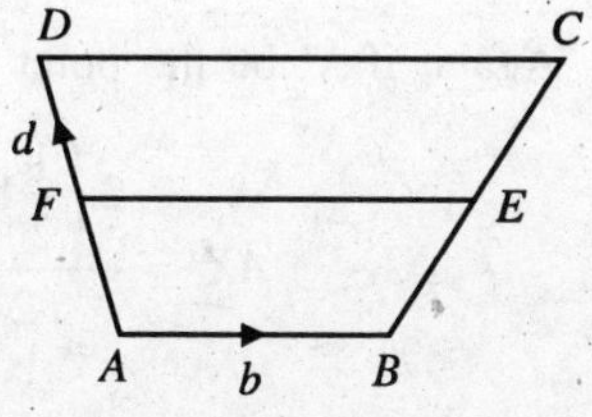

Fig. 2.6

We have

$$\overrightarrow{FE} = \overrightarrow{AE} + \overrightarrow{AF}$$

$$= \frac{1}{2}(\mathbf{b}+\mathbf{d}+t\mathbf{b}) - \frac{1}{2}\mathbf{d}$$

$$= \frac{1}{2}(1+t)\,\mathbf{b} = \frac{1}{2}(1+t)\overrightarrow{AB}$$

$\Rightarrow$ the vector $\overrightarrow{FE}$ is the product of the vector $\overrightarrow{AB}$ by a scalar $\frac{1}{2}(1+t)$

$$\Rightarrow \quad FE \parallel AB \text{ and } FE = \frac{1}{2}(1+t)\,AB.$$

Also from (*i*), we have

$$DC = t\,AB.$$

It follows that

$$AB + DC = (1 + t)\,AB = 2\,EF.$$

Example 4. *If M, N are the mid-points of the sides AB, CD of a parallelogram ABCD, prove that DM and BN cut the diagonal AC at its points of trisection which are also the points of trisection of DM and BN respectively.*

Solution. A as origin of reference. Take position vectors of B and D as $\mathbf{b}$ and $\mathbf{d}$ respectively.

ABCD is a parallelogram.

$$\Rightarrow \quad \overrightarrow{BC} = \overrightarrow{AD} = \mathbf{d}.$$

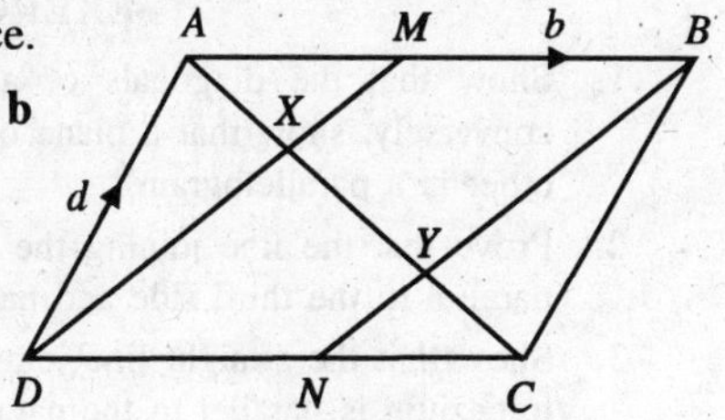

Fig. 2.7

$$\vec{AC} = \vec{AB} + \vec{BC} = \mathbf{b} + \mathbf{d}$$

If X is a point which trisects AC, then

$$\vec{AX} = \frac{1}{3}(\mathbf{b} + \mathbf{d}) \quad ...(i)$$

Again, if X' be the point which divides DM in the ratio 2 : 1, then

$$\vec{AX'} = \frac{2.\left(\frac{1}{2}\mathbf{b}\right) + 1.\mathbf{d}}{2+1} = \frac{1}{3}(\mathbf{b} + \mathbf{d})$$

$$= \vec{AX}$$

Implying that X and X' must coincide. Thus, DM cuts the diagonal AC at its point of trisection X which is also the point of trisection of DM.

Similarly other result can be established.

Example 5. *The middle points of the adjacent sides of any quadrilateral are joined; prove that the figure so formed is a parallelogram.*

Solution. $ABCD$ is a quadrilateral and E, F, G, H be the mid-points of AB, BC, CD and DA respectively. Let A be the origin of reference and **b, c, d** be the position vectors of B, C and D.

Position vectors of E, F, G, H are

$$\frac{1}{2}\mathbf{b},\ \frac{1}{2}(\mathbf{b}+\mathbf{c}),\ \frac{1}{2}(\mathbf{c}+\mathbf{d}) \text{ and } \frac{1}{2}\mathbf{d}$$

respectively.

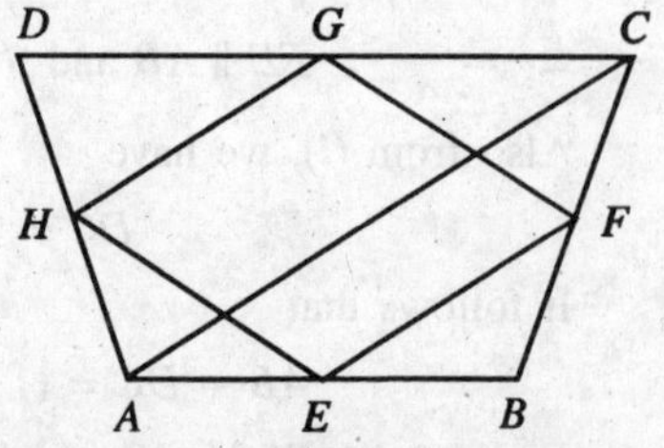

Fig. 2.8

$$\therefore \quad \vec{EF} = \frac{1}{2}(\mathbf{b}+\mathbf{c}) - \frac{1}{2}\mathbf{b} = \frac{1}{2}\mathbf{c}$$

and $$\vec{HG} = \frac{1}{2}(\mathbf{c}+\mathbf{d}) - \frac{1}{2}\mathbf{d} = \frac{1}{2}\mathbf{c}$$

$\Rightarrow$ EF is equal and parallel to HG. Hence, $EFGH$ is a parallelogram.

EXERCISES

1. Show that the diagonals of a parallelogram bisect each other. Also conversely, show that a plane quadrilateral whose diagonals bisect each other is a parallelogram.
2. Prove that the line joining the mid-points of two sides of a triangle is parallel to the third side and half of it.
3. Show that the straight line joining the mid-points of the diagonals of a trapezium is parallel to the parallel sides and is half of their difference.

4. Prove that the straight lines joining the mid-points of pairs of opposite edges of a tetrahedron are concurrent.

5. G is the centroid of a tetrahedron $ABCD$; $A'B'C'D'$ is another tetrahedron such that AA', BB', CC', and DD', are all bisected at G; show that G is also the centroid of the tetrahedron $A'B'C'D'$.

6. OP is a diagonal of the parallelopiped with coterminous edges OL, OM, ON, and OQ, OR, OS are the diagonals of the parallelopiped constructed with

$$OM, ON, OP;\ ON, OL, OP;\ OL, OM, OP$$

as coterminous edges. Finally OM is a diagonal of the parallelopiped with OQ, OR, OS as coterminous edges. Show that OM lies along OL and is five times OL.

7. Show that the four diagonals of any parallelopiped are concurrent and are bisected at the point of concurrence.

8. The diagonals of the three faces of the parallelopiped drawn from the same vertex are prolonged half their lengths; show that the three points thus obtained are coplanar with the opposite vertex.

2.4. PARAMETRIC VECTORIAL EQUATIONS

It is possible to express the position vectors of points on given lines and planes in terms of some fixed vectors and variable scalars, called *parameters*, such that

(*i*) for arbitrary values of the parameters, the resulting position vectors represent points on the locus in question, and

(*ii*) conversely, the position vector of each point on the locus arises for some suitable values of the parameters.

Parametric Vectorial Equation of a Line. To find the parametric vectorial equation of a line which passes through a given point and is parallel to a given line.

Take any point, O, as the origin of reference.

Let, $\mathbf{a}$, be the position vector of the given point, A, and let $\mathbf{b}$, be a vector parallel to the given line.

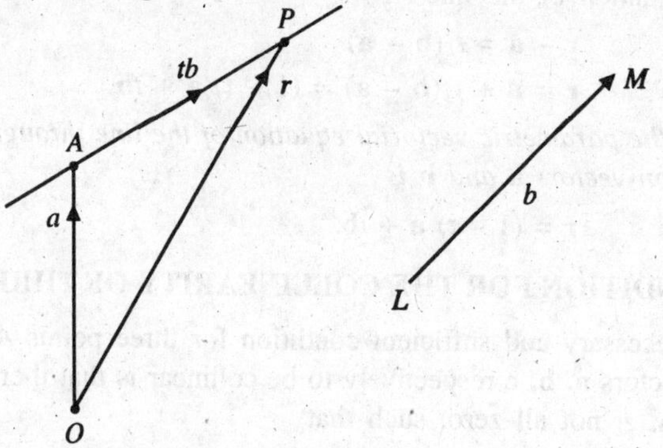

Fig. 2.9

Let, **r**, be the position vector of any point P, on the given line.

The vector $\overrightarrow{AP} = \mathbf{r} - \mathbf{a}$ and **b** being collinear, we have

$$\mathbf{r} - \mathbf{a} = t\mathbf{b} \quad \Rightarrow \quad \mathbf{r} = \mathbf{a} + t\mathbf{b}. \qquad ...(i)$$

Each point, P, of the line arises for some value of the scalar t. Also, conversely, for each value of the scalar, t,

$$\mathbf{a} + t\mathbf{b}$$

is the positive vector of a point of the line.

Hence, (i) is the required parametric vectorial equation of the given line so that the *parametric vectorial equation of the line through a point with a position vector* **a** *and is parallel to a vector,* **b**, *is*

$$\mathbf{r} = \mathbf{a} + t\mathbf{b},$$

where, t, is a scalar parameter. (Kanpur, 94)

Cor. Line through two given points. Let **a, b** be the position vectors of the two given points A, B with reference to any origin O and let **r** be the position vector of any point P on the line.

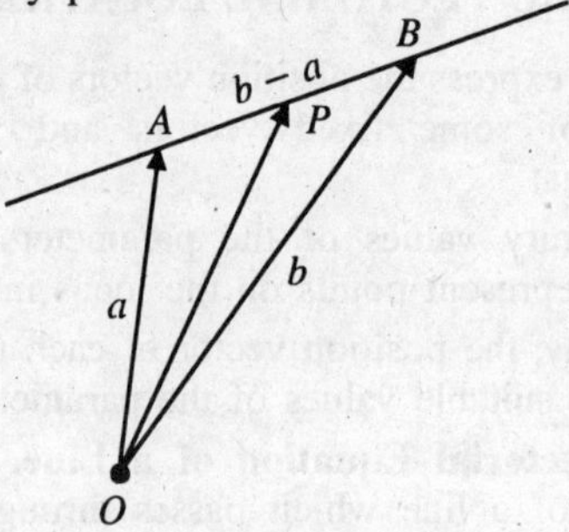

Fig. 2.10

Now the vectors $\overrightarrow{AP} = \mathbf{r} - \mathbf{a}$ and $\overrightarrow{AB} = \mathbf{b} - \mathbf{a}$ being collinear the required equation of the line AB is

$$\mathbf{r} - \mathbf{a} = t\,(\mathbf{b} - \mathbf{a})$$

$$\mathbf{r} = \mathbf{a} + t\,(\mathbf{b} - \mathbf{a}) = (1 - t)\,\mathbf{a} + t\mathbf{b}$$

Thus, *the parametric vectorial equation of the line through two points with position vectors* **a** *and* **b** *is*

$$\mathbf{r} = (1 - t)\,\mathbf{a} + t\mathbf{b}.$$

2.5. CONDITION FOR THE COLLINEARITY OF THREE POINTS

The necessary and sufficient condition for three points A, B, C with position vectors **a, b, c** respectively to be collinear is that there exist three scalars x, y, z, not all zero, such that

$$x\mathbf{a} + y\mathbf{b} + z\mathbf{c} = \mathbf{0}, \quad x + y + z = 0.$$

We have $\vec{AB} = \mathbf{b} - \mathbf{a}$, $\vec{AC} = \mathbf{c} - \mathbf{a}$.

The points A, B and C will be collinear if and only if the vectors $\mathbf{b} - \mathbf{a}$ and $\mathbf{c} - \mathbf{a}$ are collinear.

The condition is necessary. Let the points A, B, C be collinear so that the vectors $\vec{AB}$ and $\vec{AC}$ are collinear. Thus, there exists a scalar k such that

$$\mathbf{b} - \mathbf{a} = k(\mathbf{c} - \mathbf{a}) \Rightarrow (k - 1)\mathbf{a} + \mathbf{b} - k\mathbf{c} = \mathbf{0}.$$

Taking $x = k - 1$, $y = 1$, $z = -k$, we see that if the points A, B, C are collinear, then there exists scalars x, y, z not all zero such that

$$x\mathbf{a} + y\mathbf{b} + z\mathbf{c} = \mathbf{0} \text{ and } x + y + z = 0.$$

The condition is sufficient. Let there exist three scalars x, y, z are not zero such that

$$x\mathbf{a} + y\mathbf{b} + z\mathbf{c} = \mathbf{0},\ x + y + z = 0.$$

From these we obtain

$$-(y + z)\mathbf{a} + y\mathbf{b} + z\mathbf{c} = \mathbf{0}$$

$$\Rightarrow y(\mathbf{b} - \mathbf{a}) + z(\mathbf{c} - \mathbf{a}) = \mathbf{0}$$

$$\Rightarrow \mathbf{c} - \mathbf{a} = -yz^{-1}(\mathbf{b} - \mathbf{a}) \text{ if } z \neq 0.$$

Thus, the vectors $\mathbf{c} - \mathbf{a}$ and $\mathbf{b} - \mathbf{a}$ are collinear and as such the points A, B, C are collinear.

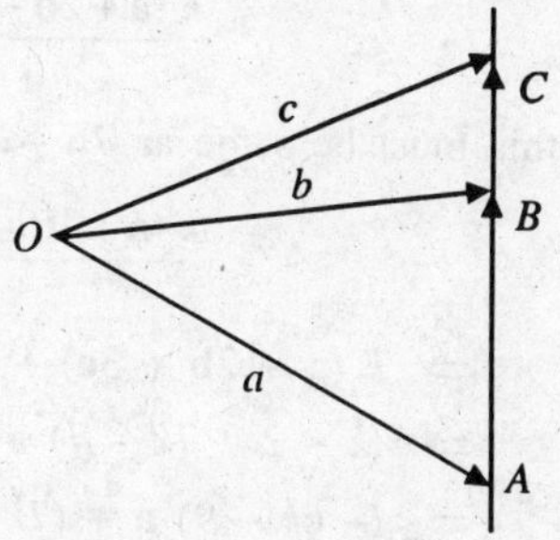

Fig. 2.11

Note. Rewriting the relations $x\mathbf{a} + y\mathbf{b} + z\mathbf{c} = \mathbf{0}$, $x + y + z = 0$ in the form $\mathbf{c} = \dfrac{x\mathbf{a} + y\mathbf{b}}{x + y}$, we may see that the point C divides AB in the ratio $y : x$.

EXAMPLES

Example 1. *Show that the three points A, B, C with position vectors*

$$-2\mathbf{a} + 3\mathbf{b} + 5\mathbf{c},\ \mathbf{a} + 2\mathbf{b} + 3\mathbf{c},\ 7\mathbf{a} - \mathbf{c}$$

are collinear. Find the ratio in which C divides AB. Obtain also the equation of the line.

Solution. Let O be the origin of reference.

We have

$$\vec{AB} = \vec{OB} - \vec{OA} = (\mathbf{a} + 2\mathbf{b} + 3\mathbf{c}) - (-2\mathbf{a} + 3\mathbf{b} + 5\mathbf{c}) = 3\mathbf{a} - \mathbf{b} - 2\mathbf{c}.$$

$$\vec{AC} = \vec{OC} - \vec{OA},$$

$$= (7\mathbf{a} - \mathbf{c}) - (-2\mathbf{a} + 3\mathbf{b} + 5\mathbf{c})$$

$$= (9\mathbf{a} - 3\mathbf{b} - 6\mathbf{c}) = 3\vec{AB}$$

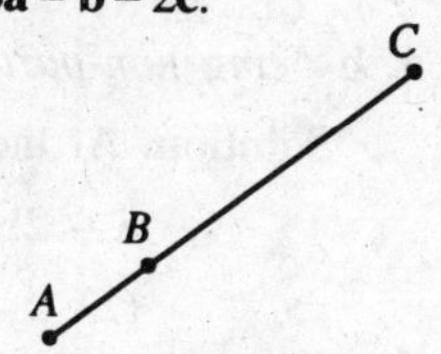

Fig. 2.12

Thus, the vectors $\vec{AB}$, $\vec{AC}$ are collinear. These vectors being also coterminous, the points A, B, C, are collinear.

Also, we have

$$2\,(-2\mathbf{a}+3\mathbf{b}+5\mathbf{c})-3\,(\mathbf{a}+2\mathbf{b}+3\mathbf{c})+(7\mathbf{a}-\mathbf{c})=0$$

where $$2-3+1=0.$$

so that the truth of the result is also seen by following the result of § 2.5.

By § 2.5 note, we may see that C divides AB in the ratio 3 : 2.

Another Method. Let the point C divide AB in the ratio k : 1.

The point dividing AB in the ratio k : 1 being

$$\frac{k\,(\mathbf{a}+2\mathbf{b}+3\mathbf{c})+1\,(-2\mathbf{a}+3\mathbf{b}+5\mathbf{c})}{k+1}$$

this must be same as $7\mathbf{a}-\mathbf{c}$ so that

$$\frac{k\,(\mathbf{a}+2\mathbf{b}+3\mathbf{c})+1\,(-2\mathbf{a}+3\mathbf{b}+5\mathbf{c})}{k+1}=7\mathbf{a}-\mathbf{c}$$

$$\Rightarrow\quad k\,(\mathbf{a}+2\mathbf{b}+3\mathbf{c})+1\,(-2\mathbf{a}+3\mathbf{b}+5\mathbf{c})=(k+1)\,(7\mathbf{a}-\mathbf{c})$$

$$\Rightarrow\quad (k-2-7k-7)\,\mathbf{a}+(2k+3)\,\mathbf{b}+(3k+5+k+1)\,\mathbf{c}=\mathbf{0}$$

$$\Rightarrow\quad (-6k-9)\,\mathbf{a}+(2k+3)\,\mathbf{b}+(4k+6)\,\mathbf{c}=\mathbf{0}$$

$$\Rightarrow\quad -6k-9=0,\;\; 2k+3=0,\;\; 4k+6=\mathbf{0}.$$

These equations are constant in k and

$$k=-\frac{3}{2}.$$

Thus, the points A, B, C are collinear and C divides AB in the ratio

$$-3:2.$$

The equation of the line through the two points A, B is

$$\mathbf{r}=-2\mathbf{a}+3\mathbf{b}+5\mathbf{c}+t\,[(\mathbf{a}+2\mathbf{b}+3\mathbf{c})-(-2\mathbf{a}+3\mathbf{b}+5\mathbf{c})]$$

$$=-2\mathbf{a}+3\mathbf{b}+5\mathbf{c}+t\,(3\mathbf{a}-\mathbf{b}-2\mathbf{c}).$$

It may be seen that the point C arises for $t=3$.

Example 2. *Find the point of intersection of the lines*

$$\mathbf{r}=\mathbf{a}-2\mathbf{b}+\lambda\,(\mathbf{b}+2\mathbf{a}),\quad \mathbf{r}=2\mathbf{a}-\mathbf{b}+\mu\,(\mathbf{a}+2\mathbf{b}),$$

a, b *being non-parallel vectors.*

Solution. At the point of intersection of the lines, we have

$$\mathbf{a}-2\mathbf{b}+\lambda\,(\mathbf{b}+2\mathbf{a})=2\mathbf{a}-\mathbf{b}+\mu\,(\mathbf{a}+2\mathbf{b})$$

$$\Rightarrow\quad (2\lambda-\mu-1)\,\mathbf{a}+(\lambda-2\mu-1)\,\mathbf{b}=\mathbf{0}.$$

As, **a, b** are non-parallel vectors, we have

$$2\lambda - \mu - 1 = 0, \quad \lambda - 2\mu - 1 = 0$$

giving $\lambda = \frac{1}{3}$ and $\mu = -\frac{1}{3}$.

Putting $\lambda = \frac{1}{3}$, in the equation of the line with λ parameter, we see that the position vector of the point of intersection, is $\frac{5}{2}(\mathbf{a} - \mathbf{b})$. It may be verified that for $\mu = -\frac{1}{3}$, we get the same point.

Example 3. *Through the middle point M of the side AD of a parallelogram ABCD, the straight line BM is drawn cutting AC at R and CD produced at Q; prove that*

$$QR = 2RB.$$

Solution. Take A as the origin of reference.

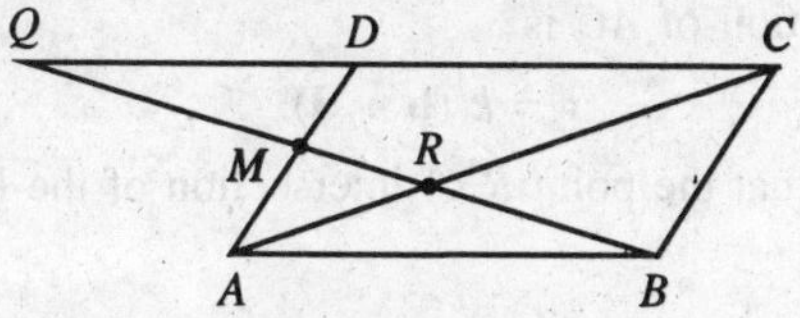

Fig. 2.13

Let **b, d** be the position vectors of the point *B* and *D* so that

$$\vec{AB} = \mathbf{b}, \; \vec{AD} = \mathbf{d}.$$

The position vector of *M* is $\frac{1}{2}\mathbf{d}$ and that of *C* is

$$= \vec{AB} + \vec{BC} = \vec{AB} + \vec{AD} = \mathbf{b} + \mathbf{d}.$$

The equation of the line *BM* joining the points, B, M with position vectors **b**, $\frac{1}{2}\mathbf{d}$ is

$$\mathbf{r} = \mathbf{b} + t\left(\frac{1}{2}\mathbf{d} - \mathbf{b}\right) \qquad ...(i)$$

and the equation of the line *DC* through the point *D* with position vector **d** and parallel to $\vec{AB} = \mathbf{b}$ is

$$\mathbf{r} = \mathbf{d} + p\mathbf{b}. \qquad ...(ii)$$

At the point of intersection of the lines (*i*) and (*ii*), we have

$$\mathbf{b}+t\left(\frac{1}{2}\mathbf{d}-\mathbf{b}\right)=\mathbf{d}+p\mathbf{b}$$

$$\Rightarrow \quad (1-t-p)\,\mathbf{b}+\left(\frac{1}{2}t-1\right)\mathbf{d}=0$$

$$\Rightarrow \quad 1-t-p=0,\ \frac{1}{2}t-1=0$$

$$\Rightarrow \quad t=2,\ \ p=-1;$$

the vectors **b, d** being non-collinear.

Substituting the value of t in (*i*) or of p in (*ii*), we see that the position vector of the point Q of intersection of the lines BM and DC is

$$\mathbf{d}-\mathbf{b}=\overrightarrow{AQ}.$$

We shall now find the position vector of the point R of intersection of the lines AC and BM.

Again the equation of AC is

$$\mathbf{r}=k\,(\mathbf{b}+\mathbf{d}).$$

We may show that the point R of intersection of the lines AC and BM is

$$\frac{1}{3}(\mathbf{b}+\mathbf{d})=\overrightarrow{AR}$$

It follows that

$$\begin{cases}\overrightarrow{QR}=\overrightarrow{AR}-\overrightarrow{AQ}=\frac{4}{3}\mathbf{b}-\frac{2}{3}\mathbf{d}=2\left(\frac{2}{3}\mathbf{b}-\frac{1}{3}\mathbf{d}\right)\\ \overrightarrow{RB}=\overrightarrow{AB}-\overrightarrow{AR}=\mathbf{b}-\frac{1}{3}(\mathbf{b}+\mathbf{d})=\frac{2}{3}\mathbf{b}-\frac{1}{3}\mathbf{d},\end{cases}$$

Thus $$\overrightarrow{QR}=2\overrightarrow{RB}\Rightarrow QR=2RB.$$

Example 4. *In a triangle ABC, D divides BC in the ratio 3 : 2 and E divides CA in the ratio 1 : 3. The lines AD and BE meet at H and CH meets AB in F. Find the ratio in which F divides AB.*

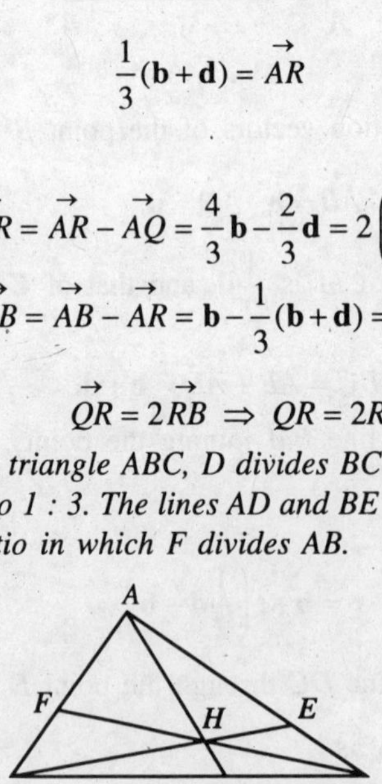

Fig. 2.14

Solution. Take C as the origin of reference. Let **a, b** be the position vectors of the points A and B respectively. Now the point D is $\frac{2}{5}\mathbf{b}$ and the point E is $\frac{1}{4}\mathbf{a}$.

The equation of AD and BE are

$$\mathbf{r} = \mathbf{a} + k\left(\frac{2}{5}\mathbf{b} - \mathbf{a}\right), \text{ and } \mathbf{r} = \mathbf{b} + t\left(\frac{1}{4}\mathbf{a} - \mathbf{b}\right) \text{ respectively.}$$

For the point of intersection H, we have

$$\mathbf{a} + k\left(\frac{2}{5}\mathbf{b} - \mathbf{a}\right) = \mathbf{b} + t\left(\frac{1}{4}\mathbf{a} - \mathbf{b}\right) \qquad \text{...}(i)$$

$$\Rightarrow \quad 1 - k = \frac{t}{4}, \; \frac{2}{5}k = 1 - t \;\Rightarrow\; k = \frac{5}{6}; \; t = \frac{2}{3}.$$

Putting these values of k and t in the equations of the lines AD and BE, we see that the point H is

$$\frac{1}{6}\mathbf{a} + \frac{1}{3}\mathbf{b}.$$

The equations of CH and AB being

$$\mathbf{r} = t\left(\frac{\mathbf{a}}{6} + \frac{\mathbf{b}}{3}\right), \text{ and } \mathbf{r} = \mathbf{a} + k(\mathbf{b} - \mathbf{a}),$$

we have for the point of intersection F,

$$t\left(\frac{\mathbf{a}}{6} + \frac{\mathbf{b}}{3}\right) = \mathbf{a} + k(\mathbf{b} - \mathbf{a})$$

$$\Rightarrow \quad \left(\frac{t}{6} - 1 + k\right) = 0, \; \left(\frac{t}{3} - k\right) = 0.$$

These give $t = 2$; $k = \frac{2}{3}$.

Thus, the point F is

$$\frac{\mathbf{a} + 2\mathbf{b}}{3},$$

so that F divides AB in the ratio 2 : 1.

Example 5. *Find the equations of the bisectors of the angles between the lines*

$$\mathbf{r} = \mathbf{a} + t\mathbf{b}, \quad \mathbf{r} = \mathbf{a} + p\mathbf{c}.$$

Solution. Let *AL*, *AM* be the given lines so that **a** is the position vector of the point *A*. Let *P*, *Q*, *R* be three points at unit distances from *A* as in the figure. Let *E*, *F* be the mid-points of *PQ*, *QR*.

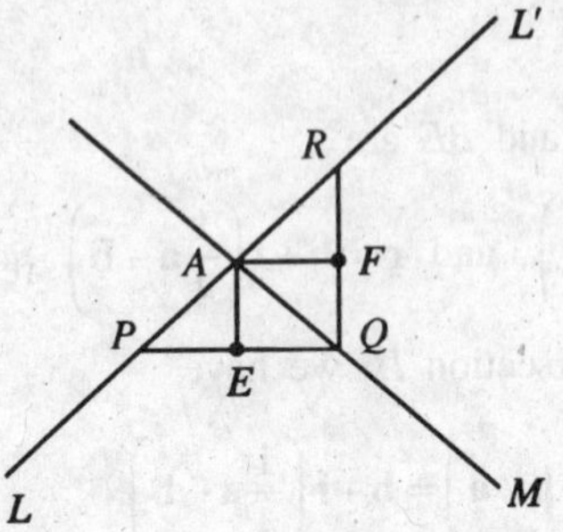

Fig. 2.15

Then *AE*, *AF* are the bisectors of the angles between the given lines. The position vectors of *P*, *Q*, *R* are

$$\mathbf{a}+\frac{\mathbf{b}}{|\mathbf{b}|},\ \mathbf{a}+\frac{\mathbf{c}}{|\mathbf{c}|},\ \mathbf{a}-\frac{\mathbf{b}}{|\mathbf{b}|}$$

respectively, implying that the position vectors of *E*, *F* are

$$\mathbf{a}+\frac{1}{2}\left(\frac{\mathbf{b}}{|\mathbf{b}|}+\frac{\mathbf{c}}{|\mathbf{c}|}\right),\ \mathbf{a}+\frac{1}{2}\left(\frac{\mathbf{c}}{|\mathbf{c}|}-\frac{\mathbf{b}}{|\mathbf{b}|}\right).$$

Thus, the required equations of the bisectors *AE*, *AF* are

$$\mathbf{r}=\mathbf{a}+t\left(\frac{\mathbf{b}}{|\mathbf{b}|}+\frac{\mathbf{c}}{|\mathbf{c}|}\right),\ \mathbf{r}=\mathbf{a}+p\left(\frac{\mathbf{c}}{|\mathbf{c}|}-\frac{\mathbf{b}}{|\mathbf{b}|}\right).$$

respectively, where *t* and *p* are scalar parameters.

Example 6. *Show that the internal (external) bisector of any angle of a triangle divides the base internally (externally) in the ratio of the sides containing the angle.*

Solution. Let the lengths of the sides *BC*, *CA*, *AB* of a triangle *ABC* be denoted by α, β, γ respectively.

Take *A* as the origin of reference. Let **b**, **c** be the position vectors of the points *B* and *C* so that

$$\overrightarrow{AB}=\mathbf{b},\ \overrightarrow{AC}=\mathbf{c}.$$

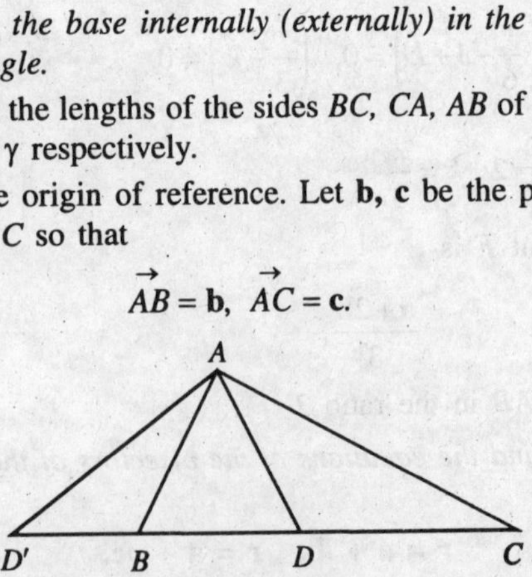

Fig. 2.16

We have

$$|\overrightarrow{AB}| = \gamma, \ |\overrightarrow{AC}| = \beta.$$

Now $\dfrac{\mathbf{b}}{\gamma}, \dfrac{\mathbf{c}}{\beta}$ are unit vectors along *AB* and *AC* so that the equation of the internal bisector *AD* is

$$\mathbf{r} = t\left(\frac{\mathbf{b}}{\gamma} + \frac{\mathbf{c}}{\beta}\right). \qquad ...(i)$$

Also the equation of the line BC joining the points B and C with position vectors **b** and **c** is

$$\mathbf{r} = \mathbf{b} + p\,(\mathbf{c} - \mathbf{b}). \qquad ...(ii)$$

At the point of intersection *D* of the lines *AD* and *BC*, we have

$$t\left(\frac{\mathbf{b}}{\gamma} + \frac{\mathbf{c}}{\beta}\right) = \mathbf{b} + p\,(\mathbf{c} - \mathbf{b})$$

$$\Rightarrow \qquad \left(\frac{t}{\gamma} - 1 + p\right) = \mathbf{b} + \left(\frac{t}{\beta} - p\right)\mathbf{c} = 0.$$

As the vector **b, c** are non-parallel, we have

$$\frac{t}{\gamma} - 1 + p = 0, \ \frac{t}{\beta} - p = 0$$

$$\Rightarrow \qquad t = \frac{\beta\gamma}{\beta + \gamma}, \ p = \frac{\gamma}{\beta + \gamma}.$$

Substituting the value of *t* in (*i*) or of *p* in (*ii*), we see that the position vector of the point of intersection *D* of *AD* and *BC* is

$$\frac{\beta\mathbf{b} + \gamma\mathbf{c}}{\beta + \gamma}$$

which divides *BC* in the ratio $\gamma : \beta$, *i.e.*, *AB* : *AC*.

Considering the external bisector *AD′*

$$\mathbf{r} = t\left(\frac{\mathbf{c}}{\beta} - \frac{\mathbf{b}}{\gamma}\right),$$

we may similarly prove the part corresponding to the external bisector.

Example 7. *ABC is a triangle; AD, AD′ are the bisectors of the angle A meeting BC in D, D′ respectively; A′ is the mid-point of DD′; B′, C′, are the points on CA and AB similarly obtained. Show that the points A′, B′, C′, are collinear.*

Let **a, b, c** be the position vectors of the vertices *A, B, C* with respect to any origin of reference *O*. Let the lengths of the sides *BC, CA, AB* of the triangle be α, β, γ.

The position vectors of *D, D′* are

$$\frac{\beta\mathbf{b}+\gamma\mathbf{c}}{\beta+\gamma},\ \frac{\beta\mathbf{b}-\gamma\mathbf{c}}{\beta-\gamma}$$

respectively, so that the position vector of the mid point *A′* of *D D′* is

$$\frac{1}{2}\left(\frac{\beta\mathbf{b}+\gamma\mathbf{c}}{\beta-\gamma}+\frac{\beta\mathbf{b}-\gamma\mathbf{c}}{\beta-\gamma}\right)=\frac{\beta^2\mathbf{b}-\gamma^2\mathbf{c}}{\beta^2-\gamma^2}=a',\ \text{say.}$$

By symmetry, the position vectors of *B′*, *C′* are

$$\frac{\gamma^2\mathbf{c}-\alpha^2\mathbf{a}}{\gamma^2-\alpha^2}=\mathbf{b}',\ \text{say, and}\ \frac{\alpha^2\mathbf{a}-\beta^2\mathbf{b}}{\alpha^2-\beta^2}=\mathbf{c}',\ \text{say.}$$

We now have

$$(\beta^2-\gamma^2)\,\mathbf{a}'+(\gamma^2-\alpha^2)\,\mathbf{b}'+(\alpha^2+\beta^2)\,\mathbf{c}'=0$$

where

$$(\beta^2-\gamma^2)+(\gamma^2-\alpha^2)+(\alpha^2+\beta^2)=0$$

It follows that the points *A′*, *B′*, *C′*, with position vectors **a′**, **b′**, **c′** are collinear (§ 2.5 Page 32).

Example 8. *Show that the internal bisectors of the angles of a triangle are concurrent.*

Solution. Take any point *O* as the origin of reference. Let **a, b, c** be the position vectors of the vertices *A, B, C* and α, β, γ the lengths of the sides of the triangle. The position vector of the point *D* where the internal bisector of the angle *A* meets *BC* is

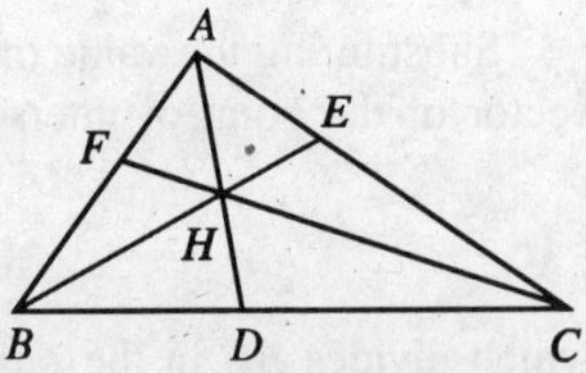

Fig. 2.17

$$\frac{\beta\mathbf{b}+\gamma\mathbf{c}}{\beta+\gamma}.$$

The position vector of the point dividing *AD* in the ratio (β + γ) / α is

$$\frac{\alpha\mathbf{a}+\beta\mathbf{b}+\gamma\mathbf{c}}{\alpha+\beta+\gamma}. \qquad ...(i)$$

By symmetry, we see that this point lies on the other two bisectors also.

Thus, the bisectors are concurrent and the point of concurrence is given by (*i*).

EXERCISES

1. Find the points of intersection of the following pairs of lines, assuming that the vectors **a** and **b** are not parallel.

(*i*) $\mathbf{r} = \gamma\,(\mathbf{b} - \mathbf{a})$, $\mathbf{r} = 2\mathbf{b} + \mu\mathbf{a}$.

(*ii*) $\mathbf{r} = \gamma\,(\mathbf{b} + \mathbf{a})$, $\mathbf{r} = \mu\,(\mathbf{b} - \mathbf{a})$.

(*iii*) $\mathbf{r} = (6\mathbf{b} - \mathbf{a}) + \gamma\,(2\mathbf{a} - \mathbf{b})$, $\mathbf{r} = (\mathbf{b} - \mathbf{a}) + \mu\,(\mathbf{b} + 3\mathbf{a})$

(*iv*) $\mathbf{r} = 2\mathbf{a} + 2\mathbf{b} + \gamma\,(\mathbf{a} - \mathbf{b})$, $\mathbf{r} = 3\mathbf{a} + \mu\,(\mathbf{b} - \mathbf{a})$.

(*v*) $\mathbf{r} = \mathbf{a} + \mu\mathbf{b}$, $\mathbf{r} = \mathbf{b} + \gamma\mathbf{a}$.

2. Test for collinearity the sets of points with the following position vectors : (Kanpur 93)

(*i*) $\mathbf{a} - 2\mathbf{b} + 3\mathbf{c}$, $2\mathbf{a} + 3\mathbf{b} - 4\mathbf{c}$, $-7\mathbf{b} + 10\mathbf{c}$.

(*ii*) $3\mathbf{a} - 4\mathbf{b} + 3\mathbf{c}$, $-4\mathbf{a} + 5\mathbf{b} - 6\mathbf{c}$, $4\mathbf{a} - 7\mathbf{b} + 6\mathbf{c}$.

(*iii*) $2\mathbf{a} + 5\mathbf{b} - 4\mathbf{c}$, $\mathbf{a} + 4\mathbf{b} - 3\mathbf{c}$, $4\mathbf{a} + 7\mathbf{b} - 6\mathbf{c}$.

(*iv*) $5\mathbf{a} + 4\mathbf{b} + 2\mathbf{c}$, $6\mathbf{a} + 2\mathbf{b} - \mathbf{c}$, $7\mathbf{a} + \mathbf{b} - \mathbf{c}$.

3. Show that the lines

$$\mathbf{r} = \mathbf{a} + t\,(\mathbf{b} + \mathbf{c}), \quad \mathbf{r} = \mathbf{b} + t\,(\mathbf{c} + \mathbf{a})$$

intersect and find also their point of intersection.

4. In a $\Delta\, OAB$, E is the mid-point of AB and F is the point in OA such that $OF = 2\,FA$. Calculate the ratios $OC : CE$ and $BC : CF$ where C is the point of intersection of OE and BF.

5. Find the position vector of the point P of intersection of the lines l_1 and l_2 with vector equations

$$r = 2\mathbf{a} + \lambda\,(\mathbf{b} - 3\mathbf{a}) \text{ and } r = 3\,(\mathbf{a} - \mathbf{b}) - \mu\,(\mathbf{a} + \mathbf{b}),$$

where **a, b** are non-parallel vectors. Q is the point in l_1 when $\lambda = 1$ and R is the point in l_2 when $\mu = 1$. If $PQSR$ is a parallelogram, give the vector equations of the lines QS and RS as also the position vector of the point S.

6. In a $\Delta\, OAB$, C is the mid-point of AB and D is the mid-point of OB. The line OC meets AD at G. Calculate the ratio in which G divides OC and AD. If E is the point in which OA meets BG, calculate the ratio in which G divides BE and show that E is the mid-point of OA.

7. The points O, A, B, X and Y are such that $\overrightarrow{OA} = \mathbf{a}$, $\overrightarrow{OB} = \mathbf{b}$, $\overrightarrow{OX} = 3\mathbf{a}$ and $\overrightarrow{OY} = 3\mathbf{b}$. Express $\overrightarrow{BX}$ and $\overrightarrow{AY}$ in terms of **a** and **b**.

The point P divides AY in the ratio $1 : 3$. Express $\overrightarrow{BP}$ in terms of **a** and **b** and show that P lies on BX.

8. In a triangle ABC, D divides BC in the ratio $2 : 1$ and E divides CA in the ratio $3 : 1$. The line DE meets the line AB in F. Find the ratio in which the point F divides AB.

9. The median AD of a triangle ABC is bisected at E and BE is produced to meet the side AC in F; show that

$$AF = \frac{1}{3}AC \text{ and } EF = \frac{1}{4}BF.$$

10. $ABCD$ is a parallelogram and L, M are the mid-points of AB and CD respectively; show that DL and BM trisect AC and are trisected by AC.

11. A line EF drawn parallel to the base BC of a triangle ABC meets AB and AC in F and E respectively; BE and CF meet in L. Show that AL bisects BC.

12. The sides CA, AB of a triangle ABC are divided internally in the same ratio at E and F; show that EF divides BC externally in the square of this ratio.

13. Points Y and Z are taken on the sides CA, AB of a triangle ABC such that $CY = YA$ and $AZ = 2\,ZB$; BY and CZ meet at P, show that $CP = 3\,PZ$. Also find the ratio in which AP divides BC.

14. A, B, C and A', B', C' are two sets of points on two skew lines such that

$AB : BC = A'B' : B'C'$.

Show that the middle points of AA', BB', CC are collinear.

2.6. PARAMETRIC VECTORIAL EQUATION OF A PLANE

To find the parametric vectorial equation of the plane which passes through a given point and is parallel to two given lines.

Let AB, AC be two lines through the point A parallel to the vector **b** and **c**.

Take any point O, as the origin of reference. Let, **a**, be the position vector of the given point A. Let, **r**, be the position vector of any point **P** on the given plane.

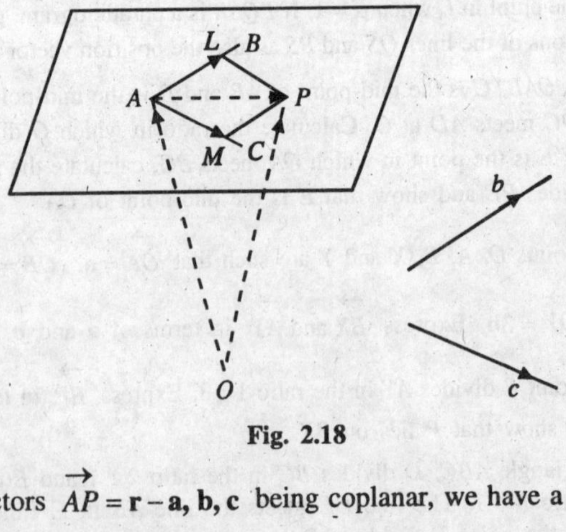

Fig. 2.18

The vectors $\overrightarrow{AP} = \mathbf{r} - \mathbf{a}, \mathbf{b}, \mathbf{c}$ being coplanar, we have a relation of the form

$$\mathbf{r} - \mathbf{a} = t\mathbf{b} + p\mathbf{c}$$

$$\Rightarrow \quad \mathbf{r} = \mathbf{a} + t\mathbf{b} + p\mathbf{c}$$

where t and p are scalars.

Here each point of the plane arises for some values of the scalars t, p. Also conversely, it is clear that for arbitrary scalar values of t and p, $\mathbf{a} + t\mathbf{b} + p\mathbf{c}$ is the position of a point on the plane.

Thus, the parametric vector equation of the plane which passes through the point with position vector **a**, *and which is parallel to the vectors* **b** *and* **c** *is*

$$\mathbf{r} = \mathbf{a} + t\mathbf{b} + p\mathbf{c},$$

where t, p are scalars parameters.

Cor. 1. Plane through two given points and parallel to a given line. If **a**, **b** are the position vectors of the given points A, B and **c** is a vector parallel to the given plane, then the vectors $\overrightarrow{AP} = r - a$, $\overrightarrow{AB} = b - a$, and **c** being coplanar, we see that

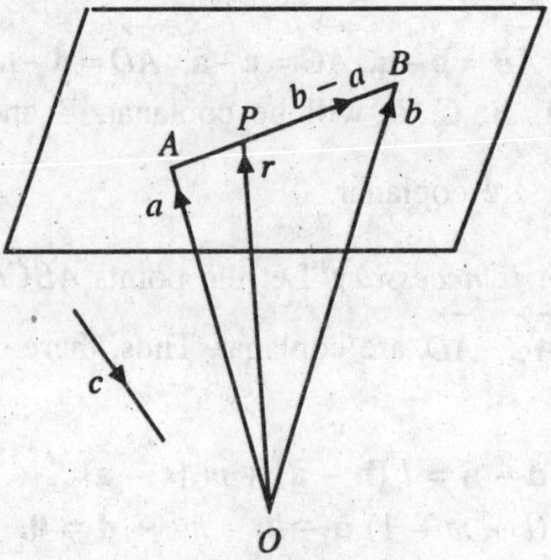

Fig. 2.19

$$\mathbf{r} = \mathbf{a} + t\,(\mathbf{b} - \mathbf{a}) + p\mathbf{c} = (1 - t)\,\mathbf{a} + t\mathbf{b} + p\mathbf{c}$$

is the required equation. (Kanpur 95)

Cor. 2. Plane through three given points. If **a, b, c** be the position vectors of three given points A, B, C, then the vectors

$$\overrightarrow{AP} = \mathbf{r} - \mathbf{a},\ \overrightarrow{AB} = \mathbf{b} - \mathbf{a},\ \overrightarrow{AC} = \mathbf{c} - \mathbf{a},$$

being coplanar, the equation of the plane is

$$\mathbf{r} = \mathbf{a} + t\,(\mathbf{b} - \mathbf{a}) + p\,(\mathbf{c} - \mathbf{a}) = (1 - t - p)\,\mathbf{a} + t\mathbf{b} + p\mathbf{c}$$

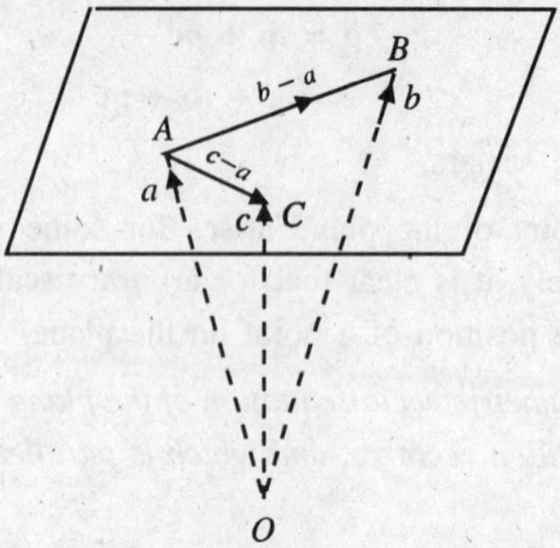

Fig. 2.20

2.7. CONDITION FOR THE COPLANARITY OF FOUR POINTS

The necessary and sufficient condition for four points A, B, C, D with position vectors **a, b, c, d** *to be coplanar is that there exists four scalars x, y, z, t not all zero, such that*

$$x\mathbf{a} + y\mathbf{b} + z\mathbf{c} + t\mathbf{d} = \mathbf{0}, \quad x + y + z + t = 0.$$

We have

$$\vec{AB} = \mathbf{b} - \mathbf{a}, \ \vec{AC} = \mathbf{c} - \mathbf{a}, \ \vec{AD} = \mathbf{d} - \mathbf{a}.$$

The points *A, B, C, D* will be coplanar, if and only if the vectors $\vec{AB}$, $\vec{AC}$ and $\vec{AD}$ are coplanar.

The condition is necessary. Let the points *ABCD* be coplanar so that the vectors $\vec{AB}, \vec{AC}, \vec{AD}$ are coplanar. Thus, there exist scalars *l, m* such that

$$\mathbf{d} - \mathbf{a} = l\,(\mathbf{b} - \mathbf{a}) + m\,(\mathbf{c} - \mathbf{a})$$

$$\Rightarrow \quad (l + m - 1)\,\mathbf{a} - l\mathbf{b} - m\mathbf{c} + \mathbf{d} = \mathbf{0}.$$

Taking $x = l + m - 1$, $y = -l$, $z = -m$, $t = 1$, we see that if the points *A, B, C, D* are coplanar, there exist scalars *x, y, z, t* not all zero such that $x\mathbf{a} + y\mathbf{b} + z\mathbf{c} + t\mathbf{d} = \mathbf{0}$ and $x + y + z + t = 0$.

The condition is sufficient. Let there exist scalars *x, y, z, t* not all zero such that

$$x\mathbf{a} + y\mathbf{b} + z\mathbf{c} + t\mathbf{d} = \mathbf{0}, \quad x + y + z + t = 0.$$

From these we obtain

$$-(y + z + t)\,\mathbf{a} + y\mathbf{b} + z\mathbf{c} + t\mathbf{d} = \mathbf{0}$$

$$\Rightarrow \quad \mathbf{d} - \mathbf{a} = (-y\,t^{-1})\,(\mathbf{b} - \mathbf{a}) + (-z\,t^{-1})\,(\mathbf{c} - \mathbf{a}) \quad \text{if } y \neq 0$$

so that **d − a, b − a, c − a** are coplanar vectors and as such *A, B, C, D* are coplanar points.

EXAMPLES

Example 1. *Show that the points with position vectors*

$$-6\mathbf{a} + 3\mathbf{b} + 2\mathbf{c}, \quad 3\mathbf{a} - 2\mathbf{b} + 4\mathbf{c}, \quad 5\mathbf{a} + 7\mathbf{b} + 3\mathbf{c}, \quad -13\mathbf{a} + 17\mathbf{b} - \mathbf{c}$$

are coplanar.

Solution. Denoting the given points by *A, B, C, D* respectively, we obtain

$$\vec{AB} = \vec{OB} - \vec{OA} = (3\mathbf{a} - 2\mathbf{b} + 4\mathbf{c}) - (-6\mathbf{a} + 3\mathbf{b} + 2\mathbf{c})$$
$$= 9\mathbf{a} - 5\mathbf{b} + 2\mathbf{c}.$$
$$\vec{AC} = 11\mathbf{a} + 4\mathbf{b} + \mathbf{c}.$$
$$\vec{AD} = -7\mathbf{a} + 14\mathbf{b} - 3\mathbf{c}.$$
$$\vec{AB} - 2\vec{AC} = -13\mathbf{a} - 13\mathbf{b}.$$
$$\vec{AD} + 3\vec{AC} = 26\mathbf{a} + 26\mathbf{b} = -2(\vec{AB} - 2\vec{AC})$$
$$\Rightarrow \quad \vec{AD} + 3\vec{AC} = -2\vec{AB} + 4\vec{AC}$$
$$\Rightarrow \quad -2\vec{AB} + \vec{AC} = \vec{AD}.$$

Thus, the vectors $\vec{AB}, \vec{AC}, \vec{AD}$ are coplanar. These three vectors, being also coterminous, we see that the four points *A, B, C, D* are coplanar.

This result may also be proved with the help of the result of § 2.7. The student may find the equation of the plane through the points *A, B, C* and show that the point *D* lies on it.

Example 2. *Show that the lines*

$$\mathbf{r} = 8\mathbf{a} - 9\mathbf{b} + 10\mathbf{c} + t\,(3\mathbf{a} - 16\mathbf{b} + 7\mathbf{c}),$$
$$\mathbf{r} = 15\mathbf{a} + 29\mathbf{b} + 5\mathbf{c} + p\,(3\mathbf{a} + 8\mathbf{b} - 5\mathbf{c}),$$

are non-coplanar; **a, b, c** *being non-coplanar vectors.*

Solution. We shall show that the two lines have no point in common.

Assuming that the lines have a point in common, we have

$$8\mathbf{a} - 9\mathbf{b} + 10\mathbf{c} + t\,(3\mathbf{a} - 16\mathbf{b} + 7\mathbf{c}),$$
$$= 15\mathbf{a} + 29\mathbf{b} + 5\mathbf{c} + p\,(3\mathbf{a} + 8\mathbf{b} - 5\mathbf{c})$$
$$\Rightarrow (8 + 3t - 15 - 3p)\,\mathbf{a} + (-9 - 16t - 29 - 8p)\,\mathbf{b}$$
$$+ (10 + 7t - 5 + 5p)\,\mathbf{c} = \mathbf{0}$$
$$\Rightarrow (-7 + 3t - 3p)\,\mathbf{a} + (-38 - 16t - 8p)\,\mathbf{b} + (5 + 7t + 5p)\,\mathbf{c} = \mathbf{0}.$$

As **a, b, c** are non-coplanar vectors, it follows that

$$-7 + 3t - 3p = 0, \quad -38 - 16t - 8p = 0, \quad 5 + 7t + 5p = 0.$$

It may be shown that these three equations in the two unknowns *t* and *p* are not consistent.

Thus, the two given lines have no point in common and hence are non-coplanar. These lines are also not parallel.

Example 3. *Assuming that the vectors* **a** *and* **c** *are not collinear, examine the following lines for intersection :*

$$\mathbf{r} = 6\mathbf{a} - \mathbf{c} + \lambda\,(2\mathbf{c} - \mathbf{a});\;\; \mathbf{r} = \mathbf{a} - \mathbf{c} + \mu\,(\mathbf{a} + 3\mathbf{c}).$$

Solution. Assuming that the lines intersect, there must exist values of λ and μ such that

$$6\mathbf{a} - \mathbf{c} + \lambda\,(2\mathbf{c} - \mathbf{a}) = \mathbf{a} - \mathbf{c} + \mu\,(\mathbf{a} + 3\mathbf{c})$$

$$\Rightarrow \quad (6 - \lambda - \mu - 1)\,\mathbf{a} + (2\lambda + 1 + 1 - 3\mu)\,\mathbf{c} = \mathbf{0}$$

$$\Rightarrow \quad (5 - \lambda - \mu)\,\mathbf{a} + (2\lambda - 3\mu)\,\mathbf{c} = \mathbf{0}$$

$$\Rightarrow \quad \left.\begin{array}{r} 5 - \lambda - \mu = 0 \\ 2\lambda - 3\mu = 0 \end{array}\right\}$$

$$\Rightarrow \quad \left.\begin{array}{r} -\lambda - \mu = -5 \\ 2\lambda - 3\mu = 0 \end{array}\right\}$$

These give $\lambda = 3$; $\mu = 2$.

Putting the values of λ and μ in the given equations, we see that the lines intersect and the point of intersection is

$$3\mathbf{a} + 5\mathbf{c}.$$

Also therefore the lines are coplanar.

Example 4. *Find the point of intersection of the line*

$$\mathbf{r} = 2\mathbf{a} + \mathbf{b} + t\,(\mathbf{b} - \mathbf{c})$$

and the plane

$$\mathbf{r} = \mathbf{a} + \lambda\,(\mathbf{b} + \mathbf{c}) + \mu\,(\mathbf{a} + 2\mathbf{b} - \mathbf{c})$$

Solution. At the point of intersection of the line and the plane, we have

$$2\mathbf{a} + \mathbf{b} + t\,(\mathbf{b} - \mathbf{c}) = \mathbf{a} + \lambda\,(\mathbf{b} + \mathbf{c}) + \mu\,(a + 2\mathbf{b} - \mathbf{c})$$

$$\Rightarrow \quad (2 - 1 - \mu)\,\mathbf{a} + (1 + t - \lambda - 2\mu)\,\mathbf{b} + (-t + \lambda + \mu)\,\mathbf{c} = \mathbf{0}$$

$$\Rightarrow \quad 1 - \mu = 0,\;\; 1 + t - \lambda - 2\mu = 0,\;\; -t - \lambda + \mu = 0$$

These give $\mu = 1$; $\lambda = 0$; $t = 1$.

Putting the value of t in the given equation of the line, we see that the position vector of the point of intersection is

$$2\mathbf{a} + 2\mathbf{b} - \mathbf{c}.$$

The same result may be obtained by putting the values of λ and μ in the equation of the plane.

Example 5. *Obtain the vector equation of the line of intersection of the planes*

$$\mathbf{r} = \mathbf{b} + \lambda_1\,(\mathbf{b} - \mathbf{a}) + \mu_1\,(\mathbf{a} + \mathbf{c});\;\; \mathbf{r} = \mathbf{c} + \lambda_2\,(\mathbf{b} - \mathbf{c}) + \mu_2\,(\mathbf{a} + \mathbf{b});$$

a, b, c *being non-coplanar vectors.*

Solution. At the point of intersection of the two planes, we have

$$\mathbf{b} + \lambda_1\,(\mathbf{b} - \mathbf{a}) + \mu_1\,(\mathbf{a} + \mathbf{c}) = \mathbf{c} + \lambda_2\,(\mathbf{b} - \mathbf{c}) + \mu_2\,(\mathbf{a} + \mathbf{b})$$

$$\Rightarrow \quad (-\lambda_1 + \mu_1 - \mu_2)\,\mathbf{a} + (1 - \lambda_1 + \lambda_2 + \mu_2)\,\mathbf{b}$$

$+ (\mu_1 - 1 + \lambda_2)$ **c** = **0**.

As **a, b, c** form a linearly independent system, we have

$$-\lambda_1 + \mu_1 - \mu_2 = 0,\ \ 1 + \lambda_1 - \lambda_2 + \mu_2 = 0,\ \ \mu_1 - 1 + \lambda_2 = 0.$$

We have 3 equations in four unknowns $\lambda_1, \mu_1, \lambda_2, \mu_2$.

We now eliminate λ_2, μ_2 and find a relation between λ_1 and μ_1. We may obtain

$$\lambda_1 = 0.$$

Writing $\lambda_1 = 0$, we may see that the line in equation is $\mathbf{r} = \mathbf{b} + \mu_1 (\mathbf{a} + \mathbf{c})$ where μ_1 is the parameter.

We also have $\lambda_2 = 1 - \mu_1$.

Writing $\lambda_2 = 1 - \mu_2$ and $\mu_2 = \mu_1$ we see that the vectorial equation of the line may be described as

$$\mathbf{r} = \mathbf{c} + (1 - \mu_1)(\mathbf{b} - \mathbf{c}) + \mu_1 (\mathbf{a} + \mathbf{b})$$

$$\Rightarrow \quad \mathbf{r} = \mathbf{b} + \mu_2 (\mathbf{a} + \mathbf{c}).$$

which is the same result as obtained by putting $\lambda_1 = 0$ in the first equation of the plane.

Example 6. *Find the vector equation of the line of intersection of the planes*

$$\mathbf{r} = \lambda_1 (\mathbf{a} + \mathbf{b}) + \mu_1 (\mathbf{a} - \mathbf{c})$$

and

$$\mathbf{r} = -2\mathbf{b} + \lambda_2 (\mathbf{a} + 2\mathbf{b} - \mathbf{c}) + \mu_2 \mathbf{a}.$$

Solution. For points on the line of intersection, we have

$$\lambda_1 (\mathbf{a} + \mathbf{b}) + \mu_1 (\mathbf{a} - \mathbf{c}) = -2\mathbf{b} + \lambda_2 (\mathbf{a} + 2\mathbf{b} - \mathbf{c}) + \mu_2 \mathbf{a}$$

$$\Rightarrow (\lambda_1 + \mu_1 - \lambda_2 - \mu_2)\,\mathbf{a} + (\lambda_1 + 2 - 2\lambda_2)\,\mathbf{b} + (-\mu_1 + \lambda_2)\,\mathbf{c} = \mathbf{0}.$$

$$\Rightarrow \left.\begin{array}{r} \lambda_1 + \mu_1 - \lambda_2 - \mu_2 = 0 \\ \lambda_1 - 2\lambda_2 + 2 = 0 \\ -\mu_1 + \lambda_2 = 0 \end{array}\right\}$$

Here we have 3 linear equations in four unknowns $\lambda_1, \lambda_2, \mu_1$ and μ_2. Solving these we get

$$\lambda_2 = k\ ;\ \lambda_1 = 2k - 2\ ;\ \mu_2 = 2k - 2. \quad \text{Here } \mu_1 = k.$$

Putting the values of l_1 and m_1 in the first equation of the plane, we have

$$\mathbf{r} = (2k - 2)(\mathbf{a} + \mathbf{b}) + k(\mathbf{a} - \mathbf{c})$$

$$= (3k - 2)\,\mathbf{a} + (2k - 2)\,\mathbf{b} - k\mathbf{c} \qquad ...(i)$$

Also putting the values of λ_2 and μ_2 in the second equation of the plane, we obtain

$$\mathbf{r} = -2b + k(\mathbf{a} + 2\mathbf{b} - \mathbf{c}) + (2k - 2)\,\mathbf{a}$$

$$= (3k - 2)\,\mathbf{a} + (2k - 2)\,\mathbf{b} - k\mathbf{c}.$$

Thus, the required vector equation of the line of the intersection of the given planes is

$$\mathbf{r} = -\,2\mathbf{a} - 2\mathbf{b} + k\,(3\mathbf{a} + 2\mathbf{b} - \mathbf{c}).$$

Example 7. *Show that four points* $\mathbf{r}_i = l_i\mathbf{a} + m_i\mathbf{b} + n_i\mathbf{c}$, $i = 1, 2, 3, 4$ *are coplanar if and only if*

$$\begin{vmatrix} l_1 & l_2 & l_3 & l_4 \\ m_1 & m_2 & m_3 & m_4 \\ n_1 & n_2 & n_3 & n_4 \\ 1 & 1 & 1 & 1 \end{vmatrix} = 0.$$

Solution. If the given points are coplanar, then there must exist a relation of the form

$$x\,(l_1\mathbf{a} + m_1\mathbf{b} + n_1\mathbf{c}) + y\,(l_2\mathbf{a} + m_2\mathbf{b} + n_2\mathbf{c}) + z\,(l_3\mathbf{a} + m_3\mathbf{b} + n_3\mathbf{c}) + t\,(l_4\mathbf{a} + m_4\mathbf{b} + n_4\mathbf{c}) = 0 \quad ..(1)$$

where $x + y + z + t = 0$ ...(2)

(1) $\Rightarrow$ $(xl_1 + yl_2 + zl_3 + tl_4)\,\mathbf{a} + (xm_1 + ym_2 + zm_3 + tm_4)\,\mathbf{b} + (xn_1 + yn_2 + zn_3 + tn_4)\,\mathbf{c} = 0$

Since **a, b, c** are non-coplanar vectors, hence above equation holds only when

$$xl_1 + yl_2 + zl_3 + tl_4 = 0 \quad ...(3)$$

$$xm_1 + ym_2 + zm_3 + tm_4 = 0 \quad ...(4)$$

$$xn_1 + yn_2 + zn_3 + tn_4 = 0 \quad ...(5)$$

Eliminating x, y, z and t between (2), (3), (4) and (5), we get the required result in the determinant form.

EXERCISES

(In the following **a, b, c** denote a set of non-coplanar vectors.)

1. $OABC$ is a tetrahedron such that

$$\overrightarrow{OA} = \mathbf{a},\ \overrightarrow{OB} = \mathbf{b},\ \overrightarrow{OC} = \mathbf{c}.$$

Find the equations of its faces.

2. $O\,A\,B\,C\,D\,E\,F\,G$ is a parallelopiped such that

$$\overrightarrow{OA} = \mathbf{a},\ \overrightarrow{OB} = \mathbf{b},\ \overrightarrow{OC} = \mathbf{c}.$$

Obtain the equations of its six faces.

3. Find the equations of the planes through the following triads of points :

(*i*) $2\mathbf{a} + 2\mathbf{b} - 2\mathbf{c}$, $3\mathbf{a} + 4\mathbf{b} + 2\mathbf{c}$, $7\mathbf{a} + 6\mathbf{c}$,

(*ii*) $\mathbf{a} + \mathbf{b} + \mathbf{c}$, $\mathbf{a} - \mathbf{b} + \mathbf{c}$, $-\,7\mathbf{a} - 3\mathbf{b} - 5\mathbf{c}$,

4. Examine for coplanarity the following sets of points :

(*i*) $-\,6\mathbf{a} + 3\mathbf{b} + 2\mathbf{c}$, $3\mathbf{a} - 2\mathbf{b} + 4\mathbf{c}$, $5\mathbf{a} + 7\mathbf{b} + 3\mathbf{c}$, $-\,13\mathbf{a} + 17\mathbf{b} -$

c.

(*ii*) $6\mathbf{a} - 4\mathbf{b} + 4\mathbf{c}, \; -\mathbf{a} - 2\mathbf{b} - 3\mathbf{c}, \; \mathbf{a} + 2\mathbf{b} - 5\mathbf{c}, \; -4\mathbf{c}$.

(*iii*) $3\mathbf{a} + 2\mathbf{b} - 5\mathbf{c}, \; 3\mathbf{a} + 8\mathbf{b} + 5\mathbf{c}, \; -3\mathbf{a} + 2\mathbf{b} + \mathbf{c}, \; \mathbf{a} + 4\mathbf{b} - 3\mathbf{c}$.

5. Show that the four points with position vectors

$$-\mathbf{a} + 4\mathbf{b} - 3\mathbf{c}, \quad 3\mathbf{a} + 2\mathbf{b} - 5\mathbf{c}, \quad -3\mathbf{a} + 8\mathbf{b} - 5\mathbf{c}, \quad -3\mathbf{a} + 2\mathbf{b} + \mathbf{c}$$

are coplanar.

6. Show that the line joining the points $6\mathbf{a} + 4\mathbf{b} + 4\mathbf{c}$, $-4\mathbf{c}$, intersects the join of the points $\mathbf{a} + 2\mathbf{b} - 5\mathbf{c}$ and $-\mathbf{a} - 2\mathbf{b} - 3\mathbf{c}$.

7. Find the position vectors of the points of intersection of the line

$$\mathbf{r} = \mathbf{a} + 2\mathbf{b} + t\,(\mathbf{a} - \mathbf{c})$$

with each of the following planes :

(*i*) $\mathbf{r} = \lambda\,(\mathbf{a} + 2\mathbf{b}) + \mu\,(2\mathbf{b} + \mathbf{c})$.

(*ii*) $\mathbf{r} = \mathbf{b} + \lambda\,(\mathbf{a} + \mathbf{c}) + \mu\,(2\mathbf{a} + \mathbf{b} - \mathbf{c})$.

(*iii*) $\mathbf{r} = 2\mathbf{a} + \lambda\,(\mathbf{a} + \mathbf{c}) + \mu\,(\mathbf{a} + \mathbf{b})$.

8. Find the point of intersection of the line

$$\mathbf{r} = 2\mathbf{a} + 3\mathbf{b} + t\mathbf{c}$$

with the plane

$$\mathbf{r} = \mathbf{a} - \mathbf{b} + p\,(\mathbf{a} + \mathbf{b} - \mathbf{c}) + k\,(\mathbf{a} + \mathbf{c} - \mathbf{b}).$$

9. Find the points of intersection, whenever they exist, of the following pairs of lines :

(*i*) $\mathbf{r} = \mathbf{a} + \mathbf{b} + \lambda_1\,(\mathbf{b} - \mathbf{c}), \quad \mathbf{r} = -(\mathbf{a} + \mathbf{c}) + \lambda_2\,(\mathbf{a} + \mathbf{b})$.

(*ii*) $\mathbf{r} = 3\,(\mathbf{a} + \mathbf{c}) + \lambda\,(\mathbf{b} - \mathbf{c}), \quad \mathbf{r} = 2\,(\mathbf{b} - \mathbf{c}) + \mu\,(\mathbf{a} + \mathbf{b})$.

(*iii*) $\mathbf{r} = \mathbf{b} - 2\mathbf{c} + \lambda\,(\mathbf{a} + \mathbf{b}), \quad \mathbf{r} = 2\mathbf{b} - \mathbf{c} + \mu\,(\mathbf{b} + \mathbf{c})$.

10. Obtain the vector equations of the lines of intersection of the following pairs of planes :

(*i*) $\mathbf{r} = \lambda\,(\mathbf{a} + \mathbf{b}) + \mu\,(\mathbf{b} - \mathbf{c})$,
$\mathbf{r} = -2\mathbf{a} + \lambda_1\,(2\mathbf{a} + \mathbf{b} - \mathbf{c}) + \mu_1\mathbf{b}$.

(*ii*) $\mathbf{r} = \mathbf{a} + \lambda_1\,(\mathbf{b} + \mathbf{c}) + \mu_1\,(\mathbf{a} - \mathbf{b})$,
$\mathbf{r} = \lambda_2\,(\mathbf{a} + \mathbf{c}) + \mu_2\,(2\mathbf{a} - \mathbf{b} + \mathbf{c})$.

(*iii*) $\mathbf{r} = \mathbf{b} + \lambda_1\,(\mathbf{a} + \mathbf{b}) + \mu_1\,(\mathbf{a} - \mathbf{c})$,
$\mathbf{r} = \lambda_2\,(\mathbf{a} + \mathbf{b}) + \mu_2\,(\mathbf{b} + \mathbf{c})$.

(*iv*) $\mathbf{r} = -2\mathbf{a} + \lambda_1\,(\mathbf{b} + \mathbf{c}) + \mu_1\,(\mathbf{a} - \mathbf{c})$,
$\mathbf{r} = \mathbf{a} + 3\mathbf{c} + \lambda_2\,(\mathbf{a} - \mathbf{b}) + \mu_2\,(\mathbf{a} + \mathbf{b} + 2\mathbf{c})$.

11. Show that the following lines intersect :

$$\mathbf{r} = (\mathbf{a} - \mathbf{b} - 10\mathbf{c}) + t\,(2\mathbf{a} - 3\mathbf{b} + 8\mathbf{c}),$$
$$\mathbf{r} = (4\mathbf{a} - 3\mathbf{b} - \mathbf{c}) + k\,(\mathbf{a} - 4\mathbf{b} + 7\mathbf{c}).$$

Find also the point of intersection.

12. Show that the lines

$$\mathbf{r} = -3\mathbf{a} + 6\mathbf{b} + t\,(-4\mathbf{a} + 3\mathbf{b} + 2\mathbf{c}),$$
$$\mathbf{r} = -2\mathbf{a} + 7\mathbf{c} + t\,(-4\mathbf{a} + \mathbf{b} + \mathbf{c}),$$

do not intersect.

13. Examine the intersection of the following pair of lines :

$$\mathbf{r} = -\mathbf{a} - 3\mathbf{b} - 5\mathbf{c} + t\,(3\mathbf{a} + 5\mathbf{b} + 7\mathbf{c}),$$

$$\mathbf{r} = 2\mathbf{a} + 4\mathbf{b} + 6\mathbf{c} + t\,(\mathbf{a} + 3\mathbf{b} + 5\mathbf{c}),$$

14. Show that a linear relation

$$x_1\mathbf{a}_1 + ... + x_i\mathbf{a}_i + ... + x_n\mathbf{a}_n = \mathbf{0}$$

connecting the position vectors

$$\mathbf{a}_1, ..., \mathbf{a}_i, ..., \mathbf{a}_n$$

of any n points

$$A_1, ..., A_i, ..., A_n$$

will be independent of the choice of origin if and only if the sum

$$x_1 + ... + x_i + ... + x_n$$

of the scalar coefficients of the vectors is zero.

[Compare with the conditions for coplanarity of four given points (§ 2.7, Page 45) and the collinearity of three given points, (see § 2.5, Page 32)]

15. Verify the condition obtained in Ex. 14 in each of the following cases :

(*i*) The point C divides the line AB in the ratio $m : n$.

(*ii*) A, B, C, D are four points such that the line AB is parallel to the line CD.

(*iii*) A, B, C, D are the vertices of a parallelogram.

(*iv*) G is the centroid of the triangle with vertices A, B, C.

(*v*) G is the centroid of the tetrahedron with vertices A, B, C, D.

SUMMARY

1. If **a, b** be the position vectors of two points A, B, then the position vector $\overrightarrow{OP}$ of the point P which divides AB in the ratio $m : n$ is

$$\frac{m\mathbf{b} + n\mathbf{a}}{m + n};$$

O being the origin of reference.

2. The parametric vectorial equation of the line

(*i*) which passes through the point with position vector **a** and which is parallel to the vector **b** is

$$\mathbf{r} = \mathbf{a} + t\mathbf{b}.$$

(*ii*) which passes through two points A, B with position vectors **a, b** is

$$\mathbf{r} = \mathbf{a} + t\,(\mathbf{b} - \mathbf{a}),$$

t is the parameter in each case.

3. The parametric vectorial equation of the plane

(*i*) which passes through a point with position vector **a** and which is parallel to the vectors **b, c** is

$$\mathbf{r} = \mathbf{a} + t\,(\mathbf{b} - \mathbf{a}) + p\,(\mathbf{c} - \mathbf{a}).$$

(*ii*) which passes through the points A, B with position vectors **a, b** and which is parallel to the vector **c** is

$$\mathbf{r} = \mathbf{a} + t\,(\mathbf{b} - \mathbf{a}) + p\mathbf{c}.$$

(*iii*) which passes through the three points A, B, C with position vectors **a, b, c** is

$$\mathbf{r} = \mathbf{a} + t(\mathbf{b} - \mathbf{a}) + p(\mathbf{c} - \mathbf{a}).$$

Here t, p are the parameters in each case.

4. (*i*) Three points A, B, C with position vectors **a, b, c** are collinear if there exist three scalars x, y, z such that

$$x\mathbf{a} + y\mathbf{b} + z\mathbf{c} = \mathbf{0} \text{ and } x + y + z = 0.$$

Also in this case the point C divides AB in the ratio $y : x$.

(*ii*) Four points A, B, C, D with position vectors **a, b, c, d** are coplanar if there exist scalars x, y, z, t such that

$$x\mathbf{a} + y\mathbf{b} + z\mathbf{c} + t\mathbf{d} = \mathbf{0}, \quad x + y + z + t = 0.$$

Also in this case the position vector of the point of intersection of the lines AB and CD is

$$\frac{x\mathbf{a} + y\mathbf{b}}{x + y} \text{ which is the same as } \frac{z\mathbf{c} + t\mathbf{d}}{z + t}.$$

OBJECTIVE QUESTIONS

For each of the following questions, four alternatives are given for the answer. Only one of them is correct. Choose the correct alternative.

1. If **a, b, c** are position vectors of the vertices of a ΔABC, then $\vec{AB} + \vec{BC} + \vec{CA} =$

(*a*) **0** (*b*) 2**a** (*c*) 2**b** (*d*) 3**c**

2. If **a, b, c** are the position vectors of the points A, B, C respectively, then $\vec{AB} + \vec{BC} + \vec{AC}$ is equal to

(*a*) **0** (*b*) 2 (**b** – **a**)

(*c*) 2 (**c** – **a**) (*d*) **a** + **b** + **c**

3. The position vectors of A and B are **a** and **b** respectively. Then the position vector of a point D dividing AB in the ratio 2 : 3 externally is

(*a*) $\frac{1}{5}(2\mathbf{a} + 3\mathbf{b})$ (*b*) $\frac{1}{5}(2\mathbf{b} + 3\mathbf{a})$

(*c*) $\frac{1}{6}(\mathbf{a} + \mathbf{b})$ (*d*) (3**a** – 2**b**)

4. Point A is **a** + 2**b**, and **a** divides AB in the ratio 2 : 3. The position vector of B is

(*a*) 2**a** – **b** (*b*) **b** – 2**a**

(*c*) **a** – 3**b** (*d*) **b**

5. P, Q, R, S have position vectors **p, q, r, s**, such that $\mathbf{p} - \mathbf{q} = 2(\mathbf{s} - \mathbf{r})$ then
(*a*) PQ and RS bisect each other
(*b*) PQ and PR bisect each other
(*c*) PQ and RS trisect each other
(*d*) QS and PR trisect each other

6. **a, b, c** are three non-zero vectors, no two of which are collinear and the vector $\mathbf{a} + \mathbf{b}$ is collinear with **c**, $\mathbf{b} + \mathbf{c}$ is collinear with **a**, then $\mathbf{a} + \mathbf{b} + \mathbf{c}$ is equal to
(*a*) **a** (*b*) **b** (*c*) **c** (*d*) None of these

7. Given that the vectors **a** and **b** are non-collinear, the value of x and y for which the vector quantity $2\mathbf{u} - \mathbf{v} = \mathbf{w}$ holds true if $\mathbf{u} = x\mathbf{a} + 2y\mathbf{b}$, $\mathbf{v} = -2y\mathbf{a} + 3x\mathbf{b}$, $\mathbf{w} = 4\mathbf{a} - 2\mathbf{b}$ are
(*a*) $x = 4/7,\ y = 6/7$ (*b*) $x = 10/7,\ y = 4/7$
(*c*) $x = 8/7,\ y = 2/7$ (*d*) $x = 2,\ y = 3$

8. If a, b, c are position vectors of three-collinear points such that $x\mathbf{a} + y\mathbf{b} + z\mathbf{c} = 0$ and atleast one scalar $x, y, z \neq 0$, then
(*a*) $x + y + z = 0$ (*b*) $x + y + z \neq 0$
(*c*) there exists no relation between x, y and z
(*d*) None of these

9. If $x\mathbf{u} + y\mathbf{v} + z\mathbf{w} = x'(\mathbf{u} + \mathbf{w}) + y'(-\mathbf{u} + \mathbf{w}) + z'(\mathbf{u} + \mathbf{v} + \mathbf{w})$, **u, v, w** being non-zero, non-coplanar vectors, then x' is equal to
(*a*) $-\frac{1}{2}x + y - \frac{1}{2}z$ (*b*) $-\frac{1}{2}x + \frac{1}{2}z$
(*c*) $-\frac{1}{2}x - y + \frac{1}{2}z$ (*d*) $x - \frac{1}{2}y + \frac{1}{2}z$

10. The points O, A, B, X and Y are such that $\overrightarrow{OA} = \mathbf{a}$, $\overrightarrow{OB} = \mathbf{b}$, $\overrightarrow{OX} = 3\mathbf{a}$, $\overrightarrow{OY} = 3\mathbf{b}$; and the point P divides AY in the ratio 1 : 3 internally, then the vector $\overrightarrow{PB}$ is equal to
(*a*) $\frac{1}{4}(\mathbf{a} - 3\mathbf{b})$ (*b*) $\frac{1}{4}(3\mathbf{a} - 2\mathbf{b})$
(*c*) $\frac{1}{4}(3\mathbf{a} - \mathbf{b})$ (*d*) $\frac{1}{4}(2\mathbf{a} - 3\mathbf{b})$

ANSWERS

1. (*a*)	**2.** (*c*)	**3.** (*d*)	**4.** (*c*)	**5.** (*d*)
6. (*d*)	**7.** (*b*)	**8.** (*a*)	**9.** (*c*)	**10.** (*c*)

APPENDIX I

1.1. VECTORIAL PROOFS OF SOME WELL-KNOWN CLASSICAL THEOREMS

1. Ceva's Theorem and its Converse. *If D, E, F are three points on the sides BC, CA, AB respectively of a triangle ABC such that the lines AD, BE and CF are concurrent, then*

$$\frac{BD}{CD}\cdot\frac{CE}{AE}\cdot\frac{AF}{BF}=-1$$

and conversely.

Let AD, BE and CF concur at H. Take any point, O, as the origin of reference. Let **a, b, c, h** be the position vectors of the points A, B, C, H respectively.

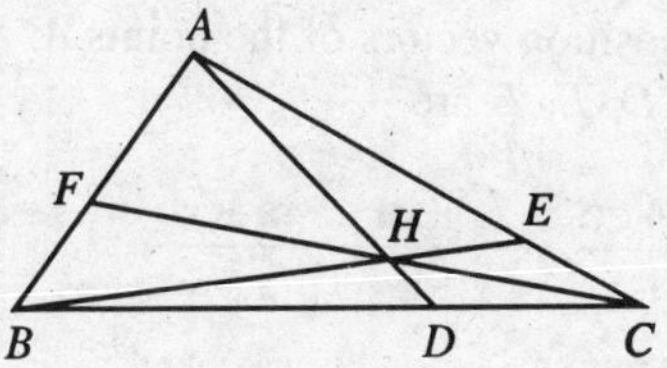

Fig. 1

These four points being coplanar, there exist four scalars x, y, z, t such that

$$x\mathbf{a} + y\mathbf{b} + z\mathbf{c} + t\mathbf{h} = \mathbf{0}, \quad x + y + z + t = 0.$$

These give

$$\frac{x\mathbf{a}+y\mathbf{b}}{x+y}=\frac{z\mathbf{c}+t}{z+t}.$$

Now $$\frac{x\mathbf{a}+y\mathbf{b}}{x+y} \text{ and } \frac{z\mathbf{c}+t\mathbf{h}}{z+t}$$

are points of the lines AB and CH respectively.

It follows that

$$(x\mathbf{a} + y\mathbf{b}) \mathbin{/} x + y$$

is the position vector of the point F of intersection of the lines AB and CH so that F divides AB in the ratio

$$AF \mathbin{/} FB = y/x \quad \Leftrightarrow \quad AF \mathbin{/} BF = -\, y/x.$$

Similarly

$$\frac{BD}{CD} = -\frac{z}{y}, \quad \frac{CE}{AE} = -\frac{x}{z}$$

$$\therefore \quad \frac{BD}{CD} \cdot \frac{CE}{AE} \cdot \frac{AF}{BF} = \left(-\frac{z}{y}\right) \cdot \left(-\frac{x}{z}\right) \cdot \left(-\frac{y}{x}\right) = -1.$$

Conversely, let D, E, F be three points on the sides BC, CA and AB respectively such that

$$\frac{BD}{CD} \cdot \frac{CE}{AE} \cdot \frac{AF}{BF} = -1. \qquad ...(i)$$

Suppose that

$$\frac{BD}{CD} = -\frac{z}{y}, \quad \frac{CE}{AE} = -\frac{x}{z} \text{ so that by } (i)$$

$$\frac{AF}{BF} = -\frac{y}{x}.$$

$$\therefore \quad \frac{BD}{DC} = \frac{z}{y}, \quad \frac{CE}{EA} = \frac{x}{z}, \quad \frac{AF}{FB} = \frac{y}{x}.$$

If **a, b, c** be the position vectors of the points A, B, C, then the position vectors of the points D, E, F are

$$\frac{y\mathbf{b} + z\mathbf{c}}{y+z}, \quad \frac{z\mathbf{c} + x\mathbf{a}}{z+x}, \quad \frac{x\mathbf{a} + y\mathbf{b}}{x+y}$$

respectively.

Therefore the position vectors of the points dividing AD, BE, CF in the ratios

$$y + z : x, \quad z + x : y, \quad x + y : z$$

are all equal to

$$\frac{x\mathbf{a} + y\mathbf{b} + z\mathbf{c}}{x + y + z}.$$

Thus, the lines AD, BE and CF are concurrent.

2. Menelau's Theorem and its Converse. *If D, E, F are three points on the sides BC, CA, AB respectively of a triangle ABC such that the points D, E, F are collinear, then*

$$\frac{BD}{CD} \cdot \frac{CE}{AE} \cdot \frac{AF}{BF} = 1.$$

and conversely.

Let BE, CF meet at H.

Let **a**, **b**, **c**, **h** be the position vectors of the four points A, B, C, H relative to any origin O of reference.

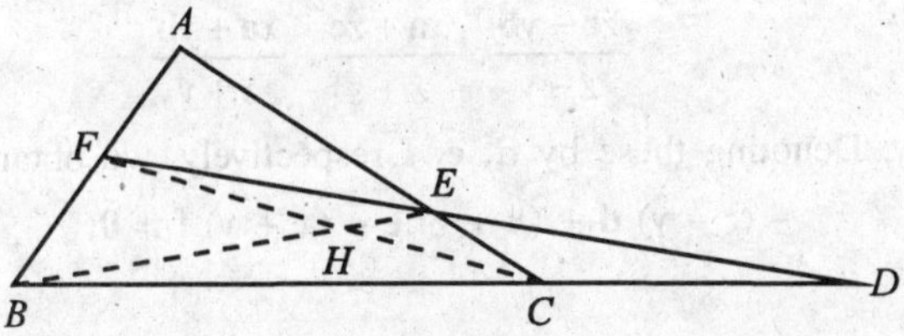

Fig. 2

These four points being coplanar, there exist four scalars x, y, z, t such that

$$x\mathbf{a} + y\mathbf{b} + z\mathbf{c} + t\mathbf{h} = \mathbf{0}, \quad x + y + z + t = 0.$$

The position vectors of the points E and F, therefore, are

$$\frac{x\mathbf{a}+z\mathbf{c}}{x+z}, \quad \frac{x\mathbf{a}+y\mathbf{b}}{x+y}$$

respectively.

We now require the position vector of the point D. Writing

$$\mathbf{e} = \frac{x\mathbf{a}+z\mathbf{c}}{x+z}, \quad \mathbf{f} = \frac{x\mathbf{a}+y\mathbf{b}}{x+y}$$

and eliminating, **a**, we have

$$(x + z)\,\mathbf{e} - (x + y)\,\mathbf{f} = z\mathbf{c} - y\mathbf{b},$$

$$\Rightarrow \quad \frac{(x+z)\,\mathbf{e}-(x+y)\,\mathbf{f}}{(x+z)-(x+y)} = \frac{z\mathbf{c}-y\mathbf{b}}{z-y}.$$

This equality shows that $(z\mathbf{c} - y\mathbf{b}) / (z - y)$ is the position vector of the point D. Thus

$$\frac{BD}{CD} = -\frac{z}{y}, \quad \frac{CE}{EA} = \frac{x}{z}, \quad \frac{AF}{FB} = \frac{y}{x}.$$

$$\Rightarrow \quad \frac{BD}{CD}\cdot\frac{CE}{AE}\cdot\frac{AF}{BF} = 1.$$

Conversely, let D, E, F be three points on the sides BC, CA, AB such that

$$\frac{BD}{CD}\cdot\frac{CE}{AE}\cdot\frac{AF}{BF} = 1.$$

Suppose that

$$\frac{BD}{CD} = \frac{z}{y}, \quad \frac{CE}{AE} = -\frac{x}{z}, \text{ so that } \frac{AF}{BF} = -\frac{y}{z}.$$

$$\Rightarrow \quad \frac{BD}{DC} = -\frac{z}{y}, \quad \frac{CE}{EA} = \frac{x}{z}, \quad \frac{AF}{EB} = \frac{y}{x}.$$

Thus, if **a, b, c** be the position vectors of the vertices *A, B, C,* then the position vectors of the points *D, E, F,* are

$$\frac{z\mathbf{c}-y\mathbf{b}}{z-y}, \quad \frac{x\mathbf{a}+z\mathbf{c}}{x+z}, \quad \frac{x\mathbf{a}+y\mathbf{b}}{x+y}$$

respectively. Denoting these by **d, e, f** respectively, we obtain

$$-(z-y)\,\mathbf{d}+(x+z)\,\mathbf{e}-(x+y)\,\mathbf{f}=\mathbf{0},$$

where

$$-(z-y)+(x+z)-(x+y)=0.$$

Thus, the points *D, E, F* are collinear.

3. Pappu's Theorem. *If A_1, A_2, A_3 and B_1, B_2, B_3 are two sets of collinear points, then the points of intersection of the pairs of lines*

$$A_1B_2,\ A_2B_1;\ A_2B_3,\ A_3B_2;\ A_3B_1,\ A_1B_2.$$

are collinear.

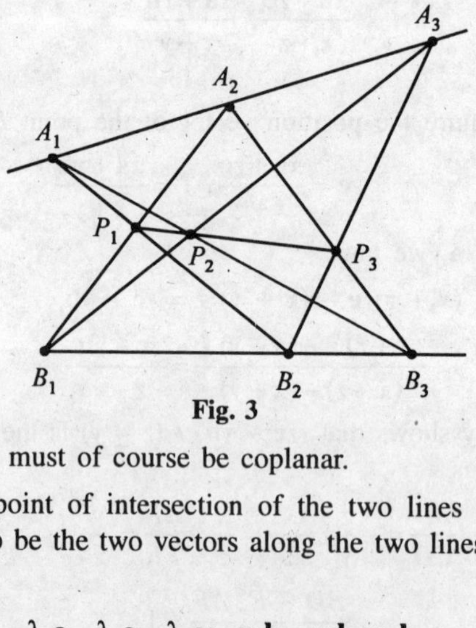

Fig. 3

The two lines must of course be coplanar.

Take *O*, the point of intersection of the two lines as the origin of reference. Let **a, b** be the two vectors along the two lines.

Let

$$\lambda_1\mathbf{a},\ \lambda_2\mathbf{a},\ \lambda_3\mathbf{a};\ \mu_1\mathbf{b},\ \mu_2\mathbf{b},\ \mu_3\mathbf{b},$$

be the position vectors of the two sets of points, A_1, A_2, A_3 and B_1, B_2, B_3 respectively.

The equations of the lines A_1B_2 and A_2B_1 are

$$\mathbf{r}=\lambda_1\mathbf{a}+t\,(\lambda_1\mathbf{a}-\mu_2\mathbf{b}), \quad \mathbf{r}=\lambda_2\mathbf{a}+p\,(\lambda_2\mathbf{a}-\mu_1\mathbf{b}).$$

At their point of intersection, we have

$$\lambda_1\mathbf{a}+t\,(\lambda_1\mathbf{a}-\mu_2\mathbf{b})=\lambda_2\mathbf{a}+p\,(\lambda_2\mathbf{a}-\mu_1\mathbf{b})$$

$$\Rightarrow \quad (\lambda_1+t\lambda_1-\lambda_2-p\lambda_2)\,\mathbf{a}+(-t\mu_2+p\mu_1)\,\mathbf{b}=\mathbf{0}$$

$$\Rightarrow \quad \lambda_1+t\lambda_1-\lambda_2-p\lambda_2=0, \quad -t\mu_2+p\mu_1=0,$$

the vectors **a, b** being non-parallel.

These give

$$p = \frac{\mu_2(\lambda_2 - \lambda_1)}{(\lambda_1\mu_1 - \lambda_2\mu_2)}.$$

Making substitution, we see that the position vector of the point of intersection of the lines A_1B_2, A_2B_1 is

$$\frac{\lambda_1\lambda_2(\mu_1 - \mu_2)}{\lambda_1\mu_1 - \lambda_2\mu_2}\mathbf{a} + \frac{\mu_1\mu_2(\lambda_1 - \lambda_2)}{\lambda_1\mu_1 - \lambda_2\mu_2}\mathbf{b} = \mathbf{p}_3, \text{ say.}$$

Changing the suffixes, 1, 2 to 2, 3 respectively and 1, 2 to 3, 1 respectively, we see that the position vectors of the other two points of intersection are

$$\frac{\lambda_2\lambda_3(\mu_2 - \mu_3)}{\lambda_2\mu_2 - \lambda_3\mu_3}\mathbf{a} + \frac{\mu_2\mu_3(\lambda_2 - \lambda_3)}{\lambda_2\mu_2 - \lambda_1\mu_1}\mathbf{b} = \mathbf{p}_1, \text{ say}$$

$$\frac{\lambda_3\lambda_1(\mu_3 - \mu_1)}{\lambda_3\mu_3 - \lambda_1\mu_1}\mathbf{a} + \frac{\mu_3\mu_1(\lambda_3 - \lambda_1)}{\lambda_3\mu_3 - \lambda_2\mu_2}\mathbf{b} = \mathbf{p}_2, \text{ say}$$

Now we may see that

$$\Sigma\lambda_1\mu_1(\lambda_2\mu_2 - \lambda_3\mu_3)\,\mathbf{p}_1 = \mathbf{0},$$

where

$$\Sigma\lambda_1\mu_1(\lambda_2\mu_2 - \lambda_3\mu_3) = 0.$$

Thus, the points are collinear.

Or directly, we may show that the vector $P_2 - P_1$ is the product of the vector $P_3 - P_1$ by a scalar.

4. Desargue's Theorem. *If ABC, $A_1B_1C_1$ are two triangles such that the three lines AA_1, BB_1 and CC_1 are concurrent, then the points of intersection of the three pairs of sides*

$$BC, B_1C_1; \quad CA, C_1A_1; \quad AB, A_1B_1$$

are collinear and conversely.

Let AA_1, BB_1 and CC_1 are concur at O. Take O as the origin of reference. Let

$$\vec{OA} = \mathbf{a}, \; \vec{OB} = \mathbf{b}, \; \vec{OC} = \mathbf{c}.$$

Then

$$\vec{OA_1} = \lambda\mathbf{a}, \; \vec{OB_1} = \mu\mathbf{b}, \; \vec{OC_1} = \nu\mathbf{c}$$

where λ, μ, ν are some scalars. The equations of BC and B_1C_1 respectively are

$$\mathbf{r} = \mathbf{b} + t\,(\mathbf{b} - \mathbf{c}), \quad \mathbf{r} = \mu\mathbf{b} + p\,(\mu\mathbf{b} - \nu\mathbf{c}),$$

so that at the point of intersection P of BC and B_1C_1, we have

$$\mathbf{b} + t\,(\mathbf{b} - \mathbf{c}) = \mu\mathbf{b} + p\,(\mu\mathbf{b} - \nu\mathbf{c}),$$

$$\Rightarrow \quad (1 + t - \mu - p\mu)\,\mathbf{b} + (-t + p\nu)\,\mathbf{c} = \mathbf{0}.$$

$$\Rightarrow \quad 1 + t - \mu - p\mu = 0, \quad -t + p\nu = 0.$$

the vectors **b** and **c** being non-parallel.

These give

$$p = (1 - \mu) / (\mu - \nu).$$

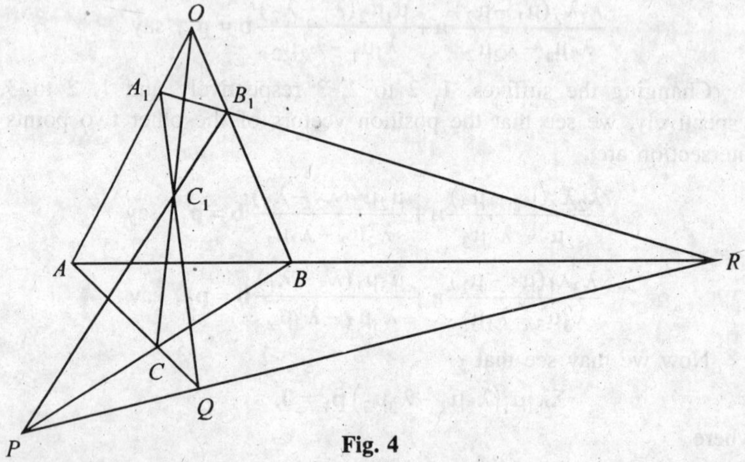

Fig. 4

Making substitution, we see that the position vector of the point P of intersection of BC, B_1C_1 is

$$\mu\mathbf{b} + \frac{1-\mu}{\mu-\nu}(\mu\mathbf{b} - \nu\mathbf{c}) = \frac{1-\nu}{\mu-\nu}\mu\mathbf{b} + \frac{1-\mu}{\nu-\mu}\nu\mathbf{c} = \mathbf{p}, \text{ say.}$$

By considerations of symmetry, the two other points of intersection Q, R are

$$\frac{1-\lambda}{\nu-\lambda}\nu\mathbf{c} + \frac{1-\nu}{\lambda-\nu}\lambda\mathbf{a} = \mathbf{q}, \text{ say.}$$

$$\frac{1-\mu}{\lambda-\mu}\lambda\mathbf{a} + \frac{1-\lambda}{\mu-\lambda}\mu\mathbf{b} = \mathbf{r}, \text{ say.}$$

We now see that

$$\Sigma\,(1 - \lambda)\,(\mu - \nu t)\,\mathbf{p} = \mathbf{0} \ \text{ where } \ \Sigma\,(1 - \lambda)\,(\mu - \nu) = 0.$$

Thus, the three points of intersection whose position vectors are **p, q, r** are collinear.

Conversely, let **a, b, c** : $\mathbf{a}_1$, $\mathbf{b}_1$, $\mathbf{c}_1$ be the position vectors of the vertices of the triangles ABC, $A_1B_1C_1$ with respect to some origin O. Let P, Q, R be the points of intersection of the pairs of lines

$$BC,\ B_1C_1;\quad CA,\ C_1A_1;\quad AB,\ A_1B_1.$$

These points are given to be collinear.

Let P divide BC in the ratio $z : y$ so that the position vector of the point P is

$$(y\mathbf{b} + z\mathbf{c}) / (y + z).$$

Let Q divide BC in the ratio $x : z$ so that the position vector of the point Q is

$$(z\mathbf{c} + x\mathbf{a}) / (z + x).$$

The position vector of the point R of intersection of PQ and AB may now be easily seen to be

$$(y\mathbf{b} - x\mathbf{a}) / (y - x).$$

Denoting these position vectors by **p, q, r** respectively, we have the relation

$$(y + z)\,\mathbf{p} - (z + x)\,\mathbf{q} - (y - x)\,\mathbf{r} = \mathbf{0} \qquad ...(1)$$

between the position vectors of the collinear points P, Q, R.

Similarly, denoting by $z_1 : y_1$ and $x_1 : y_1$ the ratios in which the points P and Q divide B_1C_1 and C_1A_1 we obtain

$$\mathbf{p} = \frac{y_1\mathbf{b} + z_1\mathbf{c}}{y_1 + z_1},\ \mathbf{q} = \frac{z_1\mathbf{c} + x_1\mathbf{a}}{z_1 + x_1},\ \mathbf{r} = \frac{y_1\mathbf{b} + x_1\mathbf{a}}{y_1 - x_1},$$

and the relation

$$(y_1 + z_1)\,\mathbf{p} - (z_1 + x_1)\,\mathbf{q} - (y_1 - x_1)\,\mathbf{r} = \mathbf{0}. \qquad ...(2)$$

From (1) and (2), we obtain

$$\frac{y+z}{y_1+z_1} = \frac{z+x}{z_1+x_1} = \frac{y-x}{y_1-x_1} = k, \text{ say.}$$

so that

$$y + z = k\,(y_1 + z_1),\ \ (z + x) = k\,(z_1 + x_1),\ \ y - x = k\,(y_1 - x_1) \qquad ...(3)$$

Also we have

$$\frac{y\mathbf{b} + z\mathbf{c}}{y + z} = \frac{y_1\mathbf{b}_1 + z_1\mathbf{c}_1}{y_1 + z_1},\ \ \frac{z\mathbf{c} + x\mathbf{a}}{z + x} = \frac{z_1\mathbf{c}_1 + x_1\mathbf{a}_1}{z_1 + x_1}$$

$$\frac{y\mathbf{b} - x\mathbf{a}}{y - x} = \frac{y_1\mathbf{b}_1 - x_1\mathbf{a}_1}{y_1 + zx_1} \qquad ...(4)$$

so that with the help of (3), we obtain

$$z\mathbf{c} - kz_1\mathbf{c}_1 = ky_1\mathbf{b}_1 - y\mathbf{b} = kx_1\mathbf{a}_1 - x\mathbf{a}$$

$$\Rightarrow \qquad \frac{z\mathbf{c} - kz_1\mathbf{c}_1}{z - kz_1} = \frac{ky_1\mathbf{b}_1 - y\mathbf{b}}{ky_1 - y} = \frac{kx_1\mathbf{a}_1 - x\mathbf{a}}{kx_1 - x} \qquad ...(5)$$

The three equal vectors in (5) being the position vectors of points on CC_1, BB_1, AA_1, we see that these lines are concurrent.

5. Complete Quadrilateral. (*i*) *The three mid-points of the three diagonals of a complete quadrilateral are collinear.*

(*ii*) *Each diagonal of a complete quadrilateral is cut harmonically by the other two.*

(*i*) Consider any four lines

$$AB,\ BC,\ CA,\ EF.$$

no three of which are concurrent. These lines intersect in pairs in **six** points called the *vertices* of the complete quadrilateral. These six vertices are divided into *three* pairs of opposite vertices, *viz.*, the intersections of the pairs of lines

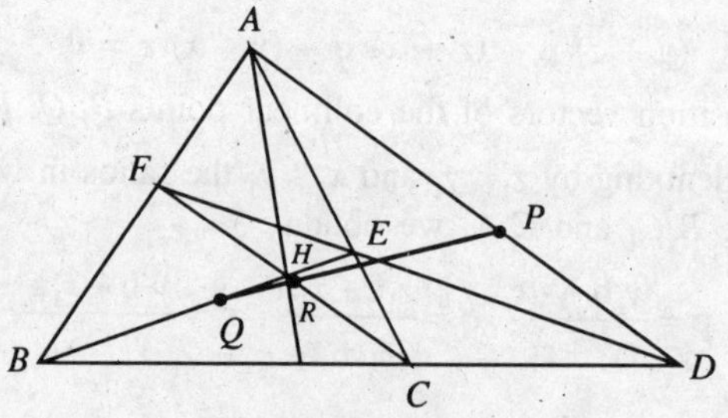

Fig. 5

$$CA,\ AB\ ;\quad BC,\ EF.$$
$$AB,\ BC\ ;\quad CA,\ EF.$$
$$BC,\ CA\ ;\quad AB,\ EF.$$

Thus

$$A,\ D\ ;\quad B\ E\ ;\quad C,\ F$$

are three pairs of opposite vertices. The lines

$$AD,\ BE,\ CF$$

joining the pairs of opposite vertices are called the three diagonals of the complete quadrilateral. We have to prove that the mid-points of *AD, BE, CF* are collinear.

Let

BE, CF meet in *H*.

With any point *O* as the origin of reference, let

$$\mathbf{a},\ \mathbf{b},\ \mathbf{c},\ \mathbf{h}$$

be the position vectors of *A, B, C, H* respectively. These points being coplanar, there exist four scalars *x, y, z, t* such that

$$x\mathbf{a} + y\mathbf{b} + z\mathbf{c} + t\mathbf{h} = \mathbf{0},\quad x + y + z + t = 0. \qquad ...(i)$$

From this we deduce that

$$\frac{z\mathbf{c}+x\mathbf{a}}{z+x}, \frac{x\mathbf{a}+y\mathbf{b}}{x+y}$$

are the position vectors of the points E and F respectively.

We now require the position vector of the point D.

Writing

$$\mathbf{e}=\frac{z\mathbf{c}+x\mathbf{a}}{z+x}, \mathbf{f}=\frac{x\mathbf{a}+y\mathbf{b}}{x+y}$$

and eliminating, **a**, we see that

$$(z + x)\,\mathbf{e} - (x + y)\,\mathbf{f} = z\mathbf{e} - y\mathbf{b}$$

$$\Rightarrow \quad \frac{(z+x)\,\mathbf{e}-(x+y)\,\mathbf{f}}{(z+x)-(x+y)}=\frac{z\mathbf{c}-y\mathbf{b}}{z-y}.$$

This equality shows that

$$\frac{z\mathbf{c}-y\mathbf{b}}{z-y}$$

is the position vector of the point, *D*, of intersection of the lines, *BC*, *EF*.

The mid-points *P*, *Q*, *R* of *AD*, *BE* and *CF* are

$$\frac{(z-y)\,\mathbf{a}-y\mathbf{b}+z\mathbf{c}}{2(z-y)}, \frac{x\mathbf{a}+(z+x)\,\mathbf{b}+z\mathbf{e}}{2(z+x)}, \frac{x\mathbf{a}+y\mathbf{b}+(x+y)\mathbf{c}}{2(x+y)}.$$

Denoting these by **p, q, r** respectively, we see that

$$x\,(z - y)\,\mathbf{p} + y\,(z + x)\,\mathbf{q} - z\,(x + y)\,\mathbf{r} = \mathbf{0}$$

where

$$x\,(z - y) + y\,(z + x) - z\,(x + y) = 0.$$

Thus, the three points *P*, *Q*, *R* with position vectors **p, q, r** are collinear.

(*ii*) The position vectors of the points *B*, *H*, *E* are

$$\mathbf{b}, \frac{x\mathbf{a}+y\mathbf{b}+z\mathbf{c}}{x+y+z}, \frac{z\mathbf{c}+x\mathbf{a}}{z+x}$$

respectively so that *H* divides *BE* in the ratio $(z + x) : y$. The point *H*′ which divides *BE* in the ratio $(z + x) : - y$ is

$$\frac{(x\mathbf{a}+z\mathbf{c})-y\mathbf{b}}{(x+z)-y}=\frac{x\mathbf{a}+(z\mathbf{c}-y\mathbf{b})}{x+(z-y)}$$

which is clearly also a point on the line *AD* dividing the same in the ratio $(z - y) : x$.

Thus, the diagonal *BE* is divided at two points of intersection *H* and *H*′ with the other two diagonals *CE*, *AD* in equal and opposite ratios. Hence, the second part.

Another Proof.

Take A as the origin of reference. Let

$$\overrightarrow{AB} = \mathbf{b}, \ \overrightarrow{AE} = \mathbf{e}.$$

It should be possible to obtain the position vector of every point in the plane of the quadrilateral in terms of **b** and **c.**

Let $\overrightarrow{AF} = \lambda\mathbf{b}, \ \overrightarrow{AC} = \mu\mathbf{e}$; $\lambda, \ \mu$ being scalars.

The equations of BC and EF are

$$\mathbf{r} = \mathbf{b} + t\,(\mu\mathbf{e} - \mathbf{b}), \ \mathbf{r} = \mathbf{e} + p\,(\mathbf{e} - \lambda\mathbf{b}).$$

At their point of intersection,

$$\mathbf{b} + t\,(\mu\mathbf{e} - \mathbf{b}) = \mathbf{e} + p\,(\mathbf{e} - \lambda\mathbf{b})$$

$$\Rightarrow \quad (1 - t + p\lambda)\,\mathbf{b} + (\mu t - 1 - p)\,\mathbf{e} = \mathbf{0}$$

$$\Rightarrow \quad 1 - t + p\lambda = 0, \ \mu t - 1 - p = 0;$$

b, e being non-parallel vectors.

These give

$$t = \frac{1-\lambda}{1-\lambda\mu}.$$

Making substitution, we see that the point D, of intersection of the lines BC and EF is

$$\mathbf{b} + \frac{1-\lambda}{1-\lambda\mu}(\mu\mathbf{e} - \mathbf{b}) = \frac{\lambda\,(1-\mu)}{1-\lambda\mu}\,\mathbf{b} + \frac{\mu\,(1-\lambda)}{1-\lambda\mu}\,\mathbf{e}.$$

Thus, the mid-points of AD, BE, CF respectively are

$$\frac{1}{2}\left[\frac{\lambda\,(1-\mu)}{1-\lambda\mu}\,\mathbf{b} + \frac{\mu\,(1-\lambda)}{1-\lambda\mu}\,\mathbf{e}\right], \ \frac{1}{2}(\mathbf{b}+\mathbf{e}), \ \frac{1}{2}(\lambda\mathbf{b} + \mu\mathbf{e}).$$

Denoting these by **p, q, r** respectively, we see that

$$(1 - \lambda\mu)\,\mathbf{p} + \lambda\mu\,\mathbf{q} + (-1)\,\mathbf{r} = \mathbf{0},$$

where

$$(1 - \lambda\mu) + \lambda\mu + (-1) = 0.$$

Thus, the points are collinear.

MISCELLANEOUS EXERCISES I

1. OAB is a given triangle such that $\vec{OA} = \mathbf{a}$, $\vec{OB} = \mathbf{b}$. Also C is a point on AB such that $AB = 2BC$. State which of the following statements are correct ?

(a) $\vec{AC} = \frac{2}{3}(\mathbf{b} - \mathbf{a})$, (b) $\vec{AC} = \frac{2}{3}(\mathbf{a} - \mathbf{b})$,

(c) $\vec{AC} = \frac{2}{3}(\mathbf{b} + \mathbf{a})$, (d) $\vec{AC} = \frac{3}{2}(\mathbf{b} - \mathbf{a})$,

(e) $\vec{AC} = \frac{3}{2}(\mathbf{a} + \mathbf{b})$.

2. O, A, B, C, D, E are five coplanar points such that

$$\vec{OA} = \mathbf{a},\ \vec{OB} = \mathbf{b},\ \vec{OC} = 2\mathbf{a} + 3\mathbf{b},\ \vec{OD} = \mathbf{a} - 2\mathbf{b},\ \vec{OE} = \mathbf{a} + \mathbf{b}.$$

Show that $\vec{OE}$, $\vec{BC}$ are parallel and $\vec{OB}$, $\vec{AD}$ are parallel.

3. E and F are the mid-points of the diagonals BD and AC of a quadrilateral $ABCD$ respectively. Show that

(i) $\vec{AB} + \vec{AD} + \vec{CB} + \vec{CD} = 4\vec{EF}$.

(ii) $\vec{AB} + \vec{BC} + \vec{CD} + \vec{AD} = 2\vec{AD}$.

4. Let $\vec{OA} = \mathbf{a}$ and $\vec{OB} = \mathbf{b}$ and $\vec{OC} = \mathbf{a} + \mathbf{b}$.

What is the type of the quadrilateral $OACB$?

5. OAB is a given triangle and P, Q, R are the mid-points of the sides OA, AB and BO respectively of the triangle. Express in terms of the vectors $\vec{OA}$ and $\vec{OB}$ the vectors $\vec{AB}, \vec{PA}, \vec{PQ}$ and $\vec{AR}$.

6. The mid-points of the sides OP and OQ of a triangle $O\ P\ Q$ are S and R respectively. Express the vectors $\vec{PQ}, \vec{SR}, \vec{SQ}$ and $\vec{RP}$ in terms of $\vec{OP}$ and $\vec{OQ}$.

Let SQ and RP meet at the point X and let

$$\frac{SX}{SQ} = m,\quad \frac{RX}{RP} = n.$$

Show that

(i) $\vec{OX} = \mathbf{a} + m(2\mathbf{b} - \mathbf{a})$ (ii) $\vec{OX} = \mathbf{b} + n(2\mathbf{a} - \mathbf{b})$

where $\vec{OP} = 2\mathbf{a}$ and $\vec{OQ} = 2\mathbf{b}$.

Deduce the X divides SQ and RP in the ratio 1 : 2.

7. The diagonals AC, BD of a quadrilateral $ABCD$ meet at a point E. Given that $\vec{AB} = \mathbf{a}$, $\vec{BC} = \mathbf{b}$, and that E is a point of trisection of each of AC and BD nearer A as well as B, express in terms of **a** and **b** the vectors $\vec{BE}$ and $\vec{DC}$.

8. The point E is a point of trisection of the straight line PQ such that $PE : EQ = 1 : 2$. Also R is any point not on the line PQ and F divides QR internally such that $QF : FR = 2 : 1$. Show that EF is parallel to PR.

9. In the triangle OAB, L is the mid-point of OA and M is a point on OB such that $\dfrac{OM}{MB} = 2$. P is the mid-point of LM, and the line AP is produced to meet OB at Q. Given that $\vec{OA} = \mathbf{a}$ and $\vec{OB} = \mathbf{b}$, find in terms of **a** and **b**, the vectors

(*i*) $\vec{OP}$, (*ii*) $\vec{AP}$.

If $\vec{AQ} = \lambda\, \vec{AP}$ and $\vec{OQ} = \mu\, \vec{OB}$, find $\lambda, \mu, \dfrac{AP}{PQ}$ and $\dfrac{OQ}{QB}$.

10. Points X and Y are taken on the sides QR and RS, respectively, of a parallelogram $PQRS$, so that $\vec{QX} = 4\vec{XR}$ and $\vec{RY} = 4\vec{YS}$. The line XY cuts the line PR at Z. Prove that $\vec{PZ} = (21/25)\vec{PR}$.

11. The diagonals of a parallelogram $ABCD$ intersect at E. The position vector with respect to an origin O of A is **a**. Also $\vec{AB} = \mathbf{p}$ and $\vec{BC} = \mathbf{q}$. Determine the position vectors of the vertices B, C, D, E with respect to O in terms of **a, p** and **q**.

12. ABC is a triangle and the position vectors of the points A, B, C relative to the origin O are **a, b, c** respectively. The point P is on the side BC such that $BP : PC = 2 : 3$, and the point Q is on the side CA such that $CQ : QA = 1 : 4$. Find the position vector of the common point R of AP and BQ. Find also the ratio in which CR divides AB.

13. In a triangle OAB, $\vec{OA} = \mathbf{a}$ and $\vec{OB} = \mathbf{b}$. Point P and Q are taken so that $\vec{AP} = \mathbf{a}$ and $\vec{BQ} = 2\mathbf{b}$. Obtain expressions for the vectors $\vec{AQ}$ and and $\vec{BP}$ in terms of **a** and **b**. The lines AQ and BP meet at R and the line OR meets AB at S. Find the ratio in which S divides AB.

14. ABC is a triangle, D is the mid-point of AB and E is the point of trisection of BC nearer C. Show that the mid-point F of CD is on AE and find also the ratio $AF : FE$.

Let DE meet AC at L. Find the ratio in which L divides AC.

15. In a triangle OAB, L is a point on the side AB and M is a point on the side OB and the lines OL and AM meet at S.

It is given that $AS = SM$ and $4\,OS = 3\,OL$, and that

$$\frac{OM}{OB} = h \text{ and } \frac{AL}{AB} = k.$$

(*i*) Express the vectors $\vec{AM}$, and $\vec{OS}$ in terms of **a**, **b** and h and the vectors $\vec{OL}$ and $\vec{OS}$ in terms of **a**, **b** and k where $\vec{OA} = \mathbf{a}$ and $\vec{OB} = \mathbf{b}$. Find h and k.

(*ii*) BS meets OA at N. Find the ratios in which L, M and N divide AB, BO and OA respectively.

16. ABC is a triangle and P, Q are the mid-points of AB, AC respectively. If $\vec{AB} = 2\mathbf{a}$ and $\vec{AC} = 2\mathbf{b}$, express in terms of **a** and **b** the vectors (*i*) $\vec{BC}$, (*ii*) $\vec{PQ}$, (*iii*) $\vec{PC}$, (*iv*) $\vec{BQ}$. What can you deduce about the line segments BC and PQ ?

17. OAB is a triangle and X is the point of the line segment AB such that $\vec{OX} = 2\,\vec{XB}$. If $\vec{OA} = \mathbf{a}$ and $\vec{OB} = \mathbf{b}$, show that $\vec{OX} = \frac{1}{3}\mathbf{a} + \frac{2}{3}\mathbf{b}$. The line OX meets the line through A parallel to OB in Y. Find Y and OX/XY.

18. If $\vec{OA} = \mathbf{a}$, $\vec{OB} = \mathbf{b}$, $\vec{OC} = \mathbf{c}$, $\vec{OD} = \mathbf{d}$ and the points P, Q, R, S are such that

$$\vec{AP} = 2\,\vec{PB},\ \vec{BQ} = 2\vec{QC},\ \vec{CR} = 2\,\vec{RD},\ \vec{DS} = 2\vec{SA}.$$

Express $\vec{PQ}$ and $\vec{SR}$ in terms of **a**, **b**, **c**, **d** and show that the condition for $P\,Q\,R\,S$ to be a parallelogram is **a** + **c** = **b** + **d**.

19. $OABCD$ is a pentagon in which the sides OA and CB are parallel, and the sides OD and AB are parallel. Also

$$\frac{OA}{CB} = 2 \text{ and } \frac{OD}{AB} = \frac{1}{3}.$$

Given that $\vec{OA} = \mathbf{a}$ and $\vec{OD} = \mathbf{d}$, express each of the vectors $\vec{AD}$, $\vec{OC}$ and $\vec{DC}$ in terms of **a** and **d**. If the diagonals OC and AD meet at X, find $\frac{OX}{XC}$ and $\frac{AX}{XD}$.

20. OAB is given triangle; X is a point on OA such that $\vec{OX} = 2\,\vec{OA}$ and Y is a point on OB, such that $\vec{OY} = \frac{2}{3}\vec{OB}$. The line XY meets the side AB of the triangle at P. Find the ratio in which P divides AB. Find also the ratio in which P divides YX.

21. In a triangle OAB, L is the point on OA such that $\frac{OL}{LA} = \frac{1}{2}$ and M is the mid-point of AB. N is the point in which the line LM meets the side OB. Find the ratio in which L divides MN.

22. Given a regular hexagon $ABCDEF$ with centre O, show that

(*i*) $\vec{OB} - \vec{OA} = \vec{OC} - \vec{OD}$,

(*ii*) $\vec{OD} + 2\,\vec{OA} = 2\,\vec{OB} + \vec{OF}$,

(*iii*) $\vec{AD} + \vec{EB} + \vec{FC} = 4\,\vec{AB}$.

23. The points O, A, B, C are the vertices of a pyramid and P, Q, R, S are the mid-points of OA, OB, BC, AC respectively.

If $\vec{OA} = \mathbf{a}$, $\vec{OB} = \mathbf{b}$, $\vec{OC} = \mathbf{c}$,

express in terms of **a, b, c** the vectors

(*i*) $\vec{OP}$, $\vec{OQ}$, $\vec{OR}$ and $\vec{OS}$.

(*ii*) $\vec{OG}$ and $\vec{OH}$ where G, H are the mid-points of PR, QS respectively.

State what *can be deduced* about PR and QS.

24. $ABCD$ is parallelogram whose diagonals intersect at E and M is the mid-point of DC. If

$$\vec{AB} = \mathbf{a} \text{ and } \vec{AD} = \mathbf{b},$$

express in terms of **a** and **b** the vectors

(*i*) $\vec{AE}$ (*ii*) $\vec{BD}$ (*iii*) $\vec{MB}$

25. In a quadrilateral $ABCD$, the point P divides DB in the ratio 1 : 2 and Q is the mid-point of AC. Prove that

$$2\vec{DC} + \vec{BC} - 2\vec{AD} - \vec{AB} = 6\vec{PQ}.$$

26. $PQRS$ is a quadrilateral and

$$\vec{PQ} = \mathbf{a}, \ \vec{QR} = \mathbf{b}, \ \vec{SP} = \mathbf{a} - \mathbf{b},$$

M is the mid-point of QR and X is the point of SM such that

$$\vec{SX} = (4/5)\,\vec{SM}.$$

(*i*) Prove that PR and PX are parallel and find the ratio of their lengths.

(*ii*) Express in terms of **a** and **b** the vectors $\vec{PR}, \vec{SM}, \vec{SX}$ and $\vec{PX}$.

(*iii*) What can be deduced about the line segments PX, XR.

27. Given that O is a point inside the triangle ABC and that D, E and F are the mid-points of BC, CA and AB respectively. Show that

(*i*) $\vec{AD} + \vec{BE} + \vec{CE} = \mathbf{0},$

(*ii*) $\vec{OD} + \vec{OE} + \vec{OF} = \vec{OA} + \vec{OB} + \vec{OC}.$

28. O, A, B and C are points such that

$$\vec{OA} = \mathbf{a}, \vec{OB} = \mathbf{b}, \vec{OC} = k\mathbf{a} + l\mathbf{b}.$$

where k and l are numbers. Express

$$\vec{AB}, \vec{AC} \text{ and } \vec{BC}$$

in terms of k, l, **a** and **b**, show that if $k + l = 1$, then AB, AC and BC are in the same direction. State the conclusion that follows from this, concerning the points A, B and C and express the ratio $AC : CB$ in terms of k and l.

X and Y are points such that

$$\vec{OX} = \frac{2}{3}\mathbf{a} \text{ and } \vec{OY} = 2\mathbf{b}.$$

Express $\vec{XY}$ in terms of **a** and **b** and hence show that the point Z, given by

$$\vec{OZ} = \vec{OX} + \frac{1}{4}\vec{XY}$$

is the mid-point of AB.

29. The triangles OAB, OCD are such that

$$\vec{OA} = \mathbf{a},\ \vec{OB} = \mathbf{b},\ \vec{OC} = -3\mathbf{a} \text{ and } \vec{OD} = -3\mathbf{b}.$$

Express in terms of **a** and **b** the vectors $\vec{AB}, \vec{BC}, \vec{AD}$ and $\vec{DC}$ and state which of them are in parallel directions.

30. In a triangle OAB, L is the mid-point of OA and M is the point on OB such that $\frac{OM}{MB} = 2$. P is the mid-point of LM and the line AP is produced to meet OB at Q.

Given that $\vec{OA} = \mathbf{a}$ and $\vec{OB} = \mathbf{b}$ find as linear combinations of **a** and **b** the vectors $\vec{OP}$ and $\vec{AP}$.

If $\vec{AQ} = h\vec{AP}$ and $\vec{OQ} = k\vec{OB}$,

find h, k, AP/PQ and OQ/QB.

31. The position vectors of points A, B, C relative to a fixed origin O are **a, b, c** respectively. If D is the mid-point of AB, and if E is the point which divides CB internally in the ratio 1 : 2, write down the position vectors of D and E in terms of **a, b, c.**

Show that the mid-point F of CD is on AE and find the ratio $AF : FE$.

32. In the triangle OAB, M is the mid-point of AB, C is a point on OM such that $OC = \frac{1}{2}CM$ and X is a point on OB such that $OX = 2XB$. The line XC is produced to meet OA at Y. Find

$$\frac{OY}{YA} \text{ and } \frac{XC}{CY}$$

33. $OABC$ is a quadrilateral in which the diagonals OB and AC meet at X. A line drawn through B parallel to CA meets OA produced at D. Show that $\frac{BD}{CA} = \frac{3}{4}$ and $\frac{AD}{OA} = \frac{1}{2}$.

34. The vertices P, Q and S of a triangle PQS have position vectors **p, q** and **s** respectively.

(*i*) Find m, the position vector of M, the mid-point of PQ, in terms of **p** and **q**.

(*ii*) Find **t**, the position vector of T on SM such that $ST : TM = 2 : 1$, in terms of **p**, **q** and **s**.

(*iii*) If the parallelogram *PQRS* is now completed. Express **r**, the position vector of the point *R* in terms of **p**, **q** and **s**.

Prove that *P*, *T* and *R* are collinear.

35. *ABC* is a triangle and *O*, any point in the plane of the same; *AO*, *BO* and *CO* meet the sides *BC*, *CA* and *AB* in *D*, *E*, *F* respectively; show that

$$\frac{OD}{AD}+\frac{OE}{BE}+\frac{OF}{CF}=1.$$

36. *ABCD* is a tetrahedron and *O* is any point; the lines joining *O* to the vertices meet the opposite faces in *P*, *Q*, *R*, *S* respectively. Prove that

$$\frac{OP}{AP}+\frac{OQ}{BQ}+\frac{OR}{CR}+\frac{OS}{DS}=1.$$

Take *O*, as origin of reference.

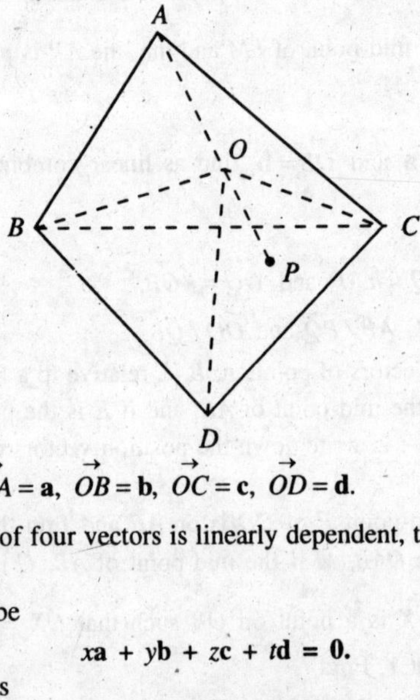

Let $\overrightarrow{OA}=\mathbf{a},\ \overrightarrow{OB}=\mathbf{b},\ \overrightarrow{OC}=\mathbf{c},\ \overrightarrow{OD}=\mathbf{d}.$

As every system of four vectors is linearly dependent, there is a linear relation connecting **a**, **b**, **c**, **d**.

Let this relation be

$$x\mathbf{a}+y\mathbf{b}+z\mathbf{c}+t\mathbf{d}=\mathbf{0}. \qquad ...(1)$$

We re-write (*i*) as

$$\frac{y\mathbf{b}+z\mathbf{c}+t\mathbf{d}}{y+z+t}=\frac{-x\mathbf{a}}{y+z+t}.$$

Now $(y\mathbf{b}+z\mathbf{c}+t\mathbf{d})\ /\ (y+z+t)$ is a point of the plane *BCD*, and $-x\mathbf{a}\ (y+z+t)$ is a point of the line *OA*. Thus, this must be the point *P* of intersection of the line *OA* with the plane *BCD*.

$$\therefore \qquad \overrightarrow{OP}=-\frac{x}{y+z+t}\mathbf{a} \quad\Rightarrow\quad \mathbf{a}=-\frac{y+z+t}{x}\overrightarrow{OP}.$$

Also $\qquad \overrightarrow{AP}=\overrightarrow{AO}+\overrightarrow{OP}$

$$=-\mathbf{a}-\frac{x}{y+z+t}\mathbf{a}=-\frac{\Sigma x}{y+z+t}\mathbf{a}$$

$$= -\frac{\Sigma x}{y+z+t} \cdot \frac{-(y+z+t)}{x} \vec{OP} = \frac{\Sigma x}{x} \vec{OP}$$

$$\Rightarrow \qquad \frac{AP}{OP} = \frac{\Sigma x}{x} \Rightarrow \frac{OP}{AP} = \frac{x}{\Sigma x}$$

Similarly

$$\frac{OQ}{BQ} = \frac{y}{\Sigma x}, \quad \frac{OR}{CR} = \frac{z}{\Sigma x}, \quad \frac{OS}{DS} = \frac{t}{\Sigma x}$$

Hence

$$\frac{OP}{AP} + \frac{OQ}{BQ} + \frac{OR}{CR} + \frac{OS}{DS} = 1.$$

37. Any plane cuts the sides *AB, BC, CD, DA* of a skew quadrilateral in *P, Q, R, S* respectively, prove that

$$\frac{AP}{PB} \cdot \frac{BQ}{QC} \cdot \frac{CR}{RD} \cdot \frac{DS}{SA} = 1.$$

Let **b, c, d** be the position vectors of the points *B, C, D* respectively with reference to *A* as origin.

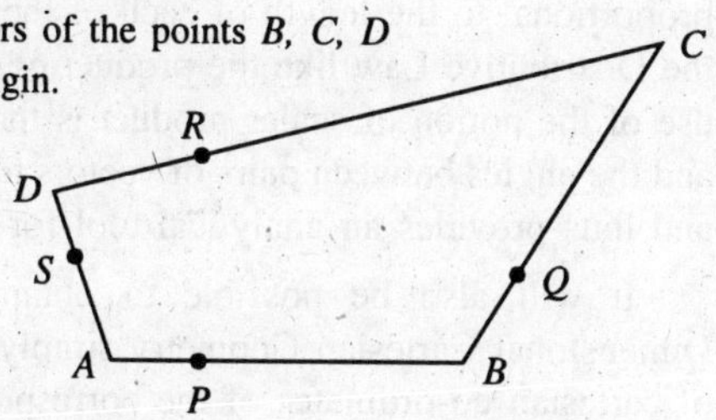

Let

$$\frac{AP}{PB} = \lambda_1, \quad \frac{BQ}{QC} = \lambda_2$$

$$\frac{CR}{DR} = \lambda_3, \quad \frac{DS}{SA} = \lambda_4,$$

so that the position vectors of the points *P, Q, R, S* are

$$\frac{\lambda_1 \mathbf{b}}{\lambda_1 + 1}, \quad \frac{\lambda_2 \mathbf{c} + \mathbf{b}}{\lambda_2 + 1}, \quad \frac{\lambda_2 \mathbf{d} + \mathbf{c}}{\lambda_3 + 1}, \quad \frac{\mathbf{d}}{\lambda_2 + 1}$$

respectively.

Denoting these by **p, q, r, s** respectively, we proceed to find a relation between the same independent of **b, c, d.** Thus

$$(\lambda_2 + 1)\,\mathbf{q} - (\lambda_3 + 1)\,\lambda_2 \mathbf{r} = \mathbf{b} - \lambda_2 \lambda_3 \mathbf{d}$$

$$= \frac{\lambda_1 + 1}{\lambda_1} \mathbf{p} - \lambda_2 \lambda_3 (\lambda_4 + 1)\,\mathbf{s}$$

$$\Rightarrow \qquad \frac{\lambda_1 + 1}{\lambda_1} \mathbf{p} - (\lambda_2 + 1)\,\mathbf{q} + (\lambda_3 + 1)\,\lambda_2 \mathbf{r} - \lambda_2 \lambda_3 (\lambda_4 + 1)\,\mathbf{s} = \mathbf{0}.$$

As, however, **p, q, r, s** are the position vectors of four coplanar points, we have

$$\frac{\lambda_1 + 1}{\lambda_1} - (\lambda_2 + 1) + (\lambda_2 + 1)\,\lambda_2 - \lambda_2 \lambda_2 (\lambda_4 + 1) = 0,$$

$$\Rightarrow \qquad \lambda_1 \lambda_2 \lambda_3 \lambda_4 = 1.$$

Hence the result.

3

Scalar Product

Introduction. The concept of the *Scalar product* of two vectors, as a result of which a scalar is associated to any given pair of vectors, will be introduced in this chapter. The justification for the use of the word *Product* lies in the fact that the so-called scalar product of two vectors is a scalar proportional to the length of each of the two factor vectors and also obeys the Distributive Law like the product of numbers. Also one very important use of the notion of scalar product is that it enables the lengths of vectors and the angles between pairs of vectors to be expressed in terms of the same and thus provides an analytical tool for the study of *Metric Geometry.*

It will also be possible to obtain various formulae of the three Dimensional Cartesian Geometry simply as a result of translation in terms of cartesian co-ordinates of the corresponding Vector notation formulae.

Some well-known properties of tetrahedra amenable to treatment by scalar products will also be obtained.

3.1. SCALAR PRODUCT

3.1.1. Scalar Product of Two Vectors

Def. *The scalar product of two vectors* **a, b** *is the scalar*

$$|\mathbf{a}|\ |\mathbf{b}|\ \cos\theta$$

where, θ, *is the angle, between the vectors* **a** *and* **b.**

Also the scalar product of any vector with the zero vector is, by definition, the scalar zero.

The scalar product of the vectors **a, b** is denoted by the symbol

$$\mathbf{a}\cdot\mathbf{b}$$

and is, on this account, also called the *Dot product* of the two vectors **a, b.**

Thus, we have, by definition,

$$\mathbf{a}\cdot\mathbf{b} = |\mathbf{a}|\ |\mathbf{b}|\ \cos\theta;$$

θ being the angle between the vectors **a, b.**

Note. It may be easily seen that the scalar product of two vectors remains unaltered when they are replaced by vectors equal to the same so that

$$\mathbf{a} = \mathbf{a}' \text{ and } \mathbf{b} = \mathbf{b}' \Rightarrow \mathbf{a} \cdot \mathbf{b} = \mathbf{a}' \cdot \mathbf{b}'.$$

Ex. D is the mid-point of the side BC of a triangle ABC; show that

$$\vec{DA} \cdot \vec{DB} + \vec{DA} \cdot \vec{DC} = 0.$$

3.1.2. Sign of the Scalar Product

Def. *If* **a**, **b** *be two non-zero vectors, then the scalar product*

$$\mathbf{a} \cdot \mathbf{b} = |\mathbf{a}| \, |\mathbf{b}| \cos \theta$$

is positive, negative or zero, according as the angle, θ, *between the vectors is acute, obtuse or right.*

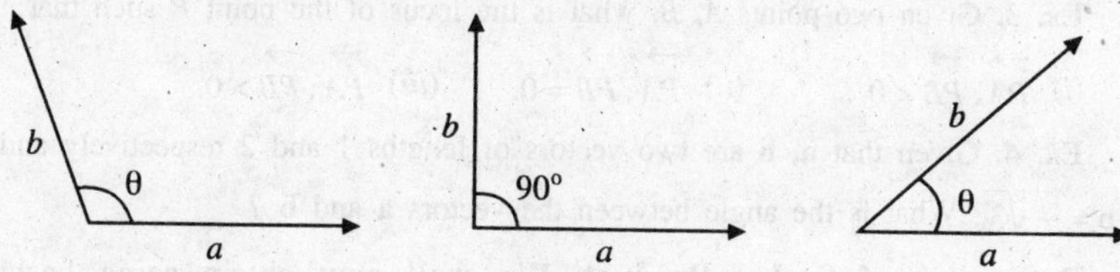

Fig. 3.1

Thus

$$\theta \text{ is acute} \Rightarrow \cos \theta > 0 \Rightarrow \mathbf{a} \cdot \mathbf{b} > 0,$$

$$\theta \text{ is right} \Rightarrow \cos \theta = 0 \Rightarrow \mathbf{a} \cdot \mathbf{b} = 0,$$

$$\theta \text{ is obtuse} \Rightarrow \cos \theta < 0 \Rightarrow \mathbf{a} \cdot \mathbf{b} < 0.$$

We notice that if **a**, **b** be two vectors, then their scalar product will be zero if and only if, either one at least of the two vectors is the zero vector or the two vectors are at right angles to each other. Thus

$$\mathbf{a} \cdot \mathbf{b} = 0 \Rightarrow \mathbf{a} = \mathbf{0} \text{ or } \mathbf{b} = \mathbf{0} \text{ or } \theta = 90^\circ;$$

θ being the angle between the vectors **a**, **b**.

It is very important to remember that the *scalar product of two non-zero vectors is zero if and only if they are at right angles to each other.*

3.1.3. Length of a Vector as a Scalar Product

If **a** be any vector, then the scalar product **a.a** of **a** with itself is given by

$$\mathbf{a} \cdot \mathbf{a} = |\mathbf{a}| \, |\mathbf{a}| \cos 0 = |\mathbf{a}|^2.$$

Thus, the length $|\mathbf{a}|$ *of any vector,* **a**, *is the non-negative square root* $\sqrt{(\mathbf{a} \cdot \mathbf{a})}$ *of the scalar product* **a.a**.

A convention. **a.a** will be denoted as $\mathbf{a}^2$, so that $\mathbf{a}^2$, is a scalar which equals the square of the length of **a**.

3.1.4. Angle between Two Vectors in terms of Scalar Products

If θ be the angle between two non-zero vectors, **a, b,** we have

$$\mathbf{a} \cdot \mathbf{b} = |\mathbf{a}| \, |\mathbf{b}| \cos\theta,$$

$$\Rightarrow \qquad \theta = \cos^{-1}\frac{\mathbf{a}\cdot\mathbf{b}}{|\mathbf{a}|\,|\mathbf{b}|} = \cos^{-1}\frac{\mathbf{a}\cdot\mathbf{b}}{(+\sqrt{\mathbf{a}\cdot\mathbf{a}})(+\sqrt{(\mathbf{b}\cdot\mathbf{b})})}.$$

Ex. 1. Is it true that

(*i*) $\mathbf{a} \cdot \mathbf{b} = \mathbf{a} \cdot \mathbf{c} \Rightarrow \mathbf{b} = \mathbf{c}$? (*ii*) $\mathbf{a} \cdot \mathbf{b} = 0 \Rightarrow \mathbf{a} \perp \mathbf{b}$?

Ex. 2. Give two points *O, A*; identify the locus of the point *P* in each of the following cases :

(*i*) $\vec{OP} \cdot \vec{OA} > 0$. (*ii*) $\vec{OP} \cdot \vec{OA} = 0$. (*iii*) $\vec{OP} \cdot \vec{OA} < 0$.

Ex. 3. Given two points *A, B,* what is the locus of the point *P* such that

(*i*) $\vec{PA} \cdot \vec{PB} < 0$. (*ii*) $\vec{PA} \cdot \vec{PB} = 0$. (*iii*) $\vec{PA} \cdot \vec{PB} > 0$.

Ex. 4. Given that **a, b** are two vectors of lengths 1 and 2 respectively and $\mathbf{a}.\mathbf{b} = -\sqrt{3}$. What is the angle between the vectors **a** and **b** ?

Properties of Scalar Product. We shall now obtain some basic properties of scalar product.

3.1.5. Commutativity

$$\mathbf{a} \cdot \mathbf{b} = \mathbf{b} \cdot \mathbf{a},$$

for every pair of vectors **a, b.**

This property is obvious from the definition.

3.1.6. $\mathbf{a} \cdot (-\mathbf{b}) = -(\mathbf{a} \cdot \mathbf{b})$; $(-\mathbf{a}) \cdot (-\mathbf{b}) = \mathbf{a} \cdot \mathbf{b}$ *for every pair of vectors,* **a, b.**

The proof is simple. [Refer Fig. 3.2]

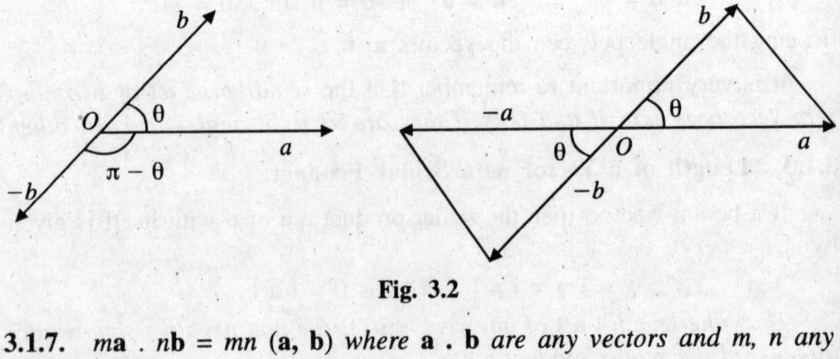

Fig. 3.2

3.1.7. *m***a** . *n***b** = *mn* (**a, b**) *where* **a** . **b** *are any vectors and m, n any scalars.*

Let *m, n* be both positive, so that the angle between *m***a** and *n***b** is the same as that between **a** and **b**.

We have

$$\begin{aligned} m\mathbf{a} \,.\, n\mathbf{b} &= |\, m\mathbf{a} \,|\; |\, n\mathbf{b} \,| \cos\theta \\ &= |\, m \,|\; |\, \mathbf{a} \,|\; |\, n \,|\; |\, \mathbf{b} \,| \cos\theta \\ &= mn \,|\, \mathbf{a} \,|\; |\, \mathbf{b} \,| \cos\theta = mn\,(\mathbf{a} \,.\, \mathbf{b}). \end{aligned}$$

Let now m be positive and n be negative so that the angle between $m\mathbf{a}$ and $n\mathbf{b}$ is $\pi - \theta$; θ being the angle between **a** and **b.**

We have

$$\begin{aligned} m\mathbf{a} \,.\, n\mathbf{b} &= |\, m\mathbf{a} \,|\; |\, n\mathbf{b} \,| \cos(\pi - \theta) \\ &= |\, m \,|\; |\, \mathbf{a} \,|\; |\, n \,|\; |\, \mathbf{b} \,| (-\cos\theta) \\ &= -mn \,|\, \mathbf{a} \,|\; |\, \mathbf{b} \,| (-\cos\theta) \\ &= mn \,|\, \mathbf{a} \,|\; |\, \mathbf{b} \,| \cos\theta = mn\,(\mathbf{a} \,.\, \mathbf{b}). \end{aligned}$$

Other cases may be similarly disposed of.

3.1.8. *If the scalar product of a vector,* **r**, *with each of three* **non-coplanar** *vectors is zero, then,* **r**, *must be the zero vector.*

This follows from the fact that no non-zero vector can be perpendicular to each of three non-coplanar vectors.

In particular, if a vector, **r**, is perpendicular to every vector, then **r** must be the zero vector.

3.1.9. *Let* $\mathbf{b} = \overrightarrow{PQ}$ *and let L, M be the feet of the perpendiculars for P and Q on the support AB of the vector* **a.**

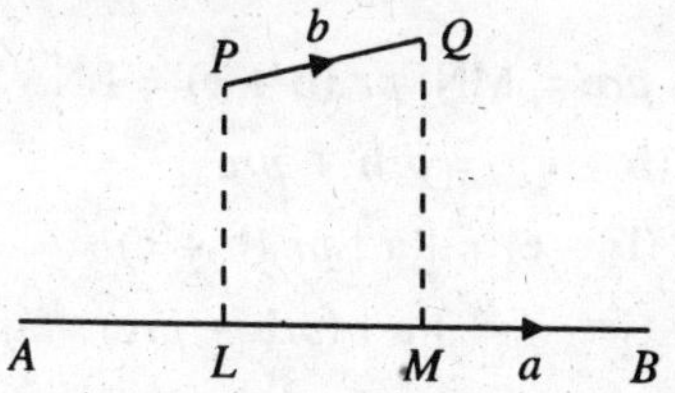

Fig. 3.3

The projection of **b** *upon* $AB = LM = |\,\mathbf{b}\,| \cos\theta$.

Thus, the scalar product **a . b** *is the product of the length of* **a**, *with the projection of* **b**, *upon* **a**, *taken with the proper sign.*

3.1.10. Distributivity

Scalar multiplication of vectors distributes the addition of vectors.

We have to prove that

$$\mathbf{a} \,.\, (\mathbf{b} + \mathbf{c}) = \mathbf{a} \,.\, \mathbf{b} + \mathbf{a} \,.\, \mathbf{c}$$

for all vectors **a, b, c.**

This property is an immediate consequence of the fact that *the projection of the sum of two vectors on any line is equal to the sum of their projections on the same line.*

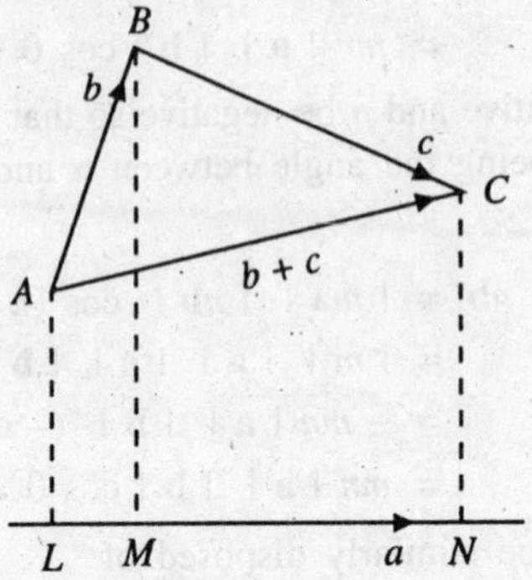

Fig. 3.4

Let

$$\vec{AB} = \mathbf{b},\ \vec{BC} = \mathbf{c} \text{ so that } \vec{AC} = \mathbf{b}+\mathbf{c}.$$

Let *L, M, N* be the feet of the perpendiculars from *A, B, C* on the line of support of the vector **a**.

Let

$$pr\mathbf{b},\ \ pr\mathbf{c},\ \ pr\ (\mathbf{b} + \mathbf{c})$$

denote the projections of **b, c** and **b + c** on **a**.

We have

$$pr\mathbf{b} = \text{LM},\ \ pr\mathbf{c} = \text{MN},\ pr\ (\mathbf{b} + \mathbf{c}) = \text{LN},$$

so that $pr\ (\mathbf{b} + \mathbf{c}) = pr\mathbf{b} + pr\mathbf{c}$

Thus

$$\begin{aligned}\mathbf{a} \cdot (\mathbf{b} + \mathbf{c}) &= |\mathbf{a}|\ pr\ (\mathbf{b} + \mathbf{c})\\ &= |\mathbf{a}|\ (pr\mathbf{b} + pr\mathbf{c})\\ &= |\mathbf{a}|\ pr\mathbf{b} + |\mathbf{a}|\ pr\mathbf{c} = \mathbf{a} \cdot \mathbf{b} + \mathbf{a} \cdot \mathbf{c}.\end{aligned}$$

It may also be shown that

$$\mathbf{a} \cdot (\mathbf{b} - \mathbf{c}) = \mathbf{a} \cdot \mathbf{b} - \mathbf{a} \cdot \mathbf{c}.$$

We have

$$\begin{aligned}\mathbf{a} \cdot (\mathbf{b} - \mathbf{c}) &= \mathbf{a} \cdot [\mathbf{b} + (-\mathbf{c})]\\ &= \mathbf{a} \cdot \mathbf{b} + \mathbf{a} \cdot (-\mathbf{c})\\ &= \mathbf{a} \cdot \mathbf{b} + [-(\mathbf{a} \cdot \mathbf{c})] = \mathbf{a} \cdot \mathbf{b} - \mathbf{a} \cdot \mathbf{c}.\end{aligned}$$

3.1.11. Some Simple Identities

(*i*) $(\mathbf{a} + \mathbf{b}) \cdot (\mathbf{a} - \mathbf{b}) = \mathbf{a} \cdot \mathbf{a} - \mathbf{b} \cdot \mathbf{b} = \mathbf{a}^2 - \mathbf{b}^2,$

(ii) $\quad (\mathbf{a} + \mathbf{b})^2 = (\mathbf{a} + \mathbf{b}) \cdot (\mathbf{a} + \mathbf{b})$

$$= \mathbf{a} \cdot \mathbf{a} + 2\mathbf{a} \cdot \mathbf{b} + \mathbf{b} \cdot \mathbf{b} = \mathbf{a}^2 + 2\mathbf{a} \cdot \mathbf{b} + \mathbf{b}^2,$$

(iii) $\quad (\mathbf{a} - \mathbf{b})^2 = (\mathbf{a} - \mathbf{b}) \cdot (\mathbf{a} - \mathbf{b})$

$$= \mathbf{a} \cdot \mathbf{a} - 2\mathbf{a} \cdot \mathbf{b} + \mathbf{b} \cdot \mathbf{b} = \mathbf{a}^2 - 2\mathbf{a} \cdot \mathbf{b} + \mathbf{b}^2.$$

These identities are simple consequences of the fact that the scalar multiplication distributes the addition of vectors and of the other results in § 19.

We may also note that

$$\mathbf{a} \cdot \mathbf{b} = \frac{1}{4}(|\mathbf{a}+\mathbf{b}|^2 - |\mathbf{a}-\mathbf{b}|^2),$$

so that the scalar product of two vectors is expressed in terms of the lengths of their sum and difference.

APPLICATIONS

EXAMPLES

Example 1. Cosine Formula for Triangles. *To prove that for any triangle ABC*

$$c^2 = b^2 + a^2 - 2ab \cos C, \qquad (\textit{Kolkata } 2005)$$

in the usual notation of plane Trigonometry.

Solution. We have

$$\vec{AC} + \vec{CB} = \vec{AB}$$

$$\Rightarrow \quad (\vec{AC} + \vec{CB}) \cdot (\vec{AC} + \vec{CB}) = \vec{AB} \cdot \vec{AB}$$

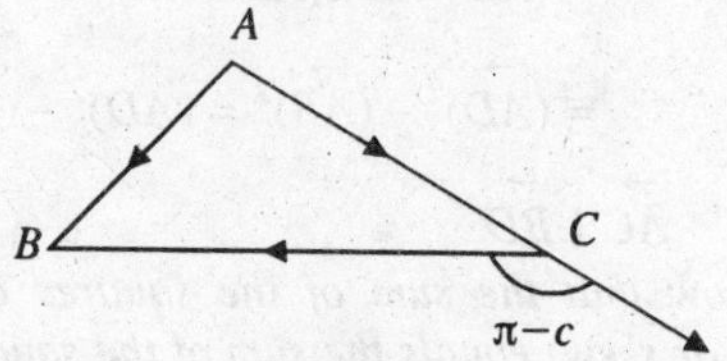

Fig 3.5

$$\Rightarrow \quad AC^2 + CB^2 + 2\vec{AC} \cdot \vec{CB} = AB^2$$

$$\Rightarrow \quad b^2 + a^2 - 2ab \cos C = c^2.$$

Example 2. Projection Formula for Triangles. *To prove that for any triangle ABC*

$$c = b \cos A + a \cos B.$$

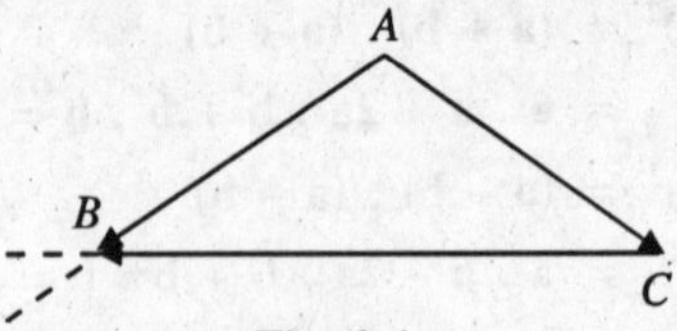

Fig. 3.6

Solution. We have

$$\vec{AC} + \vec{CB} = \vec{AB}$$

$$\Rightarrow \quad \vec{AC} \,.\, \vec{AB} + \vec{CB} \,.\, \vec{AB} = \vec{AB} \,.\, \vec{AB}$$

$$\Rightarrow \quad bc \cos A + ca \cos B = c^2$$

$$\Rightarrow \quad b \cos A + a \cos B = c.$$

Example 3. *Show that the diagonals of a rhombus are at right angles.*

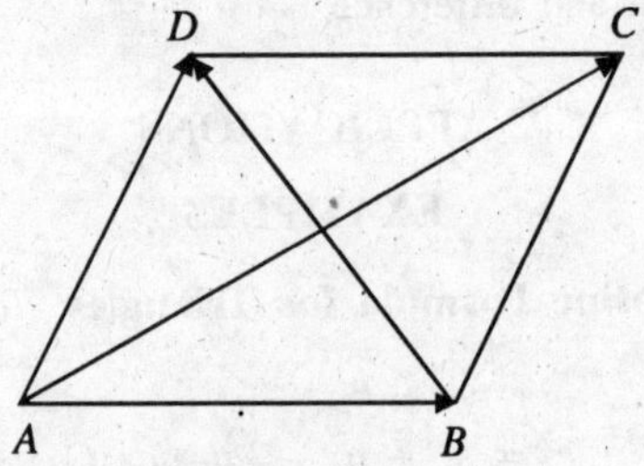

Fig. 3.7

Solution. Let *ABCD* be a rhombus.

We have

$$\vec{AC}.\vec{BD} = (\vec{AB} + \vec{BC}).(\vec{AD} - \vec{AB})$$

$$= (\vec{AB} + \vec{AD}).(\vec{AD} - \vec{AB})$$

$$= (\vec{AD})^2 - (\vec{AB})^2 = (\vec{AD})^2 - (\vec{AD})^2 = 0$$

$$\Rightarrow \quad \vec{AC} \perp \vec{BD}.$$

Example 4. *Show that the sum of the squares of the sides of any quadrilateral (plane or skew) equals the sum of the squares of its diagonals together with four times the square of the line joining their middle points.*

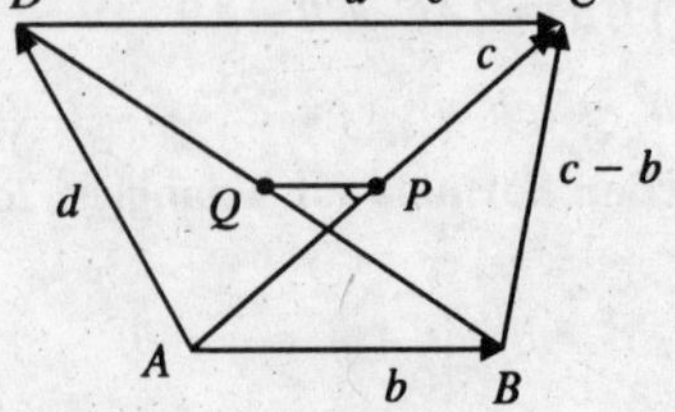

Fig. 3.8

Solution. Take the vertex A as origin.

Let **b, c, d** be the position vectors of the vertices B, C, D.

The position vectors of the mid-points P, Q of the diagonals AC, BD are $\frac{1}{2}\mathbf{c}$, $\frac{1}{2}(\mathbf{b}+\mathbf{d})$ respectively. We have

$$AB^2 + BC^2 + CD^2 + DA^2 = \mathbf{b}^2 + (\mathbf{c} - \mathbf{b})^2 + (\mathbf{d} - \mathbf{c})^2 + \mathbf{d}^2$$
$$= 2\,(\mathbf{b}^2 + \mathbf{c}^2 + \mathbf{d}^2) - 2\,(\mathbf{b.c} + \mathbf{c.d})$$

Also $$AC^2 + BD^2 + 4PQ^2 = \mathbf{c}^2 + (\mathbf{d} - \mathbf{b})^2 + 4\left[\frac{1}{2}(\mathbf{b}+\mathbf{d}) - \frac{1}{2}\mathbf{c}\right]^2$$
$$= 2\,(\mathbf{b}^2 + \mathbf{c}^2 + \mathbf{d}^2) - 2\,(\mathbf{b.c} + \mathbf{c.d}).$$

Hence the result.

Example 5. *D is the mid-point of the side BC of a triangle ABC; show that*

$$AB^2 + AC^2 = 2\,(AD^2 + BD^2).$$

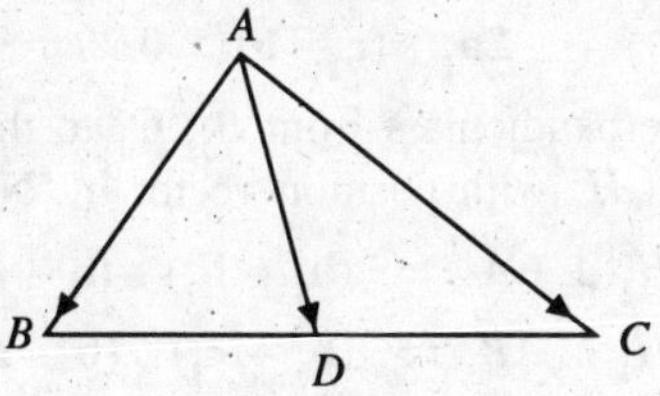

Fig. 3.9

Solution. We have

$$\vec{AB} = \vec{AD} + \vec{DB}$$

$$\Rightarrow \quad AB^2 = (\vec{AD} + \vec{DB})^2$$
$$= AD^2 + DB^2 + 2\,\vec{AD}\,.\,\vec{DB}. \qquad ...(i)$$

Also we have

$$\vec{AC} = \vec{AD} = \vec{DC}$$

$$\Rightarrow \quad AC^2 = (\vec{AD} + \vec{DC})^2$$
$$= AD^2 + DC^2 + 2\,\vec{AD}\,.\,\vec{DC}. \qquad ...(ii)$$

Adding (*i*) and (*ii*) we get

$$AB^2 + AC^2 = 2AD^2 + 2BD^2 + 2\,\vec{AD}.(\vec{DB} + \vec{DC})$$
$$= 2(DA^2 + DB^2), \text{ for } \vec{DB} + \vec{DC} = \mathbf{0}.$$

Example 6. *ABC and $A_1B_1C_1$ are two coplanar triangles such that the perpendiculars from A, B, C to the sides B_1C_1, C_1A_1, A_1B_1 of the triangle $A_1B_1C_1$ are concurrent; show that the perpendiculars from A_1, B_1, C_1 to the sides BC, CA, AB of the triangle ABC are also concurrent.*

Solution. Let the position vectors of the vertices of two triangles with respect to any origin *O* of reference be

$$\mathbf{a}, \mathbf{b}, \mathbf{c}; \mathbf{a}_1, \mathbf{b}_1, \mathbf{c}_1.$$

Let the perpendiculars from *A, B, C* to the sides of the second triangle concur at a point *H* with position vector **h**.

Now

$$AH \perp B_1C_1 \Rightarrow (\mathbf{h} - \mathbf{a}) \cdot (\mathbf{c}_1 - \mathbf{b}_1) = 0,$$

$$BH \perp C_1A_1 \Rightarrow (\mathbf{h} - \mathbf{b}) \cdot (\mathbf{a}_1 - \mathbf{c}_1) = 0,$$

$$CH \perp A_1B_1 \Rightarrow (\mathbf{h} - \mathbf{c}) \cdot (\mathbf{b}_1 - \mathbf{a}_1) = 0.$$

Adding we obtain

$$\Sigma (\mathbf{h} - \mathbf{a}) \cdot (\mathbf{c}_1 - \mathbf{b}_1) = 0$$

$$\Rightarrow \quad \Sigma \mathbf{a} \cdot (\mathbf{c}_1 - \mathbf{b}_1) = 0$$

$$\Rightarrow \quad \Sigma \mathbf{a}_1 \cdot (\mathbf{c} - \mathbf{b}) = 0. \qquad ...(i)$$

Again, let the perpendiculars from B_1, C_1 to the sides *CA*, *AB* of the triangle *ABC* meet at H_1 with position vector $\mathbf{h}_1$. Now

$$B_1H_1 \perp CA \Rightarrow (\mathbf{h}_1 - \mathbf{b}_1) \cdot (\mathbf{a} - \mathbf{c}) = 0 \qquad ...(ii)$$

$$C_1H_1 \perp AB \Rightarrow (\mathbf{h}_1 - \mathbf{c}_1) \cdot (\mathbf{b} - \mathbf{a}) = 0. \qquad ...(iii)$$

Adding (*i*), (*ii*) and (*iii*), we get

$$(\mathbf{h}_1 - \mathbf{a}_1) \cdot (\mathbf{c} - \mathbf{b}) = 0$$

$$\Rightarrow \quad A_1H_1 \perp BC.$$

Hence the result.

Example 7. *Show that the circumcentre, the centroid and the orthocentre of a triangle are collinear and the centroid divides the join of the circumcentre and the orthocentre in the ratio 1 : 2.*

Solution. Let *O, G, H* denote the circumcentre, centroid and orthocentre respectively of a triangle *ABC*.

Let **a, b, c** be the position vectors of the vertices *A, B, C* of the triangle with respect to the circumcentre *O*, as the origin of reference.

Fig. 3.10

We have

$$OA = OB = OC$$

$$\Rightarrow \quad \mathbf{a}^2 = \mathbf{b}^2 = \mathbf{c}^2 \qquad ...(i)$$

Also the position vector of G is

$$\vec{OG} = \frac{1}{3}(\mathbf{a}+\mathbf{b}+\mathbf{c}).$$

We have, by (*i*)

$$[(\mathbf{a} + \mathbf{b} + \mathbf{c}) - \mathbf{a}] \cdot (\mathbf{b} - \mathbf{c}) = 0,$$
$$[(\mathbf{a} + \mathbf{b} + \mathbf{c}) - \mathbf{b}] \cdot (\mathbf{c} - \mathbf{a}) = 0,$$
$$[(\mathbf{a} + \mathbf{b} + \mathbf{c}) - \mathbf{c}] \cdot (\mathbf{a} - \mathbf{b}) = 0.$$

so that if H' denotes the point with position vector

$$\mathbf{a} + \mathbf{b} + \mathbf{c},$$

we see that

$$H'A \perp BC, \quad H'B \perp CA, \quad H'C \perp AB.$$

so that H is the orthocentre of the triangle ABC

and we have $\quad \vec{OH} = \mathbf{a}+\mathbf{b}+\mathbf{c}.$

Thus, we have

$$\vec{OH} = 3\vec{OG}$$

$\Rightarrow$ G divides OH in the ratio 1 : 2.

Hence the result.

Example 8. *If two pairs of opposite edges of a tetrahedron are at right angles, then show that the third pair is also at right angle. Further show that for such a tetrahedron, the sum of the squares of each pair of opposite edges is the same.*

Solution. Let $OABC$ be a given tetrahedron such that

$$OA \perp BC \text{ and } OB \perp CA.$$

Let

$$\vec{OA} = \mathbf{a}, \quad \vec{OB} = \mathbf{b}, \quad \vec{OC} = \mathbf{c}.$$

Now

$$OA \perp BC \Rightarrow \vec{OA} \cdot \vec{BC} = 0$$

$\Rightarrow \quad \mathbf{a} \cdot (\mathbf{c} - \mathbf{b}) = 0 \Rightarrow \mathbf{a} \cdot \mathbf{c} = \mathbf{a} \cdot \mathbf{b} \quad$ ...(*i*)

Also

$$OB \perp CA \Rightarrow \vec{OB} \cdot \vec{CA} = 0$$

$\Rightarrow \quad \mathbf{b} \cdot (\mathbf{a} - \mathbf{c}) = 0 \Rightarrow \mathbf{b} \cdot \mathbf{a} = \mathbf{b} \cdot \mathbf{c} \quad$ ...(*ii*)

From the results (*i*) and (*ii*), we have

$$\mathbf{a} \cdot \mathbf{c} = \mathbf{a} \cdot \mathbf{b} = \mathbf{b} \cdot \mathbf{c}$$

$$\mathbf{a} \cdot \mathbf{c} = \mathbf{b} \cdot \mathbf{c} \quad \Rightarrow \quad (\mathbf{a} - \mathbf{b}) \cdot \mathbf{c} = 0. \qquad ...(iii)$$

$$\Rightarrow \qquad \vec{BA} \cdot \vec{OC} = 0 \Rightarrow OC \perp AB.$$

Again

$$OA^2 + BC^2 = (\vec{OA})^2 + (\vec{BC})^2$$
$$= \mathbf{a}^2 + (\mathbf{c} - \mathbf{b})^2 = \mathbf{a}^2 + \mathbf{b}^2 + \mathbf{c}^2 - 2\mathbf{b} \cdot \mathbf{c}.$$

Similarly, we have

$$OB^2 + CA^2 = \mathbf{a}^2 + \mathbf{b}^2 + \mathbf{c}^2 - 2\mathbf{c} \cdot \mathbf{a},$$
$$OC^2 + AB^2 = \mathbf{a}^2 + \mathbf{b}^2 + \mathbf{c}^2 - 2\mathbf{a} \cdot \mathbf{b}.$$

Thus, we have

$$OA^2 + BC^2 = OB^2 + CA^2 = OC^2 + AB^2.$$

for $\mathbf{a} \cdot \mathbf{b} = \mathbf{b} \cdot \mathbf{c} = \mathbf{c} \cdot \mathbf{a}$.

Example 9. *If the perpendiculars from two vertices B and C to the opposite faces of a tetrahedron ABCD intersect, then BC is perpendicular to AD and the perpendicular from A and D to the opposite faces also intersect.*

Solution. Take A as the origin of reference. Let

$$\vec{AB} = \mathbf{b}, \ \vec{AC} = \mathbf{c}, \ \vec{AD} = \mathbf{d}.$$

Let the perpendiculars from B and C to the opposite faces ACD and ABD meet at a point H whose position vector is $\mathbf{h}$. Thus, we have

$$\vec{BH} \perp \vec{AC} \Rightarrow (\mathbf{h} - \mathbf{b}) \cdot \mathbf{c} = 0; \ \vec{BH} \perp \vec{AD} \Rightarrow (\mathbf{h} - \mathbf{b}) \cdot \mathbf{d} = 0,$$

$$\vec{CH} \perp \vec{AB} \Rightarrow (\mathbf{h} - \mathbf{c}) \cdot \mathbf{b} = 0; \ \vec{CH} \perp \vec{AD} \Rightarrow (\mathbf{h} - \mathbf{c}) \cdot \mathbf{d} = 0.$$

These give

$$\begin{cases} \mathbf{h} \cdot \mathbf{c} = \mathbf{b} \cdot \mathbf{c}, & \mathbf{h} \cdot \mathbf{d} = \mathbf{b} \cdot \mathbf{d} \\ \mathbf{h} \cdot \mathbf{b} = \mathbf{b} \cdot \mathbf{c}; & \mathbf{h} \cdot \mathbf{d} = \mathbf{c} \cdot \mathbf{d} \end{cases}$$

$$\Rightarrow \qquad \mathbf{b} \cdot \mathbf{d} = \mathbf{c} \cdot \mathbf{d} \Rightarrow (\mathbf{b} - \mathbf{c}) \cdot \mathbf{d} = 0 \quad \Rightarrow \quad BC \perp AD.$$

Let K be the foot of the perpendicular from A to the opposite face BCD and let $\mathbf{k}$ be its position vector. As AK is perpendicular to the plane BCD, we have

$$\mathbf{k} \cdot (\mathbf{c} - \mathbf{b}) = 0, \quad \mathbf{k} \cdot (\mathbf{b} - \mathbf{d}) = 0$$

$$\Rightarrow \qquad \mathbf{k} \cdot \mathbf{c} = \mathbf{k} \cdot \mathbf{b} = \mathbf{k} \cdot \mathbf{d}.$$

We shall show that there is a point L on AK such that DL is perpendicular to the face ABC. If $\lambda\mathbf{k}$ be the position vector of this point, the two equations

$$(\lambda\mathbf{k} - \mathbf{d}) \cdot \mathbf{b} = 0, \qquad (\lambda\mathbf{k} - \mathbf{d}) \cdot \mathbf{c} = 0$$

must be consistent in relation to λ.

These equations are really the same for

$$\mathbf{k} \cdot \mathbf{b} = \mathbf{k} \cdot \mathbf{c} \quad \text{and} \quad \mathbf{b} \cdot \mathbf{d} = \mathbf{c} \cdot \mathbf{d},$$

so that λ is determined.

Hence the result.

Example 10. *Prove that the angle in a semi-circle is a right angle.*

Solution. Let O be the centre and AB the bounding diameter of the semi-circle. Let P be any point on the circumference. With O as origin, let $\overrightarrow{OA} = \mathbf{a}$, $\overrightarrow{OB} = -\mathbf{a}$ and $\overrightarrow{OP} = \mathbf{r}$.

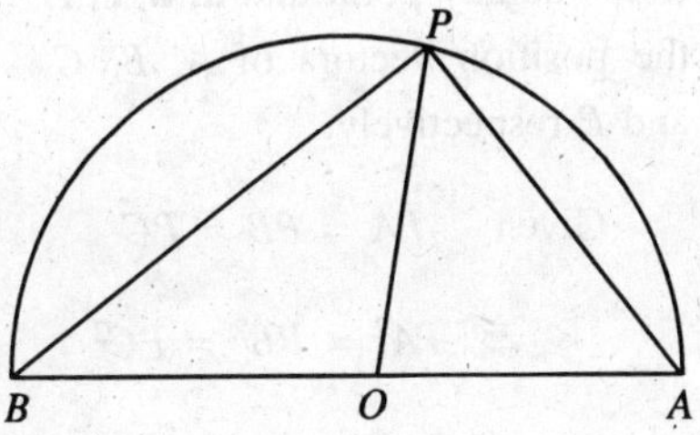

Fig. 3.11

Obviously, $OA = OB = OP$, each being equal to radius of the semi-circle.

$$\overrightarrow{AP} = \mathbf{r} - \mathbf{a} \quad \text{and} \quad \overrightarrow{BP} = \mathbf{r} - (-\mathbf{a}) = \mathbf{r} + \mathbf{a}$$

$$\therefore \quad \overrightarrow{AP} \cdot \overrightarrow{BP} = (\mathbf{r} - \mathbf{a}) \cdot (\mathbf{r} + \mathbf{a}) = \mathbf{r}^2 - \mathbf{a}^2$$

$$= OP^2 - OA^2 = 0$$

$\Rightarrow$ AP and BP are perpendicular to each other, *i.e.*, $\angle APB = 90°$.

Example 11. *The base BC of a* ΔABC *is divided at D so that mBD = nCD. Show that* $mAB^2 + nAC^2 = mBD^2 + nCD^2 + (m + n) AD^2$.

Solution. With A as origin, let the position vectors of B and C be $\mathbf{b}$ and $\mathbf{c}$.

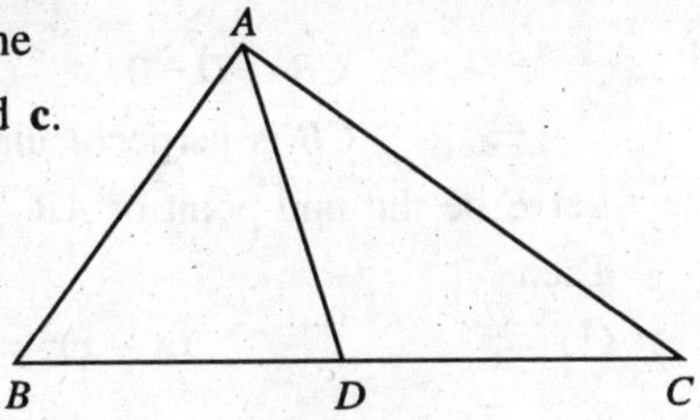

$$m\overrightarrow{BD} = n\overrightarrow{DC}$$

$$\Rightarrow \quad m\overrightarrow{DB} + n\overrightarrow{DC} = 0$$

Fig. 3.12

Now,

$$AB^2 = (\overrightarrow{AB})^2 = (\overrightarrow{AD} + \overrightarrow{DB})^2 = AD^2 + DB^2 + 2\overrightarrow{AD} \cdot \overrightarrow{DB}$$

$$AC^2 = (\overrightarrow{AC})^2 = (\overrightarrow{AD} + \overrightarrow{DC})^2 = AD^2 + DC^2 + 2\overrightarrow{AD} \cdot \overrightarrow{DC}$$

Multiplying by m and n respectively and adding, we get

$$mAB^2 + nAC^2 = (m + n) AD^2 + mBD^2 + nDC^2 + 2\overrightarrow{AD} \cdot (m\overrightarrow{DB} + n\overrightarrow{DC})$$

$$= (m + n) AD^2 + mBD^2 + nDC^2.$$

Example 12. *Prove that if a point is equidistant from the vertices of a right angled triangle, its joint to the mid-point of the hypotenuse is perpendicular to the plane of the triangle.*

Solution. ABC be a right angled triangle and let D be the mid-point of its hypotenuse BC. Let P be any point and **a, b, c, r** the position vectors of A, B, C and P respectively.

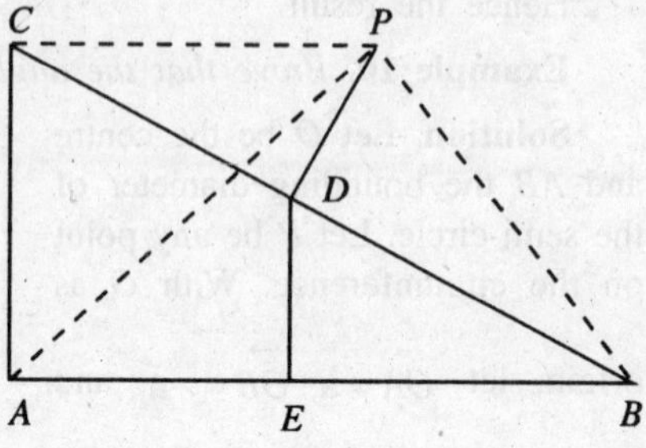

Fig. 3.13

Given $PA = PB = PC$

$\Rightarrow PA^2 = PB^2 = PC^2$

$$\Rightarrow \quad (\mathbf{a} - \mathbf{r})^2 = (\mathbf{b} - \mathbf{r})^2 = (\mathbf{c} - \mathbf{r})^2 \qquad ...(1)$$

$$(1) \Rightarrow \quad (\mathbf{b} - \mathbf{r}) \cdot (\mathbf{b} - \mathbf{r}) = (\mathbf{c} - \mathbf{r}) \cdot (\mathbf{c} - \mathbf{r})$$

$$\Rightarrow \quad \mathbf{b}^2 + \mathbf{r}^2 - 2\mathbf{b} \cdot \mathbf{r} = \mathbf{c}^2 + \mathbf{r}^2 - 2\mathbf{c} \cdot \mathbf{r}$$

$$\Rightarrow \quad \mathbf{b}^2 - \mathbf{c}^2 - 2\,(\mathbf{r} \cdot \mathbf{b} - \mathbf{r} \cdot \mathbf{c}) = 0$$

$$\Rightarrow \quad (\mathbf{b} - \mathbf{c}) \cdot (\mathbf{b} + \mathbf{c}) - 2\mathbf{r} \cdot (\mathbf{b} - \mathbf{c}) = 0$$

$$\Rightarrow \quad (\mathbf{b} - \mathbf{c}) \cdot [(\mathbf{b} + \mathbf{c}) - 2\mathbf{r}] = 0$$

$$\Rightarrow \quad 2\,(\mathbf{b} - \mathbf{c}) \cdot \left[\frac{\mathbf{b}+\mathbf{c}}{2} - \mathbf{r}\right] = 0$$

$$\Rightarrow \quad \vec{CB} \cdot \vec{PD} = 0$$

$\Rightarrow$ CB is perpendicular to PD. ...(1)

Let E be the mid-point of AB.

Then

$$(1) \Rightarrow \quad (\mathbf{a} - \mathbf{r})^2 = (\mathbf{b} - \mathbf{r})^2$$

$$\Rightarrow \quad 2\,(\mathbf{a} - \mathbf{b}) \cdot \left(\frac{\mathbf{a}+\mathbf{b}}{2} - \mathbf{r}\right) = 0$$

$$\Rightarrow \quad 2\vec{BA} \cdot \vec{PE} = 0$$

$\Rightarrow$ BA is perpendicular to PE. ...(2)

Since D and E are mid-points of BC and AB respectively, DE must be parallel to AC. Again, AC is perpendicular to BA.

Hence, DE is perpendicular to BA, *i.e.*,

$$\vec{AB} \cdot \vec{DE} = 0 \qquad ...(3)$$

Now,

$$\vec{AD} = \vec{AB} + \vec{BD}$$

$$\Rightarrow \quad \vec{AD} \cdot \vec{DP} = (\vec{AB} + \vec{BD}) \cdot (\vec{DP})$$

$$= \vec{AB} \cdot \vec{DP} + \vec{BD} \cdot \vec{DP}$$

$$= \vec{AB} \cdot \vec{DP} + \frac{1}{2}\vec{BC} \cdot \vec{DP}$$

[$\because$ D is mid-point of BC]

$$= \vec{AB} \cdot \vec{DP} \qquad \text{[by (1)]}$$

$$= \vec{AB} \cdot (\vec{DE} + \vec{EP})$$

$$= \vec{AB} \cdot \vec{DE} + \vec{AB} \cdot \vec{EP} = 0, \qquad \text{[by (2) and (3)]}$$

$\Rightarrow$ DP is perpendicular to AD. ...(4)

From (1) and (4), it follows that PD is perpendicular to both CB and AD and hence, PD must be perpendicular to the plane of $\Delta\, ABC$.

EXERCISES

1. Prove that the altitudes of a triangle are concurrent.
2. Prove that the right bisectors of the sides of a triangle are concurrent.
3. Show that the mid-point of the hypotenuse of a right angled triangle is equidistant from its vertices.
4. Show that a parallelogram whose diagonals are equal is a rectangle.
5. Show that a quadrilateral whose diagonals bisect each other at right angles is a rhombus.
6. Two medians of a triangle are equal, show that the triangle is isosceles.
7. Show that the diagonals of a trapezium having equal non-parallel sides are equal and conversely.
8. Show that the sum of the squares on the diagonals of a parallelogram is equal to the sum of the squares on its sides.
9. Show that the sum of the squares of the six edges of a tetrahedron is equal to
   (*i*) four times the sum of the squares of the lines joining the vertices to its centroid;
   (*ii*) four times the sum of the squares of the lines joining the mid-points of its opposite edges.
10. Show that the sum of the squares of the four diagonals of a parallelopiped is equal to the sum of the squares of its edges.

11. Show that the sides of a trapezium having equal non-parallel sides are equally inclined to the parallel sides.

12. A triangle OAB is right angled at O; squares $OALM$ and $OBPQ$ are constructed on the sides OA and OB externally.

    Show that the lines AP and BL intersect on the altitude through O.

13. Show that a parallelopiped with equal diagonals is a cuboid.

    Also show that a cuboid such that the angle between any two of its diagonals is the same is a cube.

14. If a straight line be perpendicular to each of two intersecting straight lines at their point of intersection, prove that it is perpendicular to every line in the plane determined by the two lines.

15. In a tetrahedron $OABC$, $OA \perp BC$, show that

$$OB^2 + CA^2 = OC^2 + AB^2.$$

16. If two opposite edges of a tetrahedron are equal in length and are at right angles to the line joining their mid-points, show that the remaining pairs of opposite edges have the same property.

17. The line joining the mid-points of two opposite edges of a tetrahedron is perpendicular to the edges; show that the remaining pairs of opposite edges are equal. Also prove the converse.

18. ABCD is a tetrahedron such that the perpendiculars AK, BL, CM and DN to the opposite faces are concurrent. Prove that

    (*i*) any two opposite edges of the tetrahedron are orthogonal.

    (*ii*) K is the orthocentre of the triangle BCD.

19. Two opposite edges AB, CD of a tetrahedron are perpendicular to each other; show that the distance between the mid-points of BD and AC is equal to the distance between the mid-points of BC, AD.

20. $ABCD$ is a tetrahedron and G is the centroid of the base BCD. Prove that

$$AB^2 + AC^2 + AD^2 = GB^2 + GC^2 + GD^2 + 3GA^2.$$

3.2. ORTHOGONAL BASES

It has been seen in Chapter 1 that a linearly independent set of three vectors can be considered as a *Base* in the sense that any given vector can be expressed as a linear combination of its members. Also we have seen that any non-coplanar triad of vectors is a linearly independent system.

In view of the concepts of the length of a vector and angle between vectors as introduced in this chapter, it is possible to consider as Bases systems such that

(*i*) *Length of each member of the system is unity;*

(*ii*) *and any two members of the system are mutually perpendicular.*

Such bases are known as **Orthogonal.**

Let **i, j, k** denote vectors of unit length along three perpendicular lines as shown in the figure. As the length of each of the vectors **i, j, k** is unity and they are mutually perpendicular. We have the relations

$$\mathbf{i}\,.\,\mathbf{i} = 1,\quad \mathbf{j}\,.\,\mathbf{j} = 1,\quad \mathbf{k}\,.\,\mathbf{k} = 1;$$

$$\mathbf{i}\,.\,\mathbf{j} = 0 = \mathbf{j}\,.\,\mathbf{i},\quad \mathbf{j}\,.\,\mathbf{k} = 0 = \mathbf{k}\,.\,\mathbf{j},\quad \mathbf{k}\,.\,\mathbf{i} = 0 = \mathbf{i}\,.\,\mathbf{k}. \qquad ...(i)$$

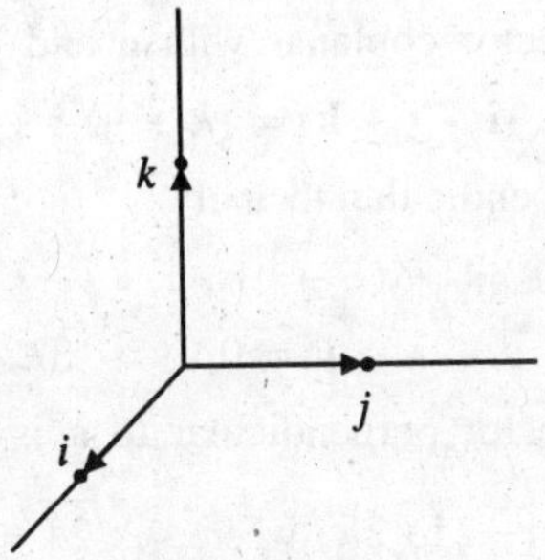

Fig. 3.14

3.2.1. Scalar Product in Terms of Components

Let two given vectors **a, b** expressed as linear combinations of the vectors **i, j, k** of an orthogonal base be

$$\mathbf{a} = a_1\mathbf{i} + a_2\mathbf{j} + a_3\mathbf{k},\quad \mathbf{b} = b_1\mathbf{i} + b_2\mathbf{j} + b_3\mathbf{k}.$$

Making use of the property that scalar multiplication distributes the sum of vectors and of the relation (*i*), we obtain

$$\mathbf{a.b} = (a_1\mathbf{i} + a_2\mathbf{j} + a_3\mathbf{k})\,.\,(b_1\mathbf{i} + b_2\mathbf{j} + b_3\mathbf{k}) = a_1b_1 + a_2b_2 + a_3b_3.$$

3.2.2. Length of a Vector and Angle between Two given Vectors

We have

$$\mathbf{a} = a_1\mathbf{i} + a_2\mathbf{j} + a_3\mathbf{k} \quad\Rightarrow\quad |a|^2 = \mathbf{a}\,.\,\mathbf{a} = a_1^2 + a_2^2 + a_3^2.$$

Also the angle θ between the vectors

$$\mathbf{a} = a_1\mathbf{i} + a_2\mathbf{j} + a_3\mathbf{k},\quad \mathbf{b} = b_1\mathbf{i} + b_2\mathbf{j} + b_3\mathbf{k}$$

is given by

$$\cos\theta = \frac{\mathbf{a}\,.\,\mathbf{b}}{|\mathbf{a}|\,|\mathbf{b}|} = \frac{a_1b_1 + a_2b_2 + a_3b_3}{\sqrt{(a_1^2 + a_2^2 + a_3^2)}\,\sqrt{(b_1^2 + b_2^2 + b_3^2)}}.$$

Cor. For the perpendicularity of the two vectors

$$a_1\mathbf{i} + a_2\mathbf{j} + a_3\mathbf{k},\quad b_1\mathbf{i} + b_2\mathbf{j} + b_3\mathbf{k},$$

we have

$$(a_1\mathbf{i} + a_2\mathbf{j} + a_3\mathbf{k})\,.\,(b_1\mathbf{i} + b_2\mathbf{j} + b_3\mathbf{k}) = \cos 90^\circ = 0$$

$$\Rightarrow \qquad a_1b_1 + a_2b_2 + a_3b_3 = 0.$$

EXAMPLES

Example 1. *Given two vectors*

$$\mathbf{a} = \mathbf{i} + \mathbf{j} - \mathbf{k}; \quad \mathbf{b} = \mathbf{i} - \mathbf{j} + \mathbf{k},$$

find a unit vector **c**, *perpendicular to the vector* **a** *and coplanar with* **a** *and* **b**. *Find also a vector* **d** *perpendicular to both* **a** *and* **c**.

Solution. Any vector coplanar with **a** and **b** is

$\lambda\,(\mathbf{i} + \mathbf{j} - \mathbf{k}) + \mu\,(\mathbf{i} - \mathbf{j} + \mathbf{k}) = (\lambda + \mu)\,\mathbf{i} + (\lambda - \mu)\,\mathbf{j} + (-\lambda + \mu)\,\mathbf{k}.$

This will be perpendicular to **a** if

$$(\lambda + \mu)\,.\,1 + (\lambda - \mu)\,.\,1 + (-\lambda + \mu)\,(-1) = 0$$

$$\Rightarrow \lambda + \mu + \lambda - \mu + \lambda - \mu = 0 \Rightarrow 3\lambda - \mu = 0 \Rightarrow \mu = 3\lambda.$$

Thus, the unit vector perpendicular to **a** is

$$\lambda\,(4\mathbf{i} - 2\mathbf{j} + 2\mathbf{k}) \text{ where } \lambda\sqrt{24} = 1 \Rightarrow \lambda = \frac{1}{\sqrt{24}}$$

Thus

$$\mathbf{c} = \frac{1}{\sqrt{24}}(4\mathbf{i} - 2\mathbf{j} + 2\mathbf{k}) = \frac{1}{\sqrt{6}}(2\mathbf{i} - \mathbf{j} + \mathbf{k}).$$

If $p\mathbf{i} + q\mathbf{j} + r\mathbf{k}$ is the required unit vector **d** then, because of its perpendicularity to both **a** and **c**, we have,

$$p + q - r = 0; \quad 2p - q + r = 0$$

which give $p = 0$, $q = r$. [By taking $\mathbf{a}\,.\,\mathbf{d} = 0$ and $\mathbf{c}\,.\,\mathbf{d} = 0$]

Thus, the required vector **d** is given by $(1/\sqrt{2})(\mathbf{j} + \mathbf{k})$.

Example 2. *Given two vectors*

$$\mathbf{a} = 2\mathbf{i} - 3\mathbf{j} + \mathbf{k}; \quad \mathbf{b} = -\mathbf{i} + 2\mathbf{j} - \mathbf{k}.$$

Find the projection of **a** *on* **b** *and that of* **b** *on* **a**.

Solution. Let θ denote the angle between the vectors **a** and **b**, so that $|\mathbf{b}|\cos\theta$ is the projection of **b** on **a** and $|\mathbf{a}|\cos\theta$ is the projection of **a** on **b**.

We have

$$|\mathbf{b}|\cos\theta = \frac{\mathbf{a}\,.\,\mathbf{b}}{|\mathbf{a}|} = \frac{-9}{\sqrt{14}},$$

and

$$|\mathbf{a}|\cos\theta = \frac{\mathbf{a}\,.\,\mathbf{b}}{|\mathbf{b}|} = \frac{-9}{\sqrt{6}}$$

Hence the results.

Example 3. *Prove that* $\cos(B - A) = \cos A \cos B + \sin A \sin B.$

Solution. Let $\angle XOL = A, \quad \angle XOM = B.$

Draw $OY \perp OX.$

Let **i, j** denote unit vectors along OX and OY.

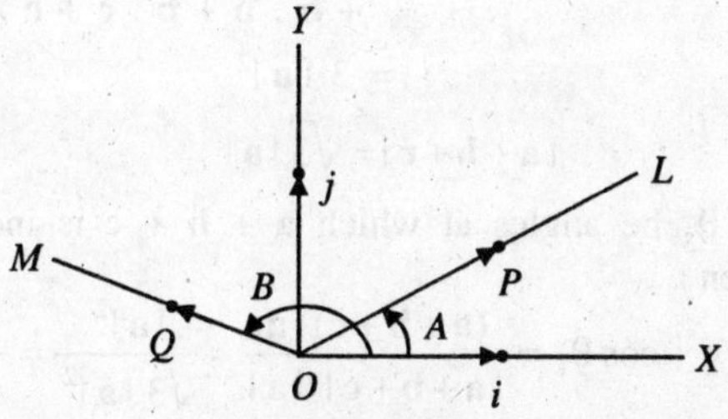

Fig. 3.15

If P and Q be points on OL and OM such that $\vec{OP}$ and $\vec{OQ}$ are unit vectors, we have

$$\vec{OP} = (\cos A)\,\mathbf{i} + (\sin A)\,\mathbf{j},\quad \vec{OQ} = (\cos B)\,\mathbf{i} + (\sin B)\,\mathbf{j}.$$

$$\Rightarrow \qquad \vec{OP} \cdot \vec{OQ} = \cos A \cos B + \sin A \sin B.$$

Also by def.

$$\vec{OP} \cdot \vec{OQ} = 1.1 \cos(B - A) = \cos(B - A).$$

Hence the result.

Example 4. *If* $\hat{\mathbf{a}}$ and $\hat{\mathbf{b}}$ *are unit vectors and* θ *is the angle between them, show that* $\sin(\theta/2) = \frac{1}{2}|\hat{\mathbf{a}} - \hat{\mathbf{b}}|$.

Solution. $\quad |\hat{\mathbf{a}} - \hat{\mathbf{b}}|^2 = (\hat{\mathbf{a}} - \hat{\mathbf{b}}).(\hat{\mathbf{a}} - \hat{\mathbf{b}})$

$$= \hat{\mathbf{a}}.\hat{\mathbf{a}} - \hat{\mathbf{a}}.\hat{\mathbf{b}} - \hat{\mathbf{b}}.\hat{\mathbf{a}} + \hat{\mathbf{b}}.\hat{\mathbf{b}}$$

$$= 1 - \cos\theta - \cos\theta + 1$$

$$= 2(1 - \cos\theta) = 4\sin^2\frac{\theta}{2}$$

$$\therefore \qquad \sin\frac{\theta}{2} = \frac{1}{2}|\hat{\mathbf{a}} - \hat{\mathbf{b}}|.$$

Example 5. *If* **a, b, c** *are mutually perpendicular vectors of equal magnitude, show that* **a + b + c** *is equally inclined to* **a, b, c.**

Solution. Given that :

$$\mathbf{b}.\mathbf{a} = \mathbf{a}.\mathbf{b} = \mathbf{b}.\mathbf{c} = \mathbf{c}.\mathbf{b} = \mathbf{a}.\mathbf{c} = \mathbf{c}.\mathbf{a} = 0 \qquad ...(1)$$

and $\qquad |\mathbf{a}| = |\mathbf{b}| = |\mathbf{c}| \qquad ...(2)$

Now,

$$|\mathbf{a} + \mathbf{b} + \mathbf{c}|^2 = (\mathbf{a} + \mathbf{b} + \mathbf{c}).(\mathbf{a} + \mathbf{b} + \mathbf{c})$$

$$= \mathbf{a} \cdot \mathbf{a} + \mathbf{a} \cdot \mathbf{b} + \mathbf{a} \cdot \mathbf{c} + \mathbf{b} \cdot \mathbf{a}$$
$$+ \mathbf{b} \cdot \mathbf{b} + \mathbf{b} \cdot \mathbf{c} + \mathbf{c} \cdot \mathbf{a} + \mathbf{c} \cdot \mathbf{b} + \mathbf{c} \cdot \mathbf{c}$$
$$= 3 \, |\mathbf{a}|^2$$

$\therefore$ $$|\mathbf{a}+\mathbf{b}+\mathbf{c}| = \sqrt{3}\, |\mathbf{a}|.$$

Let θ_1, θ_2, θ_3 be angles at which $\mathbf{a} + \mathbf{b} + \mathbf{c}$ is inclined to **a, b, c** respectively, then

$$\cos \theta_1 = \frac{(\mathbf{a}+\mathbf{b}+\mathbf{c}) \cdot \mathbf{a}}{|\mathbf{a}+\mathbf{b}+\mathbf{c}|\, |\mathbf{a}|} = \frac{|\mathbf{a}|^2}{\sqrt{3}\, |\mathbf{a}|^2} = \frac{1}{\sqrt{3}}.$$

Similarly $$\cos \theta_2 = \cos \theta_3 = \frac{1}{\sqrt{3}}.$$

$\Rightarrow$ $$\theta_1 = \theta_2 = \theta_3.$$

Example 6. *If* **a, b** *are vectors and a, b their lengths, show that*

$$\left(\frac{\mathbf{a}}{a^2} - \frac{\mathbf{b}}{b^2}\right)^2 = \left(\frac{\mathbf{a}-\mathbf{b}}{ab}\right)^2$$

Solution. $$\left(\frac{\mathbf{a}}{a^2} - \frac{\mathbf{b}}{b^2}\right)^2 = \left(\frac{\mathbf{a}}{a^2} - \frac{\mathbf{b}}{b^2}\right) \cdot \left(\frac{\mathbf{a}}{a^2} - \frac{\mathbf{b}}{b^2}\right)$$

$$= \frac{\mathbf{a} \cdot \mathbf{a}}{a^4} - \frac{2\mathbf{a} \cdot \mathbf{b}}{a^2 b^2} + \frac{\mathbf{b} \cdot \mathbf{b}}{b^4} = \frac{a^2}{a^4} - \frac{2\mathbf{a} \cdot \mathbf{b}}{a^2 b^2} + \frac{b^2}{b^4}$$

$$= \frac{1}{a^2} - \frac{2\mathbf{a} \cdot \mathbf{b}}{a^2 b^2} + \frac{1}{b^2} = \frac{b^2 - 2\mathbf{a} \cdot \mathbf{b} + a^2}{a^2 b^2}$$

$$= \frac{(\mathbf{a}-\mathbf{b}) \cdot (\mathbf{a}-\mathbf{b})}{a^2 b^2} = \frac{(\mathbf{a}-\mathbf{b})^2}{a^2 b^2}.$$

EXERCISES

1. Find the lengths of the following vectors :

$$3\mathbf{i} + 2\mathbf{j} - \mathbf{k}, \quad \sqrt{2}\mathbf{i} - \sqrt{3}\mathbf{k}, \quad 5\mathbf{i} + 4\mathbf{j} - 2\mathbf{k}.$$

2. Find the angles between the following pairs of vectors :

(*i*) $\mathbf{i} - \mathbf{j} + \mathbf{k}$, $-\mathbf{i} + \mathbf{j} + 2\mathbf{k}$, (*ii*) $3\mathbf{i} + 4\mathbf{j}$, $2\mathbf{j} - 5\mathbf{k}$,

(*iii*) $2\mathbf{i} - 3\mathbf{k}$, $\mathbf{i} + \mathbf{j} + \mathbf{k}$.

3. Find the unit vectors perpendicular to each of the following pairs of vectors :

(*i*) $\mathbf{i} - \mathbf{j} + \mathbf{k}$, $\mathbf{i} + 2\mathbf{j} - \mathbf{k}$, (*ii*) $2\mathbf{i} + \mathbf{k}$, $\mathbf{i} + \mathbf{j} + \mathbf{k}$

(*iii*) $\mathbf{i} + \mathbf{j}$, $\mathbf{i} - \mathbf{j} + \mathbf{k}$, (*iv*) $2\mathbf{i} + \mathbf{j} + \mathbf{k}$, $\mathbf{i} - 2\mathbf{j} + \mathbf{k}$.

4. Determine the unit vectors which make an angle of 60° with $\mathbf{i} - \mathbf{j}$ and an angle of 60° with $\mathbf{i} + \mathbf{k}$.

5. Given that **i, j, k** is an orthogonal base, show that

$$\mathbf{i},\ \sqrt{\frac{1}{2}}(\mathbf{j}+\mathbf{k}),\ \sqrt{\frac{1}{2}}(\mathbf{j}-\mathbf{k})$$

is also an orthonormal base.

6. Given that **i** – **j**, **i** + 2**j** are two vectors. Find a unit vector coplanar with these vectors and perpendicular to the first vector **i** – **j**. Find also the unit vector which is perpendicular to the plane of the two given vectors. Do you thus obtain an orthonormal triad ?

7. Let

(*i*) $\mathbf{a} = \mathbf{i} - \mathbf{j},\ \mathbf{b} = \mathbf{i} + \mathbf{k},$

(*ii*) $\mathbf{a} = \mathbf{i}+\mathbf{j}+\mathbf{k},\ \mathbf{b} = \mathbf{i}+\sqrt{2}\mathbf{j}-\sqrt{6}\mathbf{k}.$

Find in each case a unit vector **c** perpendicular to **a** and coplanar with **a** and **b**. Also find a unit vector **d** perpendicular to both **a** and **c**.

8. Given that

$$\mathbf{a} = \mathbf{i} + \mathbf{j} + \mathbf{k};\ \mathbf{b} = \mathbf{i} - \mathbf{j} + \mathbf{k};\quad \mathbf{c} = \mathbf{i} + \mathbf{j} - \mathbf{k},$$

evaluate

(*i*) $(\mathbf{a}\,.\,\mathbf{b}) + (\mathbf{b}\,.\,\mathbf{c}) \div (\mathbf{c}\,.\,\mathbf{a});$

(*ii*) $(\mathbf{a}\,.\,\mathbf{c})\,\mathbf{c} + (\mathbf{c}\,.\,\mathbf{b})\,\mathbf{a};$

(*iii*) $(\mathbf{a} + 2\mathbf{b})\,.\,[\mathbf{a} + (\mathbf{a}\,.\,\mathbf{c})\,\mathbf{b}].$

9. Given the vectors **a** and **b** as follows :

(*i*) $\mathbf{a} = 2\mathbf{i}-\frac{3}{2}\mathbf{j}+\frac{4}{5}\mathbf{k};\ \mathbf{b} = \mathbf{i}+\sqrt{2}\mathbf{j}+\sqrt{6}\mathbf{k}.$

(*ii*) $\mathbf{a} = \mathbf{i}+\mathbf{j}+\mathbf{k};\ \mathbf{b} = \sqrt{3\mathbf{i}+3\mathbf{j}-2\mathbf{k}}.$

Find in each case the projections of **a** on **b** and of **b** on **a**.

3.3. APPLICATIONS TO CARTESIAN GEOMETRY

Rectangular Cartesian Coordinates and Position Vectors. We shall now consider some preliminary applications to Cartesian Geometry. The major results will be obtained in the following chapter with the help of scalar triple products.

Take any point O and three mutually perpendicular straight lines OX, OY, OZ through O. Let A, B, C be three points on these lines such that

$$\overrightarrow{OA} = \mathbf{i},\ \overrightarrow{OB} = \mathbf{j},\ \overrightarrow{OC} = \mathbf{k},$$

where **i, j, k** denote mutually perpendicular unit vectors.

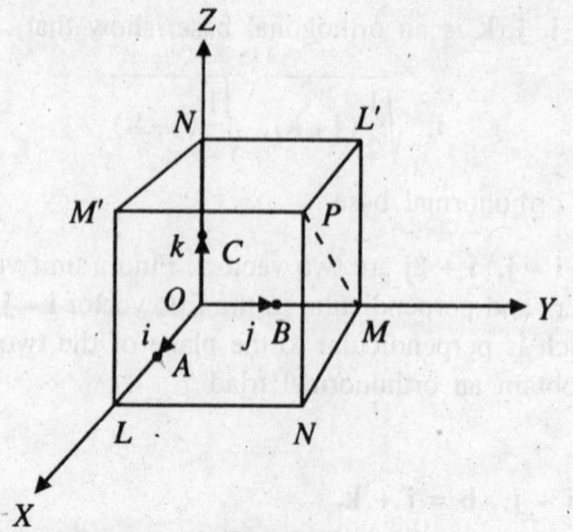

Fig. 3.16

Let P be any given point. Through P draw planes parallel to the three planes BOC, COA, AOB meeting OA, OB, OC in L, M, N respectively so that we obtain a rectangular parallelopiped having OP as a diagonal. We have

$$\mathbf{r} = \overrightarrow{OP} = \overrightarrow{OM} + \overrightarrow{MP}$$
$$= \overrightarrow{OM} + \overrightarrow{ML'} + \overrightarrow{L'P}$$
$$= \overrightarrow{OM} + \overrightarrow{ON} + \overrightarrow{OL}$$
$$= \overrightarrow{OL} + \overrightarrow{OM} + \overrightarrow{ON}.$$

There exist 3 scalars, x, y, z such that

$$\overrightarrow{OL} = x\overrightarrow{OA} = x\mathbf{i},\quad \overrightarrow{OM} = y\overrightarrow{OB} = y\mathbf{j},\quad \overrightarrow{ON} = z\overrightarrow{OC} = z\mathbf{k}.$$

Thus, we have

$$\mathbf{r} = x\mathbf{i} + y\mathbf{j} + z\mathbf{k}.$$

Here $\overrightarrow{OP} = \mathbf{r}$ has been expressed as a linear combination of three mutually perpendicular unit vectors **i, j, k.**

The numbers x, y, z are *Rectangular Cartesian Coordinates* of the point P with respect to OX, OY, OZ respectively as coordinate axes (compare 10.2 Page 16 Chapter 1).

Thus, with reference to the coordinate axes chosen as above, we see that the statement that

$x\mathbf{i} + y\mathbf{j} + z\mathbf{k}$ *is the position vector of a point* $\Leftrightarrow$ *x, y, z are the rectangular cartesian coordinates of the point.*

3.4. DISTANCE BETWEEN POINTS WHOSE RECTANGULAR CARTESIAN COORDINATES ARE

$$P\ (x_1, y_1, z_1),\quad Q\ (x_2, y_2, z_2).$$

The position vectors of the points P, Q are

$$\vec{OP} = x_1\mathbf{i} + y_1\mathbf{j} + z_1\mathbf{k},\ \vec{OQ} = x_2\mathbf{i} + y_2\mathbf{j} + z_2\mathbf{k} \text{ respectively.}$$

We have

$$\vec{PQ} = \vec{OQ} - \vec{OP} = (x_2 - x_1)\,\mathbf{i} + (y_2 - y_1)\,\mathbf{j} + (z_2 - z_1)\,\mathbf{k}$$

so that

$$PQ^2 = \vec{PQ}\,.\,\vec{PQ} = [(x_2 - x_1)\,\mathbf{i} + (y_2 - y_1)\,\mathbf{j} + (z_2 - z_1)\,\mathbf{k}].$$
$$[(x_2 - x_1)\,\mathbf{i} + (y_2 - y_1)\,\mathbf{j} + (z_2 - z_1)\,\mathbf{k}]$$
$$= (x_2 - x_1)^2 + (y_2 - y_1)^2 + (z_2 - z_1)^2.$$

3.5. DIRECTION COSINES OF A LINE

Consider any directed line and let $\vec{OA}$ denote the vector of unit length parallel to this line.

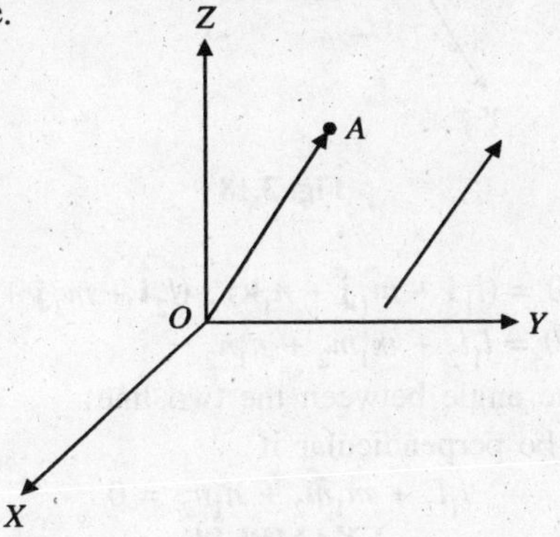

Fig. 3.17

Let $\quad \vec{OA} = l\mathbf{i} + m\mathbf{j} + n\mathbf{k}.$

We have

$$\vec{OA}\,.\,\mathbf{i} = (l\mathbf{i} + m\mathbf{j} + n\mathbf{k})\,.\,\mathbf{i} = l$$

$\Rightarrow \quad 1.1\,.\,\cos\alpha = l \quad \Rightarrow \quad l = \cos\alpha$

where α denotes the angle which the given line makes with x-axis.

Similarly we may show $m = \cos\beta$, $n = \cos\gamma$ where β and γ denote the angles which the given line makes with y-axis and z-axis respectively.

Thus, if $l\mathbf{i} + m\mathbf{j} + n\mathbf{k}$

denotes a vector of unit length parallel to a given line, then l, m, n are the *direction cosines of the line.*

Cor. *The sum of the squares of the direction cosines of any line is unity.*

We have $\quad \vec{OA}\,.\,\vec{OA} = (l\mathbf{i} + m\mathbf{j} + n\mathbf{k})\,.\,(l\mathbf{i} + m\mathbf{j} + n\mathbf{k})$

$\Rightarrow \quad 1 = l^2 + m^2 + n^2.$

3.6. ANGLE BETWEEN TWO LINES

Let $l_1, m_1, n_1; l_2, m_2, n_2$ be the direction cosines of two given lines. The vectors of unit length along the given lines being

$$l_1\mathbf{i} + m_1\mathbf{j} + n_1\mathbf{k}, \quad l_2\mathbf{i} + m_2\mathbf{j} + n_2\mathbf{k},$$

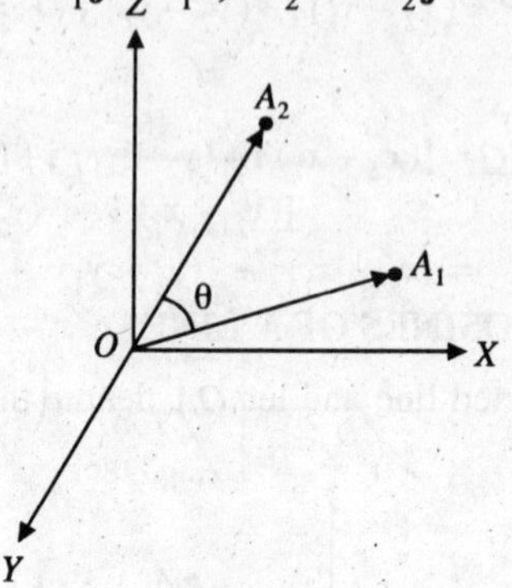

Fig. 3.18

we have

$$1.1 \,.\, \cos\theta = (l_1\mathbf{i} + m_1\mathbf{j} + n_1\mathbf{k}) \,.\, (l_2\mathbf{i} + m_2\mathbf{j} + n_2\mathbf{k})$$

$$\Rightarrow \qquad \cos\theta = l_1l_2 + m_1m_2 + n_1n_2$$

where θ denotes the angle between the two lines.

The lines will be perpendicular if

$$l_1l_2 + m_1m_2 + n_1n_2 = 0.$$

EXAMPLES

Example 1. *Find the angle between the lines AB, AC where A, B, C are the three points with rectangular cartesian coordinates*

$$(1, 2, -1), \quad (2, 0, 3), \quad (3, -1, 2)$$

respectively.

Solution. In terms of usual notation, the position vectors of the given points are

$$\vec{OA} = \mathbf{i} + 2\mathbf{j} + \mathbf{k}, \quad \vec{OB} = 2\mathbf{i} + 3\mathbf{k}, \quad \vec{OC} = 3\mathbf{i} - \mathbf{j} + 2\mathbf{k},$$

O being the origin. Thus, we have

$$\vec{AB} = \vec{OB} - \vec{OA} = \mathbf{i} - 2\mathbf{j} + 4\mathbf{k}, \quad \vec{AC} = \vec{OC} - \vec{OA} = 2\mathbf{i} - 3\mathbf{j} + 3\mathbf{k}.$$

$$\therefore \qquad \vec{AB} \,.\, \vec{AC} = (\mathbf{i} - 2\mathbf{j} + 4\mathbf{k}) \,.\, (2\mathbf{i} - 3\mathbf{j} + 3\mathbf{k}) = 20.$$

Also

$$\vec{AB}^2 = (\mathbf{i} - 2\mathbf{j} + 4\mathbf{k}) \,.\, (\mathbf{i} - 2\mathbf{j} + 4\mathbf{k}) = 21.$$

$$\vec{AC}^2 = (2\mathbf{i} - 3\mathbf{j} + 3\mathbf{k}) \,.\, (2\mathbf{i} - 3\mathbf{j} + 3\mathbf{k}) = 22.$$

If θ denote the angle between AB and AC, we have

$$\cos\theta = \frac{\vec{AB} \,.\, \vec{AC}}{|\vec{AB}|\,|\vec{AC}|} = \frac{20}{\sqrt{21}\sqrt{22}}$$

Example 2. *A line makes angles α, β, γ, δ with the diagonals of a cube; show that*

$$\cos^2\alpha + \cos^2\beta + \cos^2\gamma + \cos^2\delta = \frac{4}{3}.$$

Solution.

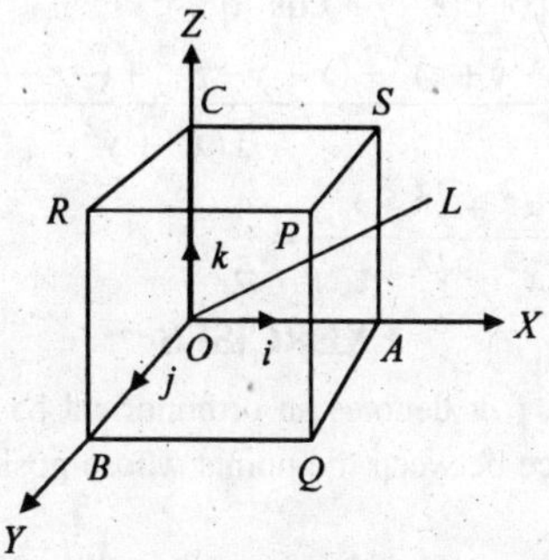

Fig. 3.19

Since the angle will remain unchanged for any size of cube, consider a unit cube. Represent the coterminous edges $\vec{OA}$, $\vec{OB}$, $\vec{OC}$ by unit vectors **i, j, k.**

Then $\vec{OA} = \mathbf{i}$, $\vec{OB} = \mathbf{j}$, $\vec{OC} = \mathbf{k}$,

$$\vec{OP} = \vec{OA} + \vec{AQ} + \vec{QP} = \mathbf{i} + \mathbf{j} + \mathbf{k}$$

$$\vec{CQ} = -\mathbf{k} + \mathbf{i} + \mathbf{j},\ \vec{AR} = -\mathbf{i} + \mathbf{k} + \mathbf{j}$$

and $$\vec{BS} = \mathbf{i} - \mathbf{j} + \mathbf{k}.$$

$\therefore$ $$OP = CQ = AR = BS = \sqrt{3}.$$

Let any line OL be given by

$$\vec{OL} = x\mathbf{i} + y\mathbf{j} + z\mathbf{k}.$$

If α be the angle between OL and OP, then

$$\vec{OL} \cdot \vec{OP} = OL \times OP \times \cos\alpha$$

$\Rightarrow$ $$(x\mathbf{i} + y\mathbf{j} + z\mathbf{k}) \cdot (\mathbf{i} + \mathbf{j} + \mathbf{k}) = \sqrt{(x^2 + y^2 + z^2)} . \sqrt{3} \cos\alpha$$

$\Rightarrow$ $$\cos\alpha = \frac{(x + y + z)}{\sqrt{3}\sqrt{(x^2 + y^2 + z^2)}}$$

Similarly,

$$\cos\beta = \frac{(x + y - z)}{\sqrt{3}\sqrt{(x^2 + y^2 + z^2)}}$$

$$\cos\gamma = \frac{(-x+y+z)}{\sqrt{3}\sqrt{(x^2+y^2+z^2)}}$$

and $$\cos\delta = \frac{(x-y+z)}{\sqrt{3}\sqrt{(x^2+y^2+z^2)}}$$

$$\therefore \quad \cos^2\alpha+\cos^2\beta+\cos^2\gamma+\cos^2\delta$$

$$= \frac{(x+y+z)^2+(x+y-z)^2+(-x+y+z)^2+(x-y+z)^2}{3(x^2+y^2+z^2)}$$

$$= \frac{4(x^2+y^2+z^2)}{3(x^2+y^2+z^2)} = \frac{4}{3}.$$

EXERCISES

(In the following **i, j, k** denotes an orthonormal base.)

1. Find the distance between the points whose position vectors are given as follows :

(*i*) 4**i** + 3**j** – 6**k**, – 2**i** + **j** – **k**; (*ii*) – 2**i** + 3**j** + 5**k**, 7**i** – **k**.

2. Find the distance between the pairs of points whose cartesian coordinates are :

(*i*) (– 1, 1, 3), (0, 5, 6); (*ii*) (2, 3, – 1), (2, 6, 2).

3. Find the angle between the lines parallel to the following pairs of vectors :

(*i*) 6**i** + 2**j** + 3**k**, 3**i** + 12 **j** + 4**k**; (*ii*) **i** + 2**j** + 3**k**, – **i** + **j** + 2**k**.

4. What are the direction cosines of the joins of the following pairs of points :

(*i*) (6, 3, 2), (5, 1, 4), (*ii*) (3, – 4, 7), (0, 2, 5).

5. Given that *P, Q, R, S* are the points (6, – 6, 0), (– 1, – 7, 6), (3, – 4, 4), (2, – 9, 2) respectively, show that $PQ \perp RS$.

6. Find the angle between the lines joining the following pairs of points :

(*i*) (0, 4, 1), (2, 3, – 1); (*ii*) (4, 5, 0), (2, 6, 2).

7. Show that the points

(5, – 1, 1), (7, – 4, 7), (1, – 6, 10), (– 1, – 3, 4)

are the vertices of a rhombus.

8. Show that the join of points (1, 2, 3), (4, 5, 7) is parallel to the join of the points (– 4, 3, – 6), (2, 9, 2).

9. Show that the points

(4, 7, 8), (2, 3, 4), (– 1, – 2, 1), (1, 2, 5)

are the vertices of a parallelogram.

10. Show that the line *AB* is perpendicular to the line *CD* if *A, B, C, D* are the points (2, 3, 4), (5, 4, – 1), (3, 6, 2), (1, 2, 0) respectively.

11. Show that the three angles of the triangle with vertices

(1, – 1, 1), (2, 3, – 1), (3, 0, 2)

are

$$\cos^{-1}\frac{2}{\sqrt{114}},\ \cos^{-1}\frac{4}{\sqrt{126}},\ \cos^{-1}\frac{-17}{\sqrt{399}}$$

respectively.

12. The points P, Q, R have rectangular cartesian coordinates $(1, 1, -1)$, $(4, 1, 2)$ and $(-2, 1, 2)$ respectively. Which of the following is true for the triangle PQR ?

A. All the sides have the same length.

B. $QR = 2\ PQ$.

C. Angle PQR is a right angle.

D. The area of the triangle is 18 square units.

13. Find the angle between two diagonals of a cube.

14. If the edges of a rectangular parallelopiped are α, β, γ, show that the angles between the four diagonals are

$$\cos^{-1} \frac{\alpha^2 \pm \beta^2 \pm \gamma^2}{\alpha^2 + \beta^2 + \gamma^2}.$$

15. With reference to mutually perpendicular axes A, B, C and D are the points $(-6, 1, 6)$, $(6, -2, 3)$, $(-2, -3, -1)$ and $(-5, -9, -7)$ respectively.

(*i*) Show that the points A, B and D lie on the surface of a sphere with centre C.

(*ii*) If DCE is a diameter of the sphere, find the coordinates of E.

(*iii*) Prove that the angle BCA is a right angle and that angle BCE equals angle ECA.

(*iv*) Show that the diameter DCE does not lie in the plane of the triangle ABC.

SUMMARY

1. The scalar product of two vectors is the product of the length of one vector and projection of other on it.

2. The square of the length of the vector **a** is

$$\mathbf{a} \cdot \mathbf{a} \text{ denoted by } |\mathbf{a}|^2.$$

3. The angle between two vectors **a** and **b** is

$$\cos^{-1} \frac{\mathbf{a} \cdot \mathbf{b}}{|\mathbf{a}|\,|\mathbf{b}|}.$$

4. If **a**, **b** are two non-zero vectors, then

$$\mathbf{a} \cdot \mathbf{b} = 0 \iff \mathbf{a} \text{ and } \mathbf{b} \text{ are perpendicular.}$$

5. Scalar multiplication distributes the addition of vectors.

6. If **i**, **j**, **k** denote mutually perpendicular vectors, each of unit length, then

$$\mathbf{i} \cdot \mathbf{i} = 1, \quad \mathbf{j} \cdot \mathbf{j} = 1, \quad \mathbf{k} \cdot \mathbf{k} = 1;$$
$$\mathbf{i} \cdot \mathbf{j} = 0, \quad \mathbf{j} \cdot \mathbf{k} = 0, \quad \mathbf{k} \cdot \mathbf{i} = 0,$$

7. A base consisting of three mutually perpendicular vectors is called *orthogonal.* In case each vector is of unit length, the base is called *orthonormal.*

8. $(a_1\mathbf{i} + a_2\mathbf{j} + a_3\mathbf{k}) \cdot (b_1\mathbf{i} + b_2\mathbf{j} + b_3\mathbf{k}) = a_1b_1 + a_2b_2 + a_3b_3.$

9. (*i*) Length of a vector $a_1\mathbf{i} + a_2\mathbf{j} + a_3\mathbf{k}$ is $\sqrt{(\mathbf{a}_1^2 + \mathbf{a}_2^2 + \mathbf{a}_3^2)}$.

(*ii*) Angle between two vectors

$a_1\mathbf{i} + a_2\mathbf{j} + a_3\mathbf{k}$ and $b_1\mathbf{i} + b_2\mathbf{j} + b_3\mathbf{k}$

is

$$\cos^{-1} \frac{a_1b_1 + a_2b_2 + a_3b_3}{\sqrt{(a_1^2 + a_2^2 + a_3^2)}\ \sqrt{(b_1^2 + b_2^2 + b_3^2)}}.$$

(*iii*) The vectors $a_1\mathbf{i} + a_2\mathbf{j} + a_3\mathbf{k}$, $b_1\mathbf{i} + b_2\mathbf{j} + b_3\mathbf{k}$ are perpendicular if and only if

$$a_1b_1 + a_2b_2 + a_3b_3 = 0.$$

10. If OX, OY, OZ be three mutually perpendicular axes and **i, j, k** are unit vectors along OX, OY, OZ respectively, then

(x, y, z) are the coordinates of P

$\Leftrightarrow$ $x\mathbf{i} + y\mathbf{j} + z\mathbf{k}$ is the position vector $\vec{OP}$ of P.

OBJECTIVE QUESTIONS

For each of the following questions, four alternatives are given for the answer. Only one of them is correct. Choose the correct alternative.

1. $a_1\mathbf{i} + a_2\mathbf{j}$ is a unit vector perpendicular to $4\mathbf{i} - 3\mathbf{j}$ if

(*a*) $a_1 = .6$, $a_2 = .8$ (*b*) $a_1 = 3$, $a_2 = 4$

(*c*) $a_1 = .8$, $a_2 = .6$ (*d*) $a_1 = 4$, $a_2 = 3$

2. If $\mathbf{a} = 2\mathbf{i} - 3\mathbf{j}$, $\mathbf{b} = 2\mathbf{j} + 3\mathbf{k}$, then $(\mathbf{a} + \mathbf{b}) . (\mathbf{a} - \mathbf{b}) =$

(*a*) 0 (*b*) – 8 (*c*) 9 (*d*) – 10

3. If $\vec{PO} + \vec{OQ} = \vec{QO} + \vec{OR}$, then P, Q, R are

(*a*) the vertices of an equilateral triangle

(*b*) the vertices of an isosceles triangle

(*c*) collinear (*d*) None of these

4. The triangle ABC is defined by the vertices A (1, – 2, 2), B (1, 4, 0) and C (– 4, 1, 1). Let M be the foot of the altitude drawn from the vertex B to side AC. Then $\vec{BM} =$

(*a*) (– 20/7, – 30/7, 10/7) (*b*) – 20, – 30, 10)

(*c*) (2, 3, – 1) (*d*) None of these

5. The vector **b**, which is collinear with the vector **a** = (2, 1, – 1) and satisfies the condition **a . b** = 3, is

(*a*) (1, 1/2, – 1/2) (*b*) (2/3, 1/3, – 1/3)

(*c*) (1/2, 1/4, – 1/4) (*d*) (1, 1, 0)

6. The angle between **a** and **b** is $\pi / 6$, then angle between 2**a** and 3**b** is

(*a*) $\pi / 3$ (*b*) $\pi / 2$ (*c*) $\pi / 6$ (*d*) None of these

7. If θ be the angle between the vectors **i + j** and **j + k**, then θ is (*a*) 0 (*b*) $\pi/4$ (*c*) $\pi/2$ (*d*) $\pi/3$

8. If **a, b, c** are three vectors such that each is inclined at an angle $\pi/3$ with the other two and $|\mathbf{a}| = 1$, $|\mathbf{b}| = 2$, $|\mathbf{c}| = 3$, then the scalar product of the vectors $2\mathbf{a} + 3\mathbf{b} - 5\mathbf{c}$ and $4\mathbf{a} - 6\mathbf{b} + 10\mathbf{c}$ is equal to

(*a*) – 334 (*b*) 188 (*c*) – 522 (*d*) – 514

9. If $\mathbf{a} + \mathbf{b} + \mathbf{c} = 0$, $|\mathbf{a}| = 3$, $|\mathbf{b}| = 5$, $|\mathbf{c}| = 7$, then the angle between **a** and **b** is

(*a*) $\pi/6$ (*b*) $2\pi/3$ (*c*) $5\pi/3$ (*d*) $\pi/3$

10. Projection of the vector $2\mathbf{i} + 3\mathbf{j} - 2\mathbf{k}$ on the vector $\mathbf{i} - 2\mathbf{j} + 3\mathbf{k}$ is

(*a*) $2/\sqrt{(14)}$ (*b*) $1/\sqrt{(14)}$ (*c*) $3/\sqrt{(14)}$ (*d*) None of these

11. The points *A* (1, 1, 2), *B* (3, 4, 2) and *C* (5, 6, 4). The exterior angle of the triangle *B* is

(*a*) $\cos^{-1}\left[-5/\sqrt{(39)}\right]$ (*b*) $\cos^{-1}\left[5/\sqrt{(39)}\right]$

(*c*) $\cos^{-1}(5/6)$ (*d*) None of these

12. If $3\mathbf{i} + 2\mathbf{j} + 8\mathbf{k}$ and $2\mathbf{i} + x\mathbf{j} + \mathbf{k}$ are at right angles then $x =$

(*a*) 7 (*b*) – 7 (*c*) 5 (*d*) – 4

13. The vectors $2\mathbf{i} + 3\mathbf{j} - 4\mathbf{k}$ and $a\mathbf{i} + b\mathbf{j} + c\mathbf{k}$ are perpendicular, when

(*a*) $a = 2$, $b = 3$, $c = -4$ (*b*) $a = 4$, $b = 4$, $c = 5$

(*c*) $a = 4$, $b = 4$, $c = -5$ (*d*) None of these

14. If $A = 2\mathbf{i} + 2\mathbf{j} + 4\mathbf{k}$, $B = -\mathbf{i} + 2\mathbf{j} + \mathbf{k}$ and $C = 3\mathbf{i} + \mathbf{j}$, then $A + tB$ is perpendicular to *c* if *t* is equal to

(*a*) 8 (*b*) 4 (*c*) 6 (*d*) 2

15. Let **a, b, c** and **d** be position vectors of four points *A, B, C* and *D* lying in a plane. If $(\mathbf{a} - \mathbf{d}) \cdot (\mathbf{b} - \mathbf{c}) = 0 = (\mathbf{b} - \mathbf{d}) \cdot (\mathbf{c} - \mathbf{a})$, then ΔABC has Δ as

(*a*) in-centre (*b*) circum-centre

(*c*) ortho-centre (*d*) centroid

16. A unit vector in *xy*-plane that makes an angle of 45° with the vector $\mathbf{i} + \mathbf{j}$ and an angle of 60° with the vector $3\mathbf{i} - 4\mathbf{j}$ is

(*a*) *i* (*b*) $(\mathbf{a} + \mathbf{j})/\sqrt{2}$

(*c*) $(\mathbf{i} - \mathbf{j})/\sqrt{2}$ (*d*) None of these

17. $(\mathbf{a} - \mathbf{i})\,\mathbf{i} + (\mathbf{a} \cdot \mathbf{j})\,\mathbf{j} + (\mathbf{a} \cdot \mathbf{k})\,\mathbf{k} =$

(*a*) **0** (*b*) **a** (*c*) 3**a** (*d*) None of these

18. Let α, β, γ be distinct real numbers. The points with position vectors $\alpha\mathbf{i} + \beta\mathbf{j} + \gamma\mathbf{k}$, $\beta\mathbf{i} + \gamma\mathbf{j} + \alpha\mathbf{k}$, $\gamma\mathbf{i} + \alpha\mathbf{j} + \beta\mathbf{k}$.

(*a*) are collinear
(*b*) form an equilateral triangle
(*c*) form an scalene triangle
(*d*) form a right angled triangle

19. If (**A** + **B**) is perpendicular to **B** and (**A** + 2**B**) is perpendicular to **A**, then

(*a*) $\mathbf{A} = \sqrt{2}\mathbf{B}$ (*b*) **A** = 2**B** (*c*) 2**A** = **B** (*d*) **A** = **B**

20. If **a**, **b**, **c** are three non-zero vectors such that **a** + **b** + **c** = 0 then the value of **a . b** + **b . c** + **c . a** is

(*a*) less than zero (*b*) equal to zero
(*c*) greater than zero (*d*) 3

21. If | **a** + **b** | = | **a** – **b** |, **a**, **b** ≠ 0, then

(*a*) **a** is ∥ to **b** (*b*) **a** is ⊥ to **b**
(*c*) | **a** | = | **b** | (*d*) None of these

22. If **x** and **y** are two unit vectors and ϕ is the angle between them, then $\frac{1}{2}|\mathbf{x} - \mathbf{y}|$ is equal to

(*a*) 0 (*b*) $\pi / 2$ (*c*) $\left|\sin \frac{1}{2}\phi\right|$ (*d*) $\left|\cos \frac{1}{2}\phi\right|$

23. The value of **b** such that the scalar product of the vector **i** + **j** + **k** with the unit vector parallel to the sum of the vector 2**i** + 4**j** – 5**k**, and *b***i** + 2**j** + 3**k** is one is

(*a*) – 2 (*b*) – 1 (*c*) 0 (*d*) 1

24. The projection of **a** = 3**i** – **j** + 5**k** on **b** = 2**i** + 3**j** + **k** is

(*a*) $8/\sqrt{(35)}$ (*b*) $8/\sqrt{(39)}$ (*c*) $8/\sqrt{(14)}$ (*d*) $\sqrt{(14)}$

25. Given two vectors **a** = 2**i** – 3**j** + 6**k** on **b** = – 2**i** + 2**j** – **k** and

$$\lambda = \frac{\text{the projection of } \mathbf{a} \text{ on } \mathbf{b}}{\text{the projection of } \mathbf{b} \text{ on } \mathbf{a}},$$

then the value of λ is

(*a*) 3/7 (*b*) 7/3 (*c*) 3 (*d*) 7

ANSWERS

1. (*a*)	**2.** (*a*)	**3.** (*c*)	**4.** (*a*)	**5.** (*a*)
6. (*c*)	**7.** (*d*)	**8.** (*a*)	**9.** (*d*)	**10.** (*a*)
11. (*b*)	**12.** (*b*)	**13.** (*b*)	**14.** (*a*)	**15.** (*c*)
16. (*d*)	**17.** (*b*)	**18.** (*b*)	**19.** (*a*)	**20.** (*a*)
21. (*b*)	**22.** (*c*)	**23.** (*d*)	**24.** (*c*)	**25.** (*b*)

4

Applications to Metric Geometry

Introduction. While in the Chapter 2, we obtained the parametric vectorial equations of planes, we shall in this chapter obtain the vectorial equations of planes in **non-parametric form.** A study of *geometry of sphere* with the help of scalar product of vectors will also be taken into account.

4.1. NORMAL FORM OF THE VECTOR EQUATION OF A PLANE.

Non-Parametric Form. Let, **n**, be the unit vector normal to given plane and let, p, be the length of the perpendicular from the origin of reference O, to the plane; p, will be always considered as positive.

Draw OK perpendicular to the plane; K being the foot of the perpendicular, We have

$$\overrightarrow{OK} = p\mathbf{n}.$$

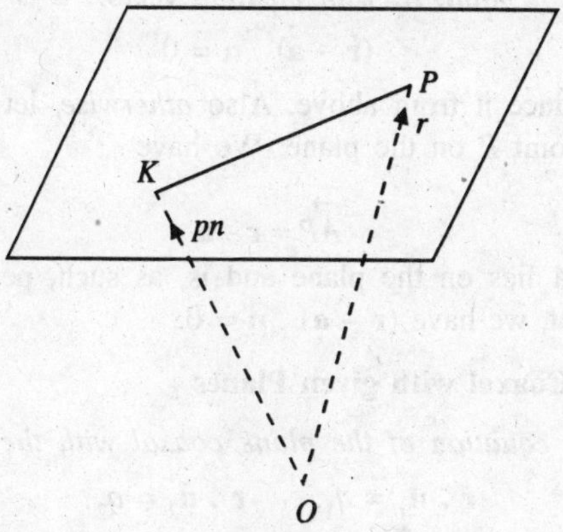

Fig. 4.1

Let, **r**, be the position vector $\overrightarrow{OP}$ of any point P on the plane. As KP lies in the plane to which OK is normal, we have

$$KP \perp OK \Rightarrow \overrightarrow{OK} \cdot \overrightarrow{KP} = 0.$$

Also we have

$$\overrightarrow{KP} = \overrightarrow{OP} - \overrightarrow{OK} = \mathbf{r} - p\mathbf{n}.$$

Thus

$$p\mathbf{n} \cdot (r - p\mathbf{n}) = 0 \qquad ...(i)$$

$$\Rightarrow \qquad \mathbf{r} \cdot \mathbf{n} = p\mathbf{n}^2 = p; \quad \mathbf{n} \text{ being a unit vector.}$$

This equation (*i*) is satisfied by the position vector of every point on the plane. Also, conversely, any point P whose position vector, $\mathbf{r}$, satisfies (*i*), is a point of the plane. Thus

$$\mathbf{r} \cdot \mathbf{n} = p$$

is the vector equation of the plane, such that $\mathbf{n}$ is the unit vector normal to the plane and p is the length of the perpendicular from the origin to the plane.

Cor. 1. The equation $\mathbf{r} \cdot \mathbf{n} = q$ *is the equation of a plane normal to the vector* $\mathbf{n}$, even if, $\mathbf{n}$, be not a unit vector, for we may rewrite it as

$$\mathbf{r} \cdot \frac{\mathbf{n}}{|\mathbf{n}|} = \frac{q}{|\mathbf{n}|}.$$

Hence, $q / |\mathbf{n}|$ is the length of the perpendicular from the origin of reference to the plane $\mathbf{r} \cdot \mathbf{n} = q$.

Cor 2. *The vectorial equation of the plane normal to the vector,* $\mathbf{n}$, *and passing through a point, A, with position vector,* $\mathbf{a}$ *is*

$$(\mathbf{r} - \mathbf{a}) \cdot \mathbf{n} = 0$$

We can deduce it from above. Also *otherwise,* let $\mathbf{r}$ be the position vector of any point P on the plane. We have

$$\overrightarrow{AP} = \mathbf{r} - \mathbf{a}.$$

The line PA lies on the plane and is, as such, perpendicular to the vector, $\mathbf{n}$, so that, we have $(\mathbf{r} - \mathbf{a}) \cdot \mathbf{n} = 0$.

4.1.1. Planes Coaxal with given Planes

To find the equation of the plane coaxal with the planes

$$\mathbf{r} \cdot \mathbf{n}_1 = q_1, \qquad \mathbf{r} \cdot \mathbf{n}_2 = q_2 \qquad ...(1)$$

and passing through the point with position vector, $\mathbf{a}$.

The equation

$$\mathbf{r} \cdot \mathbf{n}_1 - q_1 + k\,(\mathbf{r} \cdot \mathbf{n}_2 - q_2) = 0$$

$$\Rightarrow \qquad \mathbf{r} \cdot (\mathbf{n}_1 + k\mathbf{n}_2) = q_1 + kq_2 \qquad ...(2)$$

represents a plane through the line of intersection of the two given planes, whatever value k may have. The plane (2) will pass through the point, **a**, for the value of k, given by

$$\mathbf{a} \cdot (\mathbf{n}_1 + k\mathbf{n}_2) = q_1 + kq_2 \qquad ...(3)$$

$$\Rightarrow \qquad k = \frac{\mathbf{a} \cdot \mathbf{n}_1 - q_1}{q_2 - \mathbf{a} \cdot \mathbf{n}_2}.$$

Hence, the required equation is

$$(\mathbf{r} \cdot \mathbf{n}_1 - q_1) + \frac{\mathbf{a} \cdot \mathbf{n}_1 - q_1}{q_2 - \mathbf{a} \cdot \mathbf{n}_2}(\mathbf{r} \cdot \mathbf{n}_2 - q_2) = 0$$

$$\Leftrightarrow \qquad (\mathbf{r} \cdot \mathbf{n}_1 - q_1) \cdot (\mathbf{a} \cdot \mathbf{n}_2 - q_2) = (\mathbf{r} \cdot \mathbf{n}_2 - q_2) \cdot (\mathbf{a} \cdot \mathbf{n}_1 - q_1).$$

4.1.2. Angle between Two Planes

The angle θ, between the two planes

$$\mathbf{r} \cdot \mathbf{n}_1 = q_1, \qquad \mathbf{r} \cdot \mathbf{n}_2 = q_2,$$

being equal to the angle between the vectors $\mathbf{n}_1$ and $\mathbf{n}_2$ which are normal to the planes, we have

$$\theta = \cos^{-1} \frac{\mathbf{n}_1 . \mathbf{n}_2}{|\mathbf{n}_1| \, |\mathbf{n}_2|}.$$

4.1.3. Angle between a Line and a Plane

The angle θ, between any line

$$\mathbf{r} = \mathbf{a} + \mathbf{b}t$$

and any plane

$$\mathbf{r} \cdot \mathbf{n} = q,$$

being equal to the complement of the angle between the normal vector, **n**, of the plane and the direction vector, **b**, of the line, we have

$$\theta = \sin^{-1} \frac{\mathbf{n} \cdot \mathbf{b}}{|\mathbf{n}| \, |\mathbf{b}|}.$$

4.1.4. Perpendicular Distance of a Point from a Plane

Let **a** be the position vector of a given point A and

$$\mathbf{r} \cdot \mathbf{n} = q$$

be the equation of the given plane.

The equation of the line through A, normal to the plane is

$$\mathbf{r} = \mathbf{a} + t\mathbf{n}.$$

At the point of intersection of this line with the plane, we have

$$(\mathbf{a} + t\mathbf{n}) \cdot \mathbf{n} = q$$

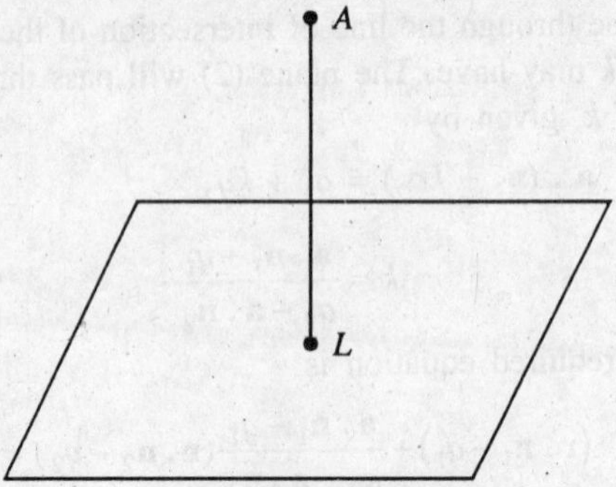

Fig. 4.2

$$\Rightarrow \qquad t = \frac{(q - \mathbf{a} \cdot \mathbf{n})}{\mathbf{n}^2}$$

Thus, the foot of the perpendicular L from the point with position vector **a** *to the given place* **r** . **n** = *q is*

$$\mathbf{a} + \frac{q - \mathbf{a} \cdot \mathbf{n}}{\mathbf{n}^2} \mathbf{n}.$$

$$\therefore \text{ the length } \quad AL = |\overrightarrow{AL}| = \left| \mathbf{a} + \frac{q - \mathbf{a} \cdot \mathbf{n}}{\mathbf{n}^2} \mathbf{n} - \mathbf{a} \right|$$

$$= \left| \frac{q - \mathbf{a} \cdot \mathbf{n}}{\mathbf{n}^2} \mathbf{n} \right| = \left| \frac{q - \mathbf{a} \cdot \mathbf{n}}{\mathbf{n}^2} \right| |\mathbf{n}|$$

$$= \frac{|q - \mathbf{a} \cdot \mathbf{n}|}{|\mathbf{n}|} \text{ for } \mathbf{n}^2 = |\mathbf{n}|^2.$$

4.1.5. Planes Bisecting the Angles between Two given Planes

Let

$$\mathbf{r} \cdot \mathbf{n}_1 = q_1 \text{ and } \mathbf{r} \cdot \mathbf{n}_2 = q_2$$

be the given planes. The perpendicular distance of any point, **r**, on either bisecting plane from the two planes being equal, we have

$$\frac{|\mathbf{r} \cdot \mathbf{n}_1 - q_1|}{|\mathbf{n}_1|} = \frac{|\mathbf{r} \cdot \mathbf{n}_2 - q_2|}{|\mathbf{n}_2|}$$

$$\Rightarrow \qquad \frac{\mathbf{r} \cdot \mathbf{n}_1 - q_1}{|\mathbf{n}_1|} = \pm \frac{\mathbf{r} \cdot \mathbf{n}_2 - q_2}{|\mathbf{n}_2|} \qquad \text{...}(i)$$

Thus, (*i*) gives the equations of the two required bisecting planes. In the standard form, these may be re-written as

$$\mathbf{r} \cdot \left(\frac{\mathbf{n}_1}{|\mathbf{n}_1|} \pm \frac{\mathbf{n}_2}{|\mathbf{n}_2|} \right) = \left(\frac{q_1}{|\mathbf{n}_1|} \pm \frac{q_2}{|\mathbf{n}_2|} \right).$$

4.1.6. Perpendicular Distance of a Point from a Line

Foot of the Perpendicular. Let, **a** be the position vector of the given point P and let

$$\mathbf{r} = \mathbf{b} + \mathbf{c}t$$

be the equation of the given line AB. If, $\mathbf{b} + \mathbf{c}t$ be the position vector of the foot L of the perpendicular, we have

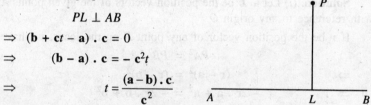

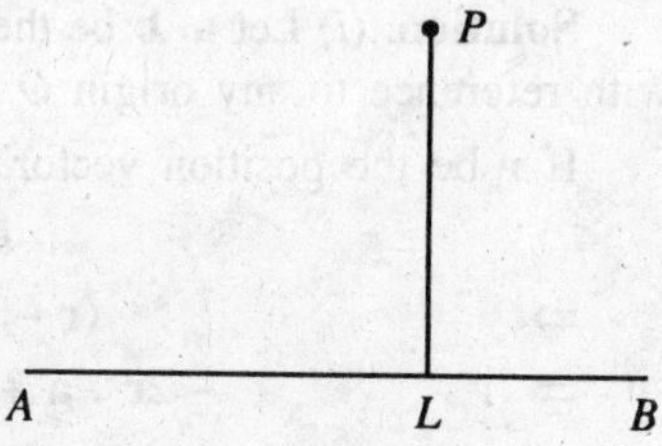

Fig. 4.3

$$PL \perp AB$$

$$\Rightarrow (\mathbf{b} + \mathbf{c}t - \mathbf{a}) \cdot \mathbf{c} = 0$$

$$\Rightarrow (\mathbf{b} - \mathbf{a}) \cdot \mathbf{c} = -\mathbf{c}^2 t$$

$$\Rightarrow t = \frac{(\mathbf{a}-\mathbf{b}) \cdot \mathbf{c}}{\mathbf{c}^2}$$

Thus, the foot of the perpendicular is

$$\mathbf{b} + \frac{(\mathbf{a}-\mathbf{b}) \cdot \mathbf{c}}{\mathbf{c}^2}\mathbf{c}.$$

Also the perpendicular distance PL

$$= |\overrightarrow{PL}| = \left| \mathbf{b} + \frac{(\mathbf{a}-\mathbf{b}) \cdot \mathbf{c}}{\mathbf{c}^2}\mathbf{c} - \mathbf{a} \right|.$$

EXAMPLES

Example 1. *Find the projection of the line*

$$\mathbf{r} = \mathbf{a} + \mathbf{b}t$$

on the plane

$$\mathbf{r} \cdot \mathbf{n} = q.$$

Solution. The equation of the line through any point, $\mathbf{a} + \mathbf{b}t$, of the given line normal to the given plane is

$$\mathbf{r} = \mathbf{a} + \mathbf{b}t + p\mathbf{n},$$

so that the projection of the point, $\mathbf{a} + \mathbf{b}t$, on the given plane is given for the value of p, satisfying

$$(\mathbf{a} + \mathbf{b}t + p\mathbf{n}) \cdot \mathbf{n} = q$$

$$\Rightarrow p = \frac{q - (\mathbf{a} + \mathbf{b}t) \cdot \mathbf{n}}{\mathbf{n}^2}$$

Thus, the projection of the point $\mathbf{a} + \mathbf{b}t$ is the point

$$\mathbf{a} + \mathbf{b}t + \frac{q - (\mathbf{a} + \mathbf{b}t) \cdot \mathbf{n}}{\mathbf{n}^2}\mathbf{n} = \left(\mathbf{a} + \frac{q - \mathbf{a} \cdot \mathbf{n}}{\mathbf{n}^2}\mathbf{n}\right) + t\left(\mathbf{b} - \frac{\mathbf{b} \cdot \mathbf{n}}{\mathbf{n}^2}\mathbf{n}\right).$$

Hence, the required projection is the line

$$\mathbf{r} = \left(\mathbf{a} + \frac{q - \mathbf{a} \cdot \mathbf{n}}{\mathbf{n}^2}\mathbf{n}\right) + t\left(\mathbf{b} - \frac{\mathbf{b} \cdot \mathbf{n}}{\mathbf{n}^2}\mathbf{n}\right).$$

Example 2. *Find the locus of a point equidistant from*
(*i*) *two given points* (*ii*) *three given points.*

Solution. (*i*) Let **a**, **b** be the position vectors of the given points *A*, *B*, with reference to any origin *O*.

If **r** be the position vector of any point *P* on the locus, we have

$$PA^2 = PB^2$$

$$\Rightarrow \quad (\mathbf{r} - \mathbf{a})^2 = (\mathbf{r} - \mathbf{b})^2$$

$$\Rightarrow \quad -2\mathbf{r} \cdot \mathbf{a} + \mathbf{a}^2 = -2\mathbf{r} \cdot \mathbf{b} + \mathbf{b}^2$$

$$\Rightarrow \quad \mathbf{r} \cdot (\mathbf{a} - \mathbf{b}) = \frac{1}{2}(\mathbf{a}^2 - \mathbf{b}^2) = \frac{1}{2}(\mathbf{a} + \mathbf{b})(\mathbf{a} - \mathbf{b})$$

$$\Rightarrow \quad \left[\mathbf{r} - \frac{1}{2}(\mathbf{a} + \mathbf{b})\right] \cdot (\mathbf{a} - \mathbf{b}) = 0.$$

Thus, the required locus is the plane bisecting the line *AB* normally.

(*ii*) If **a**, **b**, **c** be the position vectors of the given points, then the required locus is the line of intersection of the planes

$$\left[\mathbf{r} - \frac{1}{2}(\mathbf{a} + \mathbf{b})\right] \cdot (\mathbf{a} - \mathbf{b}) = 0, \quad \left[\mathbf{r} - \frac{1}{2}(\mathbf{b} + \mathbf{c})\right] \cdot (\mathbf{b} - \mathbf{c}) = 0.$$

This line passes through the circumcentre of the triangle *A*, *B*, *C* and is perpendicular to the plane of the same.

Example 3. *Show that the centre of the sphere passing through the four points with position vectors* **a, b, c, d** *is the point common to the three planes*

$$\left[\mathbf{r} - \frac{1}{2}(\mathbf{a} + \mathbf{b})\right] \cdot (\mathbf{a} - \mathbf{b}) = 0, \quad \left[\mathbf{r} - \frac{1}{2}(\mathbf{b} + \mathbf{c})\right] \cdot (\mathbf{b} - \mathbf{c}) = 0,$$

$$\left[\mathbf{r} - \frac{1}{2}(\mathbf{c} + \mathbf{a})\right] \cdot (\mathbf{c} - \mathbf{a}) = 0$$

Solution. This is an immediate consequence of the preceding result.

Example 4. *Show that the six planes through the middle point of each edge of a tetrahedron and perpendicular to the opposite edge meet in a point.*

Solution. Taking the centroid of the tetrahedron as the origin of reference, let **a, b, c, d** be the position vectors of the four vertices. We have

$$\frac{1}{4}(\mathbf{a} + \mathbf{b} + \mathbf{c} + \mathbf{d}) = \mathbf{0} \Rightarrow \mathbf{a} + \mathbf{b} + \mathbf{c} + \mathbf{d} = \mathbf{0}.$$

The equation of the plane through the mid-point of the edge AB and the perpendicular to the opposite edge CD is

$$\left[\mathbf{r}-\frac{1}{2}(\mathbf{a}+\mathbf{b})\right]\cdot(\mathbf{d}-\mathbf{c})=0$$

$$\Leftrightarrow \quad \left[\mathbf{r}+\frac{1}{2}(\mathbf{c}+\mathbf{d})\right]\cdot(\mathbf{d}-\mathbf{c})=\mathbf{0} \text{ for } \mathbf{a}+\mathbf{b}+\mathbf{c}+\mathbf{d}=\mathbf{0}.$$

$$\Leftrightarrow \quad (\mathbf{r}+\mathbf{c})^2=(\mathbf{r}+\mathbf{d})^2.$$

Thus, every point on this plane is equidistant from the points $-\mathbf{c}, -\mathbf{d}$.

Considering other planes, we see that the six planes, in question, pass through the point given by

$$(\mathbf{r}+\mathbf{a})^2=(\mathbf{r}+\mathbf{b})^2=(\mathbf{r}+\mathbf{c})^2=(\mathbf{r}+\mathbf{d})^2$$

which is the centre of the sphere through the four points

$$-\mathbf{a}, \ -\mathbf{b}, \ -\mathbf{c}, \ -\mathbf{d}.$$

This point is the centre of the circumsphere of the tetrahedron $A'B'C'D'$ where A', B', C', D' are the points such that AA', BB', CC', and DD' are bisected at the centroid of the tetrahedron $ABCD$.

Note. The tetrahedron $A'B'C'D'$ obtained above is said to be the Associate of the tetrahedron $ABCD$. Both these tetrahedra have the same centroid.

It is easy to see that the planes through the mid-point of each edge perpendicular to the opposite edge of the tetrahedron $A'B'C'D'$ pass through the circumcentre of the tetrahedron $ABCD$.

EXERCISES

1. Show that six planes bisecting the six edges of a tetrahedron perpendicularly meet in a point.
2. Find the equation of the plane through the line of intersection of the planes $\mathbf{r}\cdot\mathbf{n}_1=1$, $\mathbf{r}\cdot\mathbf{n}_2=1$ and perpendicular to the plane $\mathbf{r}\cdot\mathbf{n}_3=1$.
3. Find the reflection of the point, $\mathbf{a}$, in the plane
$$\mathbf{r}\cdot\mathbf{n}=q.$$
Also find the reflection of the line $\mathbf{r}=\mathbf{a}+t\mathbf{b}$ in the same plane.
4. Find the reflection of the point, $\mathbf{a}$, in the line
$$\mathbf{r}=\mathbf{b}+t\mathbf{c}.$$
5. Show that the square of any straight line is equal to the sum of the squares of its projections on three mutually perpendicular straight lines.
6. Prove that the square on any straight line is equal to half the sum of the squares of its projections on three mutually perpendicular planes.
7. Find the distance of the point, $\mathbf{a}$, from the plane $\mathbf{r}\cdot\mathbf{n}=q$ measured parallel to the line $\mathbf{r}=\mathbf{b}+t\mathbf{c}$.

4.2. EQUATION OF A SPHERE

To find the equation of the sphere, whose centre is **c** *and radius* **a.**

Let O be the origin and C the centre of the sphere with position vector **c.** Let the position vector of any point P on the sphere be **r.** If **a** be the radius of the sphere, then

$$CP^2 = \mathbf{a}^2 \quad \text{or} \quad (\mathbf{r} - \mathbf{c})^2 = a^2$$

$$\Rightarrow \quad \mathbf{r}^2 - 2\mathbf{r} \cdot \mathbf{c} + \mathbf{c}^2 - \mathbf{a}^2 = 0$$

$$\Rightarrow \quad \mathbf{r}^2 - 2\mathbf{r} \cdot \mathbf{c} + k = 0 \quad ...(1)$$

where $k = \mathbf{c}^2 - \mathbf{a}^2$.

Since (1) is satisfied by the position vector of every point on the surface of the sphere hence it represents the equation of the sphere.

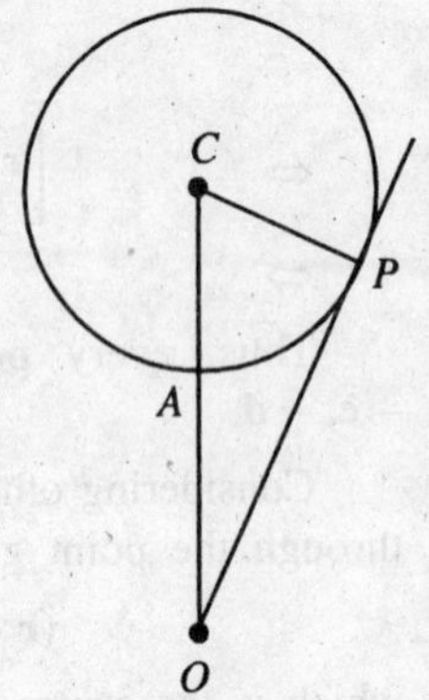

Fig. 4.4

Particular Cases :

(*i*) When the origin is the centre of the sphere, then $\mathbf{c} = 0$ and $k = \mathbf{c}^2 - \mathbf{a}^2 = -\mathbf{a}^2$. Hence, the equation of the sphere is

$$\mathbf{r}^2 = a^2.$$

(*ii*) When the origin lies on the surface of the sphere, then $|\mathbf{c}| = c = a$ and hence $k = \mathbf{c}^2 - a^2 = 0$. Therefore, the equation of the required sphere is

$$\mathbf{r}^2 - 2\mathbf{r} \cdot \mathbf{c} = 0.$$

4.2.1. Diameter Form of Equation of the Sphere

Let the position vectors of the extremities A and B of the diameter AB be **a** and **b** respectively. Let **r** be the position vector of any point P on the surface of the sphere. Then

$$\overrightarrow{AP} \cdot \overrightarrow{BP} = 0$$

$$\Rightarrow \quad (\mathbf{r} - \mathbf{a}) \cdot (\mathbf{r} - \mathbf{b}) = 0.$$

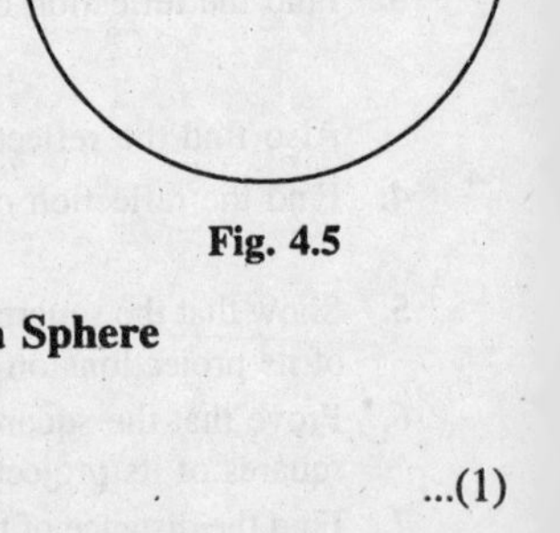

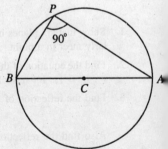

Fig. 4.5

4.2.2. Intersection of a Straight Line and a Sphere

Let

$$F(\mathbf{r}) \equiv \mathbf{r}^2 - 2\mathbf{r} \cdot \mathbf{c} + k = 0. \quad ...(1)$$

be the equation of the sphere, where $k = \mathbf{c}^2 - a^2$. Let the equation of the straight line be

$$\mathbf{r} = \mathbf{d} + t\,\hat{\mathbf{b}} \qquad \text{...(2)}$$

which passes through the point D whose position vector is $\mathbf{d}$ and is parallel to a unit vector $\hat{\mathbf{b}}$.

For the points of intersection the values of $\mathbf{r}$ satisfy (1) and (2) both. On eliminating $\mathbf{r}$ between (1) and (2), we get

$$(\mathbf{d} + t\hat{\mathbf{b}})^2 - 2\,(\mathbf{d} + t\hat{\mathbf{b}})\,.\,\mathbf{c} + k = 0$$

$$\Rightarrow \quad t^2 + 2t\hat{\mathbf{b}}\,.\,(\mathbf{d} - \mathbf{c}) + \mathbf{d}^2 - 2\mathbf{d}\,.\,\mathbf{c} + k = 0$$

$$\Rightarrow \quad t^2 + 2t\hat{\mathbf{b}}\,.\,(\mathbf{d} - \mathbf{c}) + F(\mathbf{d}) = 0 \qquad \text{...(3)}$$

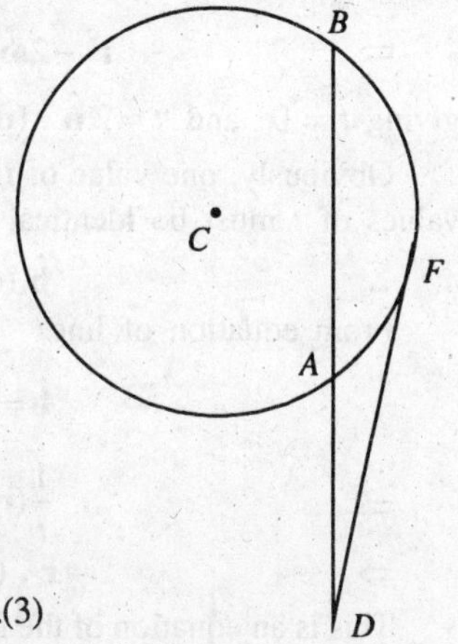

Fig. 4.6

This equation is quadratic in t and hence it gives two values t_1 and t_2 of t, when substituted in (2) give the position vectors of the two points A and B on the sphere.

Cor. 1. The points of intersection are real, coincident or imaginary according as

$$4\,[\hat{\mathbf{b}}\,.\,(\mathbf{d} - \mathbf{c})]^2 - 4F(\mathbf{d}) > 0 = 0 \text{ or } < 0.$$

Also $\quad t_1 = DA$ and $t_2 = DB$.

$\therefore \quad t_1 t_2 = DA\,.\,DB = F(\mathbf{d}) = \mathbf{d}^2 - 2\mathbf{d}\,.\,\mathbf{c} + k$

which is independent of $\hat{\mathbf{b}}$, *i.e.*, the direction of the line through the point D. Hence, for all lines through D, the product of the segment DA and DB remains unaltered.

Cor 2. If A and B tend to coincide at T, then the line becomes a tangent at T. Then

$$DT^2 = DA\,.\,DB = F(\mathbf{d})$$
$$= \mathbf{d}^2 - 2\mathbf{d}\,.\,\mathbf{c} + k.$$

This gives the *square of the length of the tangent from any point to the sphere.* It is also called the *power of the point D* with respect to the sphere.

4.2.3. Tangent Plane at a given Point

As in § 4.2.2, for the point of intersection of sphere (1) and line (2), we have

$$t^2 - 2t\hat{\mathbf{b}}\,.\,(\mathbf{d} - \mathbf{c}) + F(\mathbf{d}) = 0$$

since $\mathbf{d}$ lies on the sphere, hence

$$F(\mathbf{d}) = 0$$

$$\Rightarrow \qquad t^2 - 2t\hat{\mathbf{b}} \cdot (\mathbf{d}-\mathbf{c}) = 0$$

giving $t = 0$ and $t = 2\hat{\mathbf{b}} \cdot (\mathbf{d}-\mathbf{c})$

Obviously, one value of t is zero. If (2) touches (1) [§ 4.2.2], both the values of t must be identical

$$\therefore \qquad \hat{\mathbf{b}}.(\mathbf{d}-\mathbf{c}) = 0 \qquad ...(3)$$

From equation of line

$$\hat{\mathbf{b}} = \frac{1}{t}(\mathbf{r}-\mathbf{d})$$

$$\Rightarrow \qquad \frac{1}{t}(\mathbf{r}-\mathbf{d})\cdot(\mathbf{d}-\mathbf{c}) = 0$$

$$\Rightarrow \qquad \mathbf{r} \cdot (\mathbf{d}-\mathbf{c}) = \mathbf{d} \cdot (\mathbf{d}-\mathbf{c}) \qquad ...(4)$$

This is an equation of the form $\mathbf{r} \cdot \mathbf{n} = q$. So (4) represents a plane which is perpendicular to the radius through the point $\mathbf{d}$.

From (4),

$$\mathbf{r} \cdot \mathbf{d} - \mathbf{r} \cdot \mathbf{c} - \mathbf{d}^2 + \mathbf{d} \cdot \mathbf{c} = 0$$

or
$$\mathbf{r} \cdot \mathbf{d} - \mathbf{r} \cdot \mathbf{c} - \mathbf{d}^2 + \mathbf{d} \cdot \mathbf{c} + (\mathbf{d}^2 - 2\mathbf{d} \cdot \mathbf{c} + k) = 0$$

or
$$\mathbf{r} \cdot \mathbf{d} - \mathbf{c} \cdot (\mathbf{r} + \mathbf{d}) + k = 0 \qquad ...(5)$$

This is the required equation to the tangent plane to the sphere (1) at the given point $\mathbf{d}$.

4.2.4. Condition for Tangency of any Plane to a Sphere

Let

$$\mathbf{r}^2 - 2\mathbf{r} \cdot \mathbf{c} + k = 0 \qquad ...(1)$$

be the given sphere whose centre is $\mathbf{c}$ and radius a.

Let the given plane be

$$\mathbf{r} \cdot \mathbf{n} = q \qquad ...(2)$$

Since the tangent plane at any point is perpendicular to the radius through that point, the square of the perpendicular from the centre to the tangent plane must be equal to the square of the radius. Hence, the required condition will be

$$\left(\frac{q - \mathbf{c} \cdot \mathbf{n}}{|\mathbf{n}|}\right)^2 = a^2 = \mathbf{c}^2 - k.$$

4.2.5. Condition of Orthogonality

Let the two spheres be

$$\mathbf{r}^2 - 2\mathbf{r} \cdot \mathbf{c}_1 + k_1 = 0; \quad k_1 = \mathbf{c}_1^2 - a_1^2 \qquad ...(1)$$

$$\mathbf{r}^2 - 2\mathbf{r} \cdot \mathbf{c}_2 + k_2 = 0; \quad k_2 = \mathbf{c}_2^2 - a_2^2 \qquad ...(2)$$

These spheres will intersect orthogonally if the tangent plane to any of them at the point of intersection passes through the centre of the other. So the square of the distance between the centres must be equal to the sum of the squares at their radii. Thus,

$$(\mathbf{c}_1 - \mathbf{c}_2)^2 = a_1^2 + a_2^2 = \mathbf{c}_1^2 - k_1 + \mathbf{c}_2^2 - k_2$$

$$\Rightarrow \quad 2\mathbf{c}_1 \cdot \mathbf{c}_2 = k_1 + k_2.$$

4.2.6. Polar Plane

Def. The polar plane of a given point with respect to a sphere is the locus of the points the tangent planes at which pass through the given point.

4.2.6.1. To find the polar plane of a point

Let

$$\mathbf{r}^2 - 2\mathbf{r} \cdot \mathbf{c} + k = 0$$

be the the given sphere and **a** the given point. Let **d** be any point on the sphere the tangent plane at which pass through **a**. Tangent plane at **d** to the sphere is

$$\mathbf{r} \cdot \mathbf{d} - \mathbf{c} \cdot (\mathbf{r} + \mathbf{d}) + k = 0$$

It passes through **a**, hence

$$\mathbf{a} \cdot \mathbf{d} - \mathbf{c} \cdot (\mathbf{a} + \mathbf{d}) + k = 0$$

∴ Locus of **d** is

$$\mathbf{a} \cdot \mathbf{r} - \mathbf{c} \cdot (\mathbf{a} + \mathbf{r}) + k = 0$$

or

$$\mathbf{r} \cdot \mathbf{a} - \mathbf{c} \cdot (\mathbf{r} + \mathbf{a}) + k = 0.$$

4.2.7. Diametral Plane

Def. The locus of points which bisect a system of parallel chords of a given sphere is called the diametral plane, all chords being parallel to a given line.

4.2.7.1. To find the diametral plane of a sphere

Let

$$F(\mathbf{r}) = \mathbf{r}^2 - 2\mathbf{r} \cdot \mathbf{c} + k = 0 \qquad ...(1)$$

be the given sphere and let a chord parallel to a given unit vector $\hat{\mathbf{b}}$ and passing through the point **d** be

$$\mathbf{r} = \mathbf{d} + t\hat{\mathbf{b}} \qquad ...(2)$$

As in § 4.2.2, the points of intersection of (1) and (2) are given by

$$t^2 - 2t\mathbf{b} \cdot (\hat{\mathbf{d}} - \mathbf{c}) + F(\mathbf{d}) = 0 \qquad ...(3)$$

If d be the middle point of the chord (2), then the two values of t must be equal and opposite in sign, *i.e.*, the sum of the roots of (3) must be zero.

Thus, $\hat{\mathbf{b}} \cdot (\mathbf{d} - \mathbf{c}) = 0$

∴ the locus of the point **d** is

$$\hat{\mathbf{b}}.(\mathbf{r}-\mathbf{c})=0.$$

This is the required diametral plane.

4.2.8. Radical Plane

Def. The locus of a point, which moves so that the squares of the lengths of the tangents drawn from it to the given spheres are equal (or, powers of the point w.r.t. two spheres are equal), is called the radical plane of the two spheres.

4.2.8.1. To obtain radical plane of two spheres

Let $\mathbf{r}^2 - 2\mathbf{r} \cdot \mathbf{c}_1 + k_1 = 0$

and $\mathbf{r}^2 - 2\mathbf{r} \cdot \mathbf{c}_2 + k_2 = 0$

be the two given spheres. Let **d** be the position vector of a point which moves so that the square of the tangents drawn from it to spheres (1) are equal. Then

$$\mathbf{d}^2 - 2\mathbf{d} \cdot \mathbf{c}_1 + k_1 = \mathbf{d}^2 - 2\mathbf{d} \cdot \mathbf{c}_2 + k_2$$

or $$2\mathbf{d} \cdot (\mathbf{c}_1 - \mathbf{c}_2) = k_1 - k_2$$

$\therefore$ Locus of **d** is

$$2\mathbf{r} \cdot (\mathbf{c}_1 - \mathbf{c}_2) = k_1 - k_2.$$

Remark. The radical plane is perpendicular to $\mathbf{c}_1 - \mathbf{c}_2$. Thus, the radical plane of two spheres is perpendicular to the line joining the centres of the given spheres.

4.2.8.2. Coaxal System of Spheres

Def. A system of spheres every pair of which has the same radical plane is called a coaxal system of spheres.

4.2.8.3. Equation of a System of Coaxal Spheres

To show that the equation $\mathbf{r}^2 - 2\lambda\, \mathbf{r} \cdot \mathbf{c} + k = 0$, where λ is a parameter and k is constant represents a system of coaxal spheres.

Consider two members of the system of spheres corresponding to values λ_1 and λ_2 of the parameter λ,

$$\mathbf{r}^2 - 2\lambda_1\, \mathbf{r} \cdot \mathbf{c} + k = 0$$

and $$\mathbf{r}^2 - 2\lambda_2\, \mathbf{r} \cdot \mathbf{c} + k = 0.$$

The radical plane of these spheres is

$$2\,(\lambda_1 - \lambda_2)\, \mathbf{r} \cdot \mathbf{c} = 0$$

$\Rightarrow$ $$\mathbf{r} \cdot \mathbf{c} = 0, \qquad (\because \lambda_1 \neq \lambda_2)$$

This is independent of λ. Hence, whatever be the value of λ, the radical plane of every two members will be the same. Thus, $\mathbf{r}^2 - 2\lambda\, \mathbf{r} \cdot \mathbf{c} + d = 0$, represents a coaxal system of spheres.

Cor. The member of coaxal system of zero radius are called *limiting points.* The radius of the sphere of the system is $\sqrt{\{(\lambda c)^2 - k\}}$. Equating to zero, we get $\lambda = \pm \sqrt{(k/c)}$. Hence, the limiting points are $\pm \sqrt{(k/c)}\,\mathbf{c}$.

EXAMPLES

Example 1. *Show that the diameter subtends a right angle at any point on the surface of the sphere.*

Solution. Let ACB be the diameter of a sphere whose centre is the origin C and whose radius is a.

Let P be any point on the surface of the sphere. Equation of the sphere is

$$\mathbf{r}^2 = \mathbf{a}^2$$

$$\Rightarrow \quad (\mathbf{r} - \mathbf{a})(\mathbf{r} + \mathbf{a}) = 0$$

$$\Rightarrow \quad \vec{AP} \cdot \vec{BP} = 0$$

showing that $\angle APB = 90^\circ$.

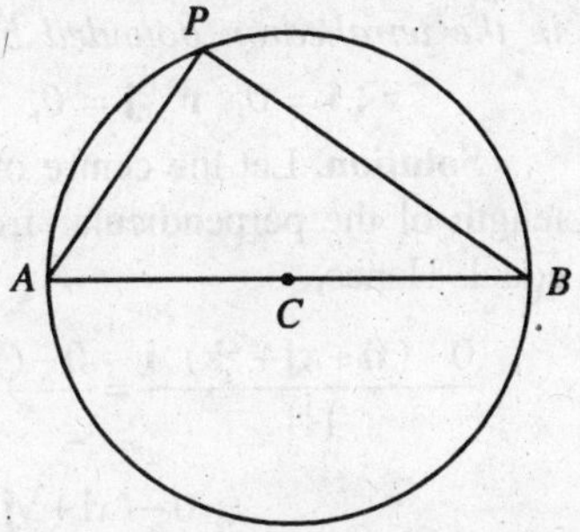

Fig. 4.7

Example 2. *A straight line is drawn from a point O to meet a fixed sphere in P. A point Q taken in OP such that OP : OQ is a fixed ratio. Show that the locus of Q is a sphere.*

Solution. With O as origin, let the fixed sphere be

$$\mathbf{r}^2 - 2\mathbf{r} \cdot \mathbf{c} + k = 0$$

Let $\vec{OQ} = \mathbf{r}'$ be the position vector Q and let $OP : OQ = n$.

$$\Rightarrow \quad \vec{OP} = n\vec{OQ} \Rightarrow \mathbf{r} = n\mathbf{r}'$$

Since P lies on the sphere, we must have

$$n^2 \mathbf{r}'^2 - 2n\mathbf{r}' \cdot \mathbf{c} + k = 0$$

or $$\mathbf{r}'^2 - 2\mathbf{r}' : (1/n)\,\mathbf{c} + k/n^2 = 0$$

$\therefore$ Locus of $\mathbf{r}'$ is

$$\mathbf{r}^2 - 2\mathbf{r} \cdot (1/n)\,\mathbf{c} + k/n^2 = 0,$$

which is a sphere.

Example 3. *Show that the sphere which cuts the spheres $F_1(\mathbf{r}) = 0$ and $F_2(\mathbf{r}) = 0$ orthogonally, also cuts $F_1(\mathbf{r}) - \lambda F_2(\mathbf{r}) = 0$ orthogonally.*

Solution. Let

$$F_1(\mathbf{r}) = \mathbf{r}^2 - 2\mathbf{r} \cdot \mathbf{c}_1 + k_1 = 0 \quad ...(1)$$

and $$F_2(\mathbf{r}) = \mathbf{r}^2 - 2\mathbf{r} \cdot \mathbf{c}_2 + k_2 = 0 \quad ...(2)$$

be the given spheres. Then the sphere $F_1(\mathbf{r}) - F_2(\mathbf{r}) = 0$ is

$$\mathbf{r}^2 - 2\mathbf{r} \cdot \mathbf{c}_1 + k_1 - \lambda(\mathbf{r}^2 - 2\mathbf{r} \cdot \mathbf{c}_2 + k_2) = 0$$

or $$\mathbf{r}^2 - 2\mathbf{r}\cdot\frac{\mathbf{c}_1 - \lambda\mathbf{c}_2}{1-\lambda} + \frac{k_1 - \lambda k_2}{1-\lambda} = 0 \quad ...(3)$$

Let the sphere $\mathbf{r}^2 - 2\mathbf{r} \cdot \mathbf{c} + k = 0$ cuts (1) and (2) orthogonally, then

$$2\mathbf{c} \cdot \mathbf{c}_1 = k + k_1 \quad ...(4)$$

and $$2\mathbf{c} \cdot \mathbf{c}_2 = k + k_2 \quad ...(5)$$

Multiplying (5) by λ and subtracting from (4), we get

$$2\mathbf{c} \cdot (\mathbf{c}_1 - \lambda\mathbf{c}_2) = k(1-\lambda) + k_1 - \lambda k_2$$

or $$2\mathbf{c}\cdot\frac{\mathbf{c}_1 - \lambda\mathbf{c}_2}{1-\lambda} = k + \frac{k_1 - \lambda k_2}{1-\lambda}$$

showing that the given sphere cuts (3) orthogonally.

Example 4. *Find the co-ordinates of the centre of the sphere inscribed in the tetrahedron bounded by the planes*

$$\mathbf{r} \cdot \mathbf{i} = 0,\ \mathbf{r} \cdot \mathbf{j} = 0,\ \mathbf{r} \cdot \mathbf{k} = 0 \text{ and } \mathbf{r} \cdot (\mathbf{i} + \mathbf{j} + \mathbf{k}) = a.$$

Solution. Let the centre of the given sphere be $x\mathbf{i} + y\mathbf{j} + z\mathbf{k}$. Then the length of the perpendicular from the centre on the given planes must be equal. Hence,

$$\frac{0-(x\mathbf{i}+y\mathbf{j}+z\mathbf{k})\cdot\mathbf{i}}{|\mathbf{i}|} = \frac{0-(x\mathbf{i}+y\mathbf{j}+z\mathbf{k})\cdot\mathbf{j}}{|\mathbf{j}|}$$

$$= \frac{0-(x\mathbf{i}+y\mathbf{j}+z\mathbf{k})\cdot\mathbf{k}}{|\mathbf{k}|} = \frac{a-(x\mathbf{i}+y\mathbf{j}+z\mathbf{k}),(\mathbf{i}+\mathbf{j}+\mathbf{k})}{|\mathbf{i}+\mathbf{j}+\mathbf{k}|}$$

$$\Rightarrow \frac{x}{1} = \frac{y}{1} = \frac{z}{1} = \frac{a-(x+y+z)}{\sqrt{3}} = \frac{x+y+z+a-(x+y+z)}{1+1+1+\sqrt{3}} = \frac{a}{3+\sqrt{3}}$$

$$\therefore \quad x = y = z = \frac{a}{3+\sqrt{3}} = \frac{a}{6}\left(3-\sqrt{3}\right).$$

Example 5. *Prove that any straight line drawn from a point O to intersect a sphere is cut harmonically by the surface and the polar plane of O.*

Solution. Let

$$\mathbf{r}^2 - 2\mathbf{r} \cdot \mathbf{c} + k = 0$$

be the given sphere with O as origin. Polar plane of O w.r.t. the sphere is

$$O^2 - (\mathbf{r} + O) \cdot \mathbf{c} + k = 0$$

or $$\mathbf{r} \cdot \mathbf{c} = k.$$

Let the equation of the given line through O is

$$\mathbf{r} = t\hat{\mathbf{b}}$$

For the points of intersection of sphere and line, we have

$$t^2 - 2t(\hat{\mathbf{b}} . \mathbf{c}) + k = 0$$

This is quadratic in t and so it gives two values t_1, t_2 which represent the distances OP and OQ respectively.

$$\therefore \quad \frac{1}{OP} + \frac{1}{OQ} = \frac{t_1 + t_2}{t_1 t_2} = \frac{2\hat{\mathbf{b}} . \mathbf{c}}{k}.$$

Again, for the points of intersection of polar plane and the sphere, we have

$$t\hat{\mathbf{b}} . \mathbf{c} = k, \text{ so that } t = OR$$

$$= \frac{k}{\hat{\mathbf{b}} . \mathbf{c}} \quad \text{or} \quad \frac{1}{OR} = \frac{\hat{\mathbf{b}} . \mathbf{c}}{k}$$

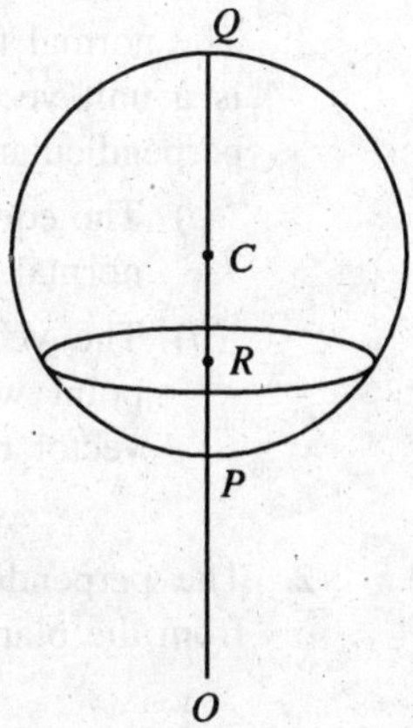

Fig. 4.8

Hence, $$\frac{1}{OP} + \frac{1}{OQ} = \frac{2}{OR}.$$

$\Rightarrow$ OP, OR, OQ are in $H.P.$

Hence, the result.

EXERCISES

1. Prove that the locus of a point, the sum of the squares of whose distances from n given points is constant, is a sphere.
2. Prove that the equation of the sphere circumscribing the tetrahedron bounded by the planes
$$\mathbf{r} . \mathbf{i} = 0, \ \mathbf{r} . \mathbf{j} = 0, \ \mathbf{r} . \mathbf{k} = 0$$
and $\mathbf{r} . (\mathbf{i} + \mathbf{j} + \mathbf{k}) = a$, is
$$\mathbf{r} . \{\mathbf{r} - a(\mathbf{i} + \mathbf{j} + \mathbf{k})\} = 0.$$
3. If the plane of the point H passes through G, then show that the polar plane of G passes through H.
4. If from any point on the surface of a sphere, straight lines are drawn to the extremities of any diameter of a concentric sphere, the sum of the squares on these lines is constant.
5. If the line joining the centre O of a sphere to any point P meets the polar plane of P in Q, prove that $OP . OQ = a^2$, where a is the radius of the sphere.
6. Find the locus of a point whose powers with respect to two spheres are equal.
7. Prove by vector method that the triangle inscribed in a semi-circle is a right angle.
8. Show that the locus of the straight lines which intersect the sphere $\mathbf{r}^2 - 2\mathbf{r} . \mathbf{c} + k = 0$ and bisected at a given point $\mathbf{d}$ is $(\mathbf{r} - \mathbf{d}) . (\mathbf{d} - \mathbf{c}) = 0$.

SUMMARY

1. The normal form of the equation of a plane is $\mathbf{r} \cdot \mathbf{n} = p$. Here $\mathbf{n}$ is a unit vector normal to the plane and p is the length of the perpendicular from the origin to the plane.

(*i*) The equation $\mathbf{r} \cdot \mathbf{n} = q$, represents a plane, $\mathbf{n}$ being a vector normal to the plane.

(*ii*) The vector equation of the plane which passes through a point with position vector $\mathbf{a}$ and which is normal to the vector $\mathbf{n}$ is

$$(\mathbf{r} - \mathbf{a}) \cdot \mathbf{n} = 0.$$

2. The perpendicular distance of the point with position vector a from the plane $\mathbf{r} \cdot \mathbf{n} = q$ is

$$\frac{|q - \mathbf{a} \cdot \mathbf{n}|}{|\mathbf{n}|}.$$

3. Equation of the sphere, whose centre is $\mathbf{c}$ and radius $\mathbf{a}$ is

$$\mathbf{r}^2 - 2\mathbf{r} \cdot \mathbf{c} + k = 0, \quad \text{where} \quad k = \mathbf{c}^2 - a^2.$$

(*i*) when the origin is the centre, the equation of sphere is

$$\mathbf{r}^2 = a^2$$

(*ii*) when the origin lies on the surface of the sphere, the equation of the sphere is

$$\mathbf{r}^2 - 2\mathbf{r} \cdot \mathbf{c} = 0.$$

(*iii*) Equation of the sphere on the join of two given points $\mathbf{a}$ and $\mathbf{b}$ as diameter is

$$(\mathbf{r} - \mathbf{a}) \cdot (\mathbf{r} - \mathbf{b}) = 0.$$

4. Equation of the tangent plane to the sphere $\mathbf{r}^2 - 2\mathbf{r} \cdot \mathbf{c} + k = 0$ at the point $\mathbf{d}$ is

$$\mathbf{r} \cdot \mathbf{d} - \mathbf{c} \cdot (\mathbf{r} + \mathbf{d}) + k = 0.$$

5. Condition that a plane $\mathbf{r} \cdot \mathbf{n} = q$ will touch the sphere $\mathbf{r}^2 - 2\mathbf{r} \cdot \mathbf{c} + k = 0$ is

$$\left(\frac{q - \mathbf{c} \cdot \mathbf{n}}{|\mathbf{n}|}\right)^2 = a^2 = \mathbf{c}^2 - k.$$

6. Two spheres $\mathbf{r}^2 - 2\mathbf{r} \cdot \mathbf{c}_1 + k_1 = 0$ and $\mathbf{r}^2 - 2\mathbf{r} \cdot \mathbf{c}_2 + k_2 = 0$ will intersect orthogonally if $2\mathbf{c}_1 \cdot \mathbf{c}_2 = k_1 + k_2$.

7. Polar plane of a point $\mathbf{a}$ with respect to the sphere $\mathbf{r}^2 - 2\mathbf{r} \cdot \mathbf{c} + k = 0$ is

$$\mathbf{r} \cdot \mathbf{a} - c \cdot (\mathbf{r} + \mathbf{a}) + k = 0.$$

8. Diameteral plane of a sphere $\mathbf{r}^2 - 2\mathbf{r} \cdot \mathbf{c} + k = 0$, parallel to a given unit vector $\hat{\mathbf{b}}$ is

$$\hat{\mathbf{b}} \cdot (\mathbf{r} - \mathbf{c}) = 0.$$

9. The radical plane of two spheres can be obtained by subtracting the equation of one sphere from the other after making the coefficients of $\mathbf{r}^2$ in each equation unity.

10. Equation $\mathbf{r}^2 - 2\lambda\mathbf{r} \cdot \mathbf{c} + k = 0$, where λ is a parameter and k is constant represents a system of coaxal spheres.

The members of coaxal system of zero radius are called limiting points. For the coaxal system $\mathbf{r}^2 - 2\lambda\mathbf{r} \cdot \mathbf{c} + k = 0$, the limiting points are $\pm\sqrt{(k/c)}\,\mathbf{c}$.

OBJECTIVE QUESTIONS

For each of the following questions, four alternatives are given for the answer. Only one of them is correct. Choose the correct alternative.

1. The angle between two planes $\mathbf{r} \cdot \mathbf{n} = q$ and $\mathbf{r} \cdot \mathbf{n}' = q'$ is

(*a*) $\sin^{-1}\left(\dfrac{\mathbf{n} \cdot \mathbf{n}'}{nn'}\right)$ (*b*) $\cos^{-1}\left(\dfrac{\mathbf{n} \cdot \mathbf{n}'}{nn'}\right)$

(*c*) $\tan^{-1}\left(\dfrac{\mathbf{n} \cdot \mathbf{n}'}{nn'}\right)$ (*d*) None of these

2. The intercept made by the plane $\mathbf{r} \cdot \mathbf{n} = q$ on the x-axis is equal to

(*a*) $q / (\mathbf{i} \cdot \mathbf{n})$ (*b*) $(\mathbf{i} \cdot \mathbf{n}) / q$

(*c*) $q\,(\mathbf{i} \cdot \mathbf{n})$ (*d*) None of these

3. Angle between a line $\mathbf{r} = \mathbf{a} + \mathbf{b}t$ and the plane $\mathbf{r} \cdot \mathbf{n} = q$ is

(*a*) $\sin^{-1} \dfrac{\mathbf{n} \cdot \mathbf{b}}{|\mathbf{n}|\,|\mathbf{b}|}$ (*b*) $\cos^{-1} \dfrac{\mathbf{n} \cdot \mathbf{b}}{|\mathbf{n}|\,|\mathbf{b}|}$

(*c*) $\tan^{-1} \dfrac{\mathbf{n} \cdot \mathbf{b}}{|\mathbf{n}|\,|\mathbf{b}|}$ (*d*) None of these

4. The equation of a sphere whose centre is the origin and radius $\mathbf{a}$ is

(*a*) $\mathbf{r}^2 - 2\mathbf{r} \cdot \mathbf{a} = 0$ (*b*) $\mathbf{r} = \mathbf{a}$

(*c*) $\mathbf{r}^2 = \mathbf{a}^2$ (*d*) $\mathbf{r} \cdot \mathbf{a} = k$

5. The two spheres $\mathbf{r}^2 - 2\mathbf{r} \cdot \mathbf{c}_1 + k_1 = 0$ and $\mathbf{r}^2 - 2\mathbf{r} \cdot \mathbf{c}_2 + k_2 = 0$ will cut orthogonally if

(*a*) $\mathbf{c}_1 \cdot \mathbf{c}_2 = k_1 + k_2$ (*b*) $2\mathbf{c}_1 \cdot \mathbf{c}_2 = k_1 + k_2$

(*c*) $\mathbf{c}_1 + \mathbf{c}_2 = k_1k_2$ (*d*) $\mathbf{c}_1 + \mathbf{c}_2 = 2k_1k_2$

ANSWERS

1. (*b*) **2.** (*a*) **3.** (*a*) **4.** (*c*) **5.** (*b*)

5

Vector Product and Scalar Triple Product

Introduction. The concept of the *Vector product* of two vectors, as a result of which a well-defined vector is associated to a given ordered pair of vectors, will be introduced in this chapter. Like Scalar products, Vector products are also proportional to the lengths of the factor vectors and obey Distributive Laws. It should be, however, remembered that quite a number of the properties of Scalar and Vector products are different from those of ordinary products of numbers.

The notion of vector products greatly facilitates the study of a type of metrical relations such as those involving areas and volumes. Also in those contexts where we require to deal with vectors perpendicular to two given vectors, vector products play an important part.

It will also be seen in the last chapter that characterisation of forces by vectors is possible with the help of Vector products.

In addition to Scalar and Vector products, *Mixed product*

$$\mathbf{a} \times \mathbf{b} \cdot \mathbf{c},$$

known as *Scalar triple product,* also plays an important part. We shall also consider this in this chapter.

The volume of a parallelopiped is given as a scalar triple product and the vanishing of a scalar triple product provides a very neat form of condition for three vectors to be coplanar and four points to be coplanar.

The part of the Algebra of vectors developed in this chapter will also provide important tools for the solution of Vector Equations.

5.1. RIGHT-HANDED AND LEFT-HANDED VECTOR TRIADS

A screw experiences a motion of translation when it is rotated and, as such, we can distinguish a screw as being right-handed or left-handed by the direction of its translation when it is rotated in any given manner. Also we shall now associate a screw with any given ordered vector triad and thus be enabled to distinguish any given vector triad also as being right-handed or left-handed.

Let **a, b, c** be three vectors, given in this order. We shall suppose that they are coinitial and no two are collinear.

Suppose, that, **a**, points towards the right of the reader and, **b**, towards the upper part of the paper.

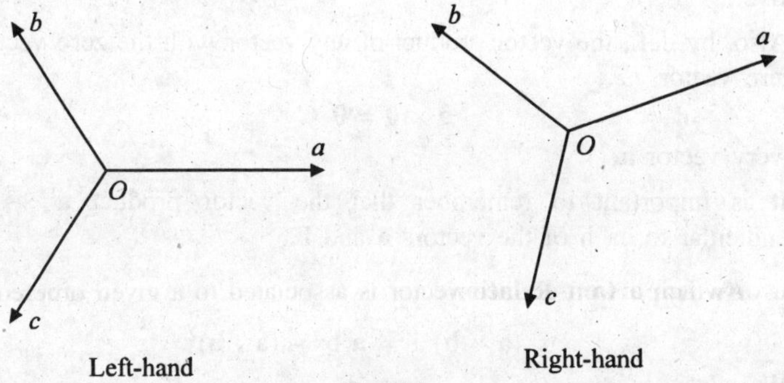

Fig. 5.1

If now a right-handed (left-handed) screw be rotated from **a**, towards **b**, through an angle < 180°, then it undergoes translation in a direction pointing towards the reader (away from the reader). On this account, *an ordered vector triad*

a, b, c

is said to be right-handed or left-handed according as the right-handed screw is translated along **c** *or opposite to* **c**, *when it is rotated from* **a** *towards* **b**, *through an angle less than 180°.*

It is easily seen that a cyclic permutation of the vectors of a triad does not change the character of the triad from the point of view of its right-handedness and left-handedness. Thus, if **a, b, c** is a right-handed (left-handed) triad, then

b, c, a and **c, a, b**

are also right-handed (left-handed), but

a, c, b; b, a, c; c, b, a

are left-handed (right-handed).

5.2. VECTOR PRODUCT OR CROSS PRODUCT

Def. *The vector product, denoted by*

$$\mathbf{a} \times \mathbf{b}$$

of two vectors, **a, b** *taken in this order, is the vector* **c**, *where*

(*i*) $|\mathbf{c}| = |\mathbf{a}|\,|\mathbf{b}| \sin\theta,$

θ *being the angle between the vectors,* **a, b** *and* $0 \le \theta \le 180°$.

(*ii*) *the support of* **c** *is perpendicular to the supports of* **a** *and* **b**.

(*iii*) *the sense of* **c** *is such that the triad*

$$\mathbf{a}, \mathbf{b}, \mathbf{c}$$

forms a right-handed system.

As $0 \leq \theta \leq 180°$, $\sin \theta$ cannot be negative so that $|\mathbf{c}|$ will not be negative.

Also, by def., the vector product of any vector with the zero vector is the zero vector, *i.e.*,

$$\mathbf{a} \times \mathbf{0} = \mathbf{0}$$

for every vector **a**.

It is important to remember that the vector product $\mathbf{a} \times \mathbf{b}$ is perpendicular to each of the vectors **a** and **b**.

5.2.1. An Important Relation

$$(\mathbf{a} \times \mathbf{b})^2 = \mathbf{a}^2\mathbf{b}^2 - (\mathbf{a} \cdot \mathbf{b})^2.$$

We have

$$\begin{aligned}(\mathbf{a} \times \mathbf{b})^2 &= (\mathbf{a} \times \mathbf{b}) \cdot (\mathbf{a} \times \mathbf{b}) \\ &= (|\mathbf{a}|\ |\mathbf{b}| \sin \theta)^2 \\ &= \mathbf{a}^2\mathbf{b}^2 \sin^2 \theta \\ &= \mathbf{a}^2\mathbf{b}^2 (1 - \cos^2 \theta) \\ &= \mathbf{a}^2\mathbf{b}^2 - (|\mathbf{a}|\ |\mathbf{b}| \cos \theta)^2 \\ &= \mathbf{a}^2\mathbf{b}^2 - (\mathbf{a} \cdot \mathbf{b})^2\end{aligned}$$

This relation expresses the length of the vector $\mathbf{a} \times \mathbf{b}$ in terms of the scalar products

$$\mathbf{a} \cdot \mathbf{a}, \quad \mathbf{b} \cdot \mathbf{b}, \quad \mathbf{a} \cdot \mathbf{b}.$$

5.2.2. Formulation of Vector Product in Terms of Scalar Products

The vector product $\mathbf{a} \times \mathbf{b}$ is the vector **c**, such that

(*i*) $|\mathbf{c}| = \sqrt{[\mathbf{a}^2\mathbf{b}^2 - (\mathbf{a} \cdot \mathbf{b})^2]}$

(*ii*) $\mathbf{c} \cdot \mathbf{a} = 0, \quad \mathbf{c} \cdot \mathbf{b} = 0.$

(*iii*) **a, b, c** form a right-handed system.

5.3. SOME PROPERTIES OF VECTOR PRODUCTS

5.3.1. (*i*) *The vector product of two parallel vectors is the zero vector*, for, in this case, $\theta = 0$ or $180°$, so that $\sin \theta = 0$ and as such $|\mathbf{c}| = 0$. Thus, $\mathbf{c} = 0$.

5.3.2. *The vector multiplication is not commutative.* In fact

$$\mathbf{a} \times \mathbf{b} = -\,\mathbf{b} \times \mathbf{a}.$$

This follows from the fact that while the magnitude and the support of $\mathbf{b} \times \mathbf{a}$ is the same as those of $\mathbf{a} \times \mathbf{b}$, their senses are different.

5.3.3.

$$-\mathbf{a} \times \mathbf{b} = -(\mathbf{a} \times \mathbf{b}), \quad \mathbf{a} \times (-\mathbf{b}) = -(\mathbf{a} \times \mathbf{b}),$$
$$(-\mathbf{a}) \times (-\mathbf{b}) = \mathbf{a} \times \mathbf{b}.$$

Generally,

$$m\mathbf{a} \times n\mathbf{b} = mn\,(\mathbf{a} \times \mathbf{b}) = \mathbf{a} \times mn\mathbf{b},$$

where m and n are any scalars, positive or negative.

These results are immediate consequences of the definition of Vector product.

5.3.4. Relations between the mutually perpendicular unit vectors i, j, k.

If **i, j, k** are mutually perpendicular unit vectors forming a right-handed system, then, as may be easily seen, from the definition of vector products, we have

$$\mathbf{i} \times \mathbf{i} = \mathbf{0}, \quad \mathbf{j} \times \mathbf{j} = \mathbf{0}, \quad \mathbf{k} \times \mathbf{k} = \mathbf{0},$$
$$\mathbf{i} \times \mathbf{j} = \mathbf{k}, \quad \mathbf{j} \times \mathbf{i} = -\mathbf{k};$$
$$\mathbf{j} \times \mathbf{k} = \mathbf{i}, \quad \mathbf{k} \times \mathbf{j} = -\mathbf{i};$$
$$\mathbf{k} \times \mathbf{i} = \mathbf{j}, \quad \mathbf{i} \times \mathbf{k} = -\mathbf{j}.$$

Note. Later on, we shall show that the Distributive Law also holds for vector multiplications, *i.e.*,

$$\mathbf{a} \times (\mathbf{b} + \mathbf{c}) = (\mathbf{a} \times \mathbf{b}) + (\mathbf{a} \times \mathbf{c}) \text{ for all vectors } \mathbf{a}, \mathbf{b}, \mathbf{c}.$$

After establishing the Distributive Law, we shall be able to express

$$(a_1\mathbf{i} + a_2\mathbf{j} + a_3\mathbf{k}) \times (b_1\mathbf{i} + b_2\mathbf{j} + b_3\mathbf{k})$$

as a linear combination of the vectors, **i, j, k.**

5.3.5. Vector Product as a Determinant

Let $\qquad \mathbf{a} = a_1\mathbf{i} + a_2\mathbf{j} + a_3\mathbf{k}, \ \mathbf{b} = b_1\mathbf{i} + b_2\mathbf{j} + b_3\mathbf{k}$

$$\begin{aligned}\mathbf{a} \times \mathbf{b} = {} & a_1b_1\ \mathbf{i} \times \mathbf{i} + a_1b_2\ \mathbf{i} \times \mathbf{j} + a_1b_3\ \mathbf{i} \times \mathbf{k} \\ & + a_2b_1\ \mathbf{j} \times \mathbf{i} + a_2b_2\ \mathbf{j} \times \mathbf{j} + a_2b_3\ \mathbf{j} \times \mathbf{k} \\ & + a_3b_1\ \mathbf{k} \times \mathbf{i} + a_3b_2\ \mathbf{k} \times \mathbf{j} + a_3b_3\ \mathbf{k} \times \mathbf{k}\end{aligned}$$

By § 5.3.4, it becomes

$$\mathbf{a} \times \mathbf{b} = (a_2b_3 - a_3b_2)\ \mathbf{i} + (a_3b_1 - a_1b_3)\ \mathbf{j} + (a_1b_2 - a_2b_1)\ \mathbf{k}$$

This expression can be re-written in the determinant form as

$$\mathbf{a} \times \mathbf{b} = \begin{vmatrix} \mathbf{i} & \mathbf{j} & \mathbf{k} \\ a_1 & a_2 & a_3 \\ b_1 & b_2 & b_3 \end{vmatrix}$$

5.3.6. Angle between Two Vectors a and b

From § 5.2, we have

$$|\mathbf{a} \times \mathbf{b}| = |\mathbf{a}|\ |\mathbf{b}| \sin\theta$$

$$\Rightarrow \quad \sin\theta = \frac{|\mathbf{a}\times\mathbf{b}|}{|\mathbf{a}|\,|\mathbf{b}|}$$

$$= \left|\hat{\mathbf{a}}\times\hat{\mathbf{b}}\right|$$

In terms of components, it can easily be shown that

$$\sin\theta = \frac{\sqrt{\{(a_2b_3-a_3b_2)^2+(a_3b_1-a_1b_3)^2+(a_1b_2-a_2b_1)^2\}}}{\sqrt{(a_1^2+a_2^2+a_3^2)}+\sqrt{(b_1^2+b_2^2+b_3^2)}}$$

5.3.7. General Form of Vector Product

Let **a** and **b** be any two vectors and *l, m, n* be three non-coplanar vectors.

Let $\mathbf{a} = a_1\mathbf{l} + a_2\mathbf{m} + a_3\mathbf{n}$ and $\mathbf{b} = b_1\mathbf{l} + b_2\mathbf{m} + b_3\mathbf{n}$. Then

$$\mathbf{a}\times\mathbf{b} = (a_1\mathbf{l} + a_2\mathbf{m} + a_3\mathbf{n}) \times (b_1\mathbf{l} + b_2\mathbf{m} + b_3\mathbf{n})$$

$$= (a_1b_2 - a_2b_1)\,\mathbf{l}\times\mathbf{m} - (a_1b_3 - a_3b_1)\,\mathbf{n}\times\mathbf{l} + (a_2b_3 - a_3b_2)\,\mathbf{m}\times\mathbf{n}$$

$$= \begin{vmatrix} \mathbf{m}\times\mathbf{n} & \mathbf{n}\times\mathbf{l} & \mathbf{l}\times\mathbf{m} \\ a_1 & a_2 & a_3 \\ b_1 & b_2 & b_3 \end{vmatrix}$$

EXAMPLES

Example 1. *Determine a unit vector perpendicular to the plane of* **a** *and* **b**, *where* **a** = *4***i** + *3***j** – **k** *and* **b** = 2**i** – *6***j** – *3***k**. *Also obtain sine of the angle between* **a** *and* **b**.

Solution.
$$\mathbf{a}\times\mathbf{b} = \begin{vmatrix} \mathbf{i} & \mathbf{j} & \mathbf{k} \\ 4 & 3 & -1 \\ 2 & -6 & -3 \end{vmatrix}$$

$$= -15\mathbf{i} + 10\mathbf{j} + 30\mathbf{k}.$$

As **a** × **b** is the vector perpendicular to the plane of **a** and **b**, hence a unit vector perpendicular to the plane of **a** and **b** is

$$= \frac{\mathbf{a}\times\mathbf{b}}{|\mathbf{a}\times\mathbf{b}|} = \frac{-15\mathbf{i}+10\mathbf{j}-30\mathbf{k}}{\sqrt{(-15)^2+(10)^2+(-30)^2}}$$

$$-\frac{3}{7}\mathbf{i}+\frac{2}{7}\mathbf{j}+\frac{6}{7}\mathbf{k}.$$

If θ be the angle between the vectors, then

$$\sin\theta = \frac{|\mathbf{a}\times\mathbf{b}|}{|\mathbf{a}|\,|\mathbf{b}|} = \frac{\sqrt{\{(-15)^2+10^2+(-30)^2\}}}{\sqrt{\{4^2+3^2+(-1)^2\}}\,\sqrt{\{2^2+(-6)^2+(-3)^2\}}}$$

$$= \frac{35}{\sqrt{26}\,\sqrt{(49)}} = \frac{5\sqrt{26}}{26}.$$

Example 2. *If* **a** × **b** = **c** × **d** *and* **a** × **c** = **b** × **d**, *show that* **a** – **d** *and* **b** – **c** *are parallel vectors.*

Solution. $(\mathbf{a}-\mathbf{d})\times(\mathbf{b}-\mathbf{c}) = \mathbf{a}\times\mathbf{b}-\mathbf{a}\times\mathbf{c}-\mathbf{d}\times\mathbf{b}+\mathbf{d}\times\mathbf{c}$

$= (\mathbf{a}\times\mathbf{b}-\mathbf{c}\times\mathbf{d})-(\mathbf{a}\times\mathbf{c}-\mathbf{b}\times\mathbf{d})$

$= 0$ [as given]

∴ (**a** – **d**) and (**b** – **c**) are parallel vectors.

Example 3. *Prove by vector method that*

$$\sin(\alpha-\beta) = \sin\alpha\cos\beta - \cos\alpha\sin\beta.$$

Solution. Let *OX* and *OY* be a set of rectangular co-ordinate axes with *O* as origin. Let *OA* and *OB* subtend angles **a** and **b** with *OX*. Let **k** be the usual unit vector perpendicular to the plane of the paper such that, **i, j, k** form a right-handed system.

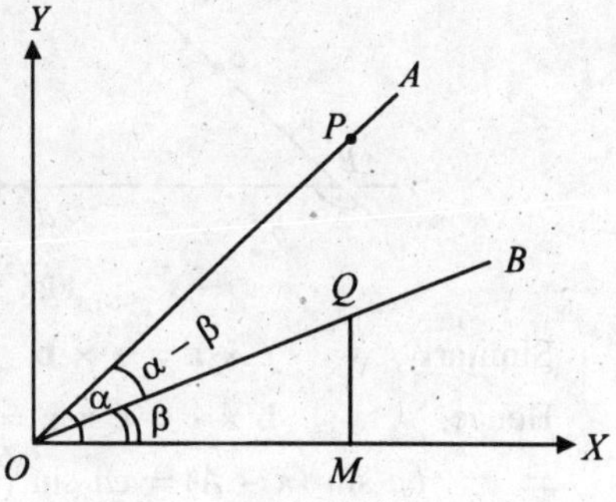

Fig. 5.2.

Let $OP = \hat{\mathbf{p}}$ and $OQ = \hat{\mathbf{q}}$ be the unit vectors along *OA* and *OB* respectively. Then, we have

$$\hat{\mathbf{q}}\times\hat{\mathbf{p}} = 1.1.\sin(\alpha-\beta)\,k$$

$$= \sin(\alpha-\beta)\,k. \quad \text{...(1)}$$

Let *QM* be perpendicular to *OX*. We have

$$OM = OQ\cos\beta = \cos\beta \text{ and } MQ = OQ\sin\beta = \sin\beta$$

$$\left[\because OQ = \left|\overrightarrow{OQ}\right| = |\hat{\mathbf{q}}| = 1\right]$$

∴ $\hat{\mathbf{q}} = (\cos\beta)\,\mathbf{i} + (\sin\beta)\,\mathbf{j}$

Similarly, $\hat{\mathbf{p}} = (\cos\alpha)\,\mathbf{i} + (\sin\alpha)\,\mathbf{j}$

∴ $\hat{\mathbf{q}}\times\hat{\mathbf{p}} = [(\cos\beta)\,\mathbf{i}+(\sin\beta)\,\mathbf{j}]\times[(\cos\alpha)\,\mathbf{i}+(\sin\alpha)\,\mathbf{j}]$

$= (\sin\alpha\cos\beta - \cos\alpha\sin\beta)\,\hat{\mathbf{k}}.$

Hence, from (1), we get

$$\sin(\alpha - \beta) = \sin\alpha \cos\beta - \cos\alpha \sin\beta.$$

Example 4. *If* **a, b, c** *be three vectors such that* **a + b + c** *= 0, prove that* **a × b = b × c = c × a** *and deduce the sine rule*

$$\frac{\sin A}{a} = \frac{\sin B}{b} = \frac{\sin C}{c}. \qquad \text{(Kolkata, 99)}$$

Solution. Let $\overrightarrow{BC}, \overrightarrow{CA}, \overrightarrow{AB}$ represent the vectors **a, b, c** respectively. Then, we have

$$\mathbf{a} + \mathbf{b} + \mathbf{c} = 0,$$

$$\Rightarrow \quad \mathbf{c} = -(\mathbf{a} + \mathbf{b})$$

$$\Rightarrow \quad \mathbf{b} \times \mathbf{c} = \mathbf{b} \times (-\mathbf{a} - \mathbf{b})$$

$$= -\mathbf{b} \times \mathbf{a} = \mathbf{a} \times \mathbf{b}$$

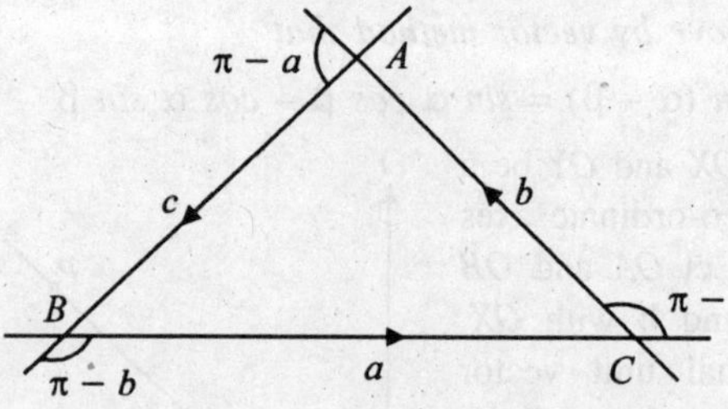

Fig. 5.3.

Similarly, $\mathbf{c} \times \mathbf{a} = \mathbf{a} \times \mathbf{b}$.

Hence, $\mathbf{b} \times \mathbf{c} = \mathbf{c} \times \mathbf{a} = \mathbf{a} \times \mathbf{b}$

$$\Rightarrow \quad bc \sin(\pi - A) = ca \sin(\pi - B)$$

$$= ab \sin(\pi - C)$$

$$\Rightarrow \quad bc \sin A = ca \sin B = ab \sin C$$

$$\Rightarrow \quad \frac{\sin A}{a} = \frac{\sin B}{b} = \frac{\sin C}{c}.$$

Example 5. *Given that* **a . b = a . c, a × b = a × c** *and* **a** *is a non-zero vector. Show that* **b = c.**

Solution. $\mathbf{a} \cdot \mathbf{b} = \mathbf{a} \cdot \mathbf{c}$ and $\mathbf{a} \neq \mathbf{0}$

$\Rightarrow$ $\mathbf{a} \cdot (\mathbf{b} - \mathbf{c}) = 0$

$\Rightarrow$ $\mathbf{b} - \mathbf{c} = 0$ or **a** is perpendicular to $(\mathbf{b} - \mathbf{c})$ ...(1)

Again $\mathbf{a} \times \mathbf{b} = \mathbf{a} \times \mathbf{c}$ and $\mathbf{a} \neq \mathbf{0}$

$\Rightarrow$ $\mathbf{a} \times (\mathbf{b} - \mathbf{c}) = \mathbf{0}$

$\Rightarrow$ $\mathbf{b} - \mathbf{c} = 0$ or **a** is parallel to $(\mathbf{b} - \mathbf{c})$ ...(2)

Since the two relations hold simultaneously. Both **a** is perpendicular to (**b** – **c**) and **a** is parallel to (**b** – **c**) is an absurd result. Hence, the only possibility is **b** – **c** = 0 so that **b** = **c**.

EXERCISES

1. Prove that the unit vector perpendicular to each of the vectors 3**i** + **j** + 2**k** and 2**i** – 2**j** + 4**k** is (**i** – **j** – **k**) / $\sqrt{(3)}$ and the sine of the angle between them in $2/\sqrt{(7)}$.
2. If $\hat{\mathbf{a}}, \hat{\mathbf{b}}, \hat{\mathbf{c}}$ be unit vectors such that $\hat{\mathbf{a}}$ and $\hat{\mathbf{b}}$ are perpendicular and $\hat{\mathbf{c}}$ is inclined at an angle θ to both $\hat{\mathbf{a}}$ and $\hat{\mathbf{b}}$.
   Show that $\mathbf{c} = \alpha\ (\mathbf{a} + \mathbf{b}) + \beta\ (\mathbf{a} \times \mathbf{b})$.
   where $\alpha = -\cos\theta$ and $\beta = -\cos 2\theta$.
3. If **a** . **b** = 0 and **a** × **b** = 0 simultaneously, prove that atleast one of the vectors **a** and **b** must be a null vector.
4. Show that the vector **a** × (**b** × **a**) is perpendicular to **a** and coplanar with **a** and **b**.
5. Show that **a** × (**b** × **c**) is coplanar with **b** and **c**.
6. **a, b, c** are three vectors such that
   $$\mathbf{a} \times \mathbf{b} = \mathbf{c},\quad \mathbf{b} \times \mathbf{c} = \mathbf{a};$$
   show that the three vectors **a, b, c** are orthogonal in pairs and
   $$|\mathbf{b}| = 1,\quad |\mathbf{c}| = |\mathbf{a}|.$$
7. **a, b** are two vectors. Also **c, d** are two vectors coplanar with **a** and **b** and perpendicular to **a** and **b** respectively. Show that
   $$(\mathbf{a} \times \mathbf{b}) \times \mathbf{c} = (\mathbf{a}\ .\ \mathbf{c})\ \mathbf{b} - (\mathbf{b}\ .\ \mathbf{c})\ \mathbf{a},\quad (\mathbf{a} \times \mathbf{b}) \times \mathbf{d} = (\mathbf{a}\ .\ \mathbf{d})\ \mathbf{b} - (\mathbf{b}\ .\ \mathbf{d})\ \mathbf{a}$$
8. Show that **a** . (**b** × **c**) = 0 if and only if the vectors **a, b, c** are coplanar.

5.4. INTERPRETATION OF VECTOR PRODUCT AS VECTOR AREA

We shall now show how it is possible to represent a plane area bounded by a closed curve which does not cross itself by a vector. For this purpose it is necessary to distinguish between the two senses in which the curve may be described.

A plane area bounded by a closed curve which does not cross itself is represented by a vector, **c**, defined as follows :

(*i*) The number of units of the length of **c** is equal to the number of units of the given area.

(*ii*) The support of **c** is perpendicular to the plane of the area.

(*iii*) The sense of **c** is such that the direction of description of the boundary of the curve and the sense of **c** correspond to a right-handed screw.

The sense of **c** will be reversed, if we reverse the direction of description of the boundary of the area.

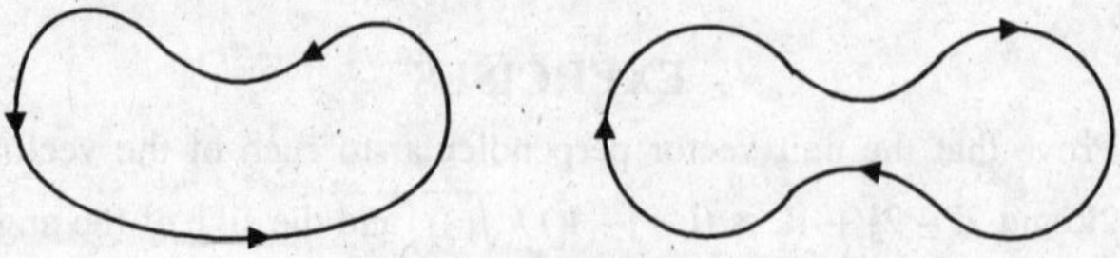

Fig. 5.4.

5.4.1. Area of a Triangle as Vector Product

Let $\vec{OA} = \mathbf{a}, \ \vec{OB} = \mathbf{b}.$

Consider the triangle OAB and the vector representing the vector area.

$$\Delta \vec{OAB}.$$

We shall see that

$$\Delta \vec{OAB} = \frac{1}{2} \mathbf{a} \times \mathbf{b}. \qquad \text{...}(i)$$

The magnitude of each side of (*i*) is $\frac{1}{2} OA \, . \, OB \sin\theta$ and the support of each is also the same. Also according to our conventions the senses of the vectors on the two sides of (*i*) are the same.

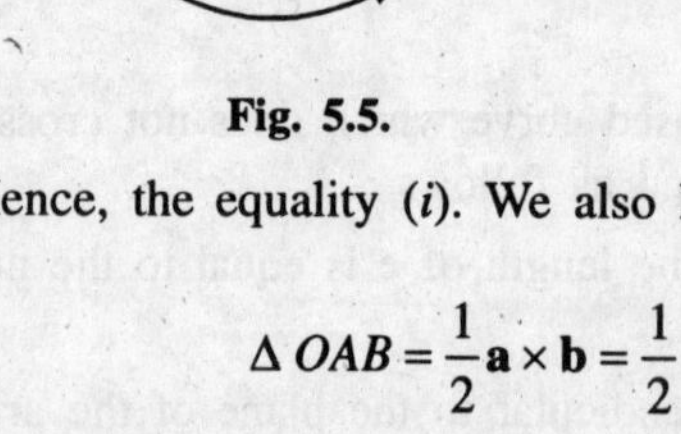

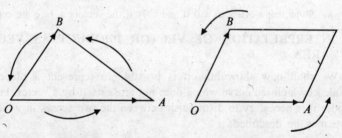

Fig. 5.5.

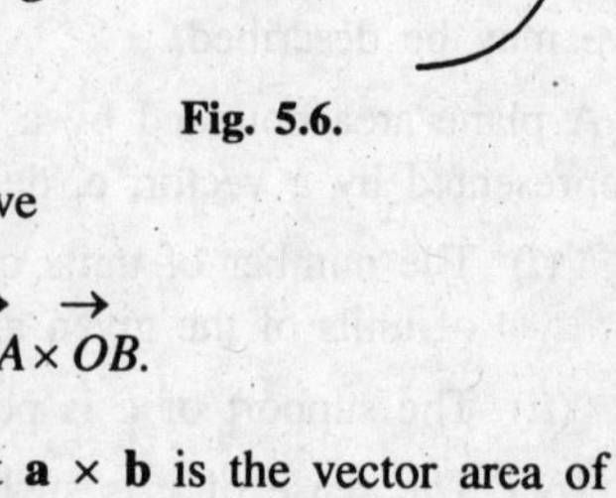

Fig. 5.6.

Hence, the equality (*i*). We also have

$$\Delta \, OAB = \frac{1}{2} \mathbf{a} \times \mathbf{b} = \frac{1}{2} \vec{OA} \times \vec{OB}.$$

It may also be now easily seen that **a** × **b** is the vector area of the parallelogram constructed with $\vec{OA}$ and $\vec{OB}$ as adjacent sides (Fig. 5.6).

5.5. SCALAR TRIPLE PRODUCT (*Patna 2003, Avadh 2005*)

The use of scalar triple product will enable us to prove that vector product distributes the sum of vectors. Also the concept of scalar triple product plays a useful part in the applications of vectors of Geometry.

Let **a, b, c** be any three vectors. Consider the expression

$$\mathbf{a} \times \mathbf{b} \cdot \mathbf{c}$$

which is the scalar product of the vectors $\mathbf{a} \times \mathbf{b}$ and **c**

Let

$$\vec{OA} = \mathbf{a},\ \vec{OB} = \mathbf{b},\ \vec{OC} = \mathbf{c}.$$

The figures show only the three faces of the parallelopiped through the point *O*.

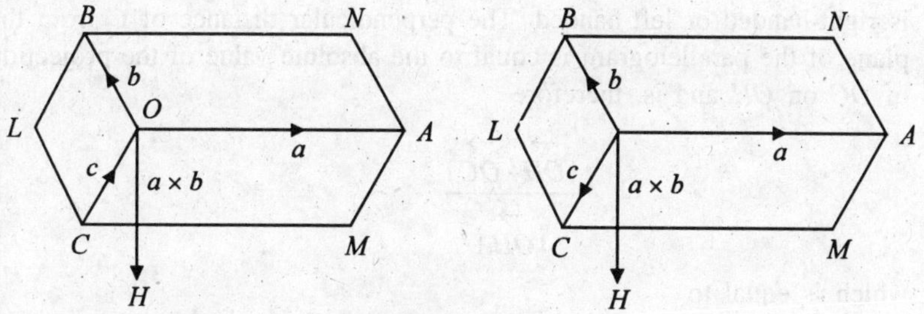

Fig. 5.7.

Firstly suppose that the three vectors are coplanar. The vector $\mathbf{a} \times \mathbf{b}$, being perpendicular to the plane of the vectors, is perpendicular to the vector, **c** and, as such

$$\mathbf{a} \times \mathbf{b} \cdot \mathbf{c} = 0.$$

Next suppose that the three vectors are not coplanar so that the lines *OA, OB* and *OC* do not lie in the same plane.

Complete the parallelopiped with *OA, OB* and *OC* as coterminous edges. Let *V*, be the volume of this parallelopiped. We shall regard, *V*, as necessarily positive.

We shall now prove the following :

$$V = [\mathbf{a} \times \mathbf{b} \cdot \mathbf{c}] \quad \text{or} \quad V = -[\mathbf{a} \times \mathbf{b} \cdot \mathbf{c}]$$

according as the ordered vector triad **a, b, c** *is right-handed or left-handed.*

The vector $\mathbf{a} \times \mathbf{b}$ denotes the vector area of the parallelogram with *OA* and *OB* as adjacent sides.

Let

$$\vec{OH} = \mathbf{a} \times \mathbf{b}.$$

The volume of the parallelopiped is the product of the area of this parallelogram with the perpendicular distance of C from its plane.

The vector triad

$$\mathbf{a},\ \mathbf{b},\ \mathbf{a} \times \mathbf{b}$$

is, by definition, right-handed.

The angle between

$$\mathbf{a} \times \mathbf{b} \text{ and } \mathbf{c}$$

i.e., between

$$\vec{OH} \text{ and } \vec{OC}$$

is acute or obtuse according as the vector triad

$$\mathbf{a},\ \mathbf{b},\ \mathbf{c}$$

is right-handed or left-handed. The perpendicular distance of C from the plane of the parallelogram is equal to the absolute value of the projection of OC on OH and is, therefore

$$\frac{|\vec{OH} \cdot \vec{OC}|}{|\vec{OH}|}$$

which is equal to

$$\frac{\vec{OH} \cdot \vec{OC}}{|\vec{OH}|} \quad \text{or} \quad \frac{-\vec{OH} \cdot \vec{OC}}{|\vec{OH}|}$$

according as the angle between $\vec{OH}$ and $\vec{OC}$ is acute or obtuse.

Thus, if **a, b, c** is a right-handed system, we have

$$V = \frac{\vec{OH} \cdot \vec{OC}}{|\vec{OH}|}\, |\vec{OH}| = \vec{OH} \cdot \vec{OC} = \mathbf{a} \times \mathbf{b} \,.\, \mathbf{c}$$

and if **a, b, c** is a left-handed system, then

$$V = \frac{\vec{OH} \cdot \vec{OC}}{|\vec{OH}|}\, |\vec{OH}| = -\vec{OH} \cdot \vec{OC} = -\,\mathbf{a} \times \mathbf{b} \cdot \mathbf{c}.$$

The foregoing investigation also shows that

$$\mathbf{a} \times \mathbf{b} \,.\, \mathbf{c}$$

is positive or negative according as **a, b, c** is a right or left-handed system.

Def. *The expression* $\mathbf{a} \times \mathbf{b} \cdot \mathbf{c}$ *is called a scalar triple or mixed product of the three vectors* **a, b, c** *taken in this order.*

5.5.1. Different Scalar Triple Products with the same Three Vectors

Let **a, b, c** be three given vectors. We can permute the three given vectors in six different ways. Also each manner of writing down the three vectors gives rise to two scalar triple products depending upon the positions of dot and cross. Thus, we have the following *twelve* scalar triple products :

$$\mathbf{a} \times \mathbf{b} \cdot \mathbf{c} \quad \mathbf{b} \times \mathbf{c} \cdot \mathbf{a} \quad \mathbf{c} \times \mathbf{a} \cdot \mathbf{b} \quad \mathbf{a} \cdot \mathbf{b} \times \mathbf{a} \quad \mathbf{b} \cdot \mathbf{c} \times \mathbf{a} \quad \mathbf{c} \cdot \mathbf{a} \times \mathbf{b}$$

$$\mathbf{a} \times \mathbf{c} \cdot \mathbf{b} \quad \mathbf{b} \times \mathbf{a} \cdot \mathbf{c} \quad \mathbf{c} \times \mathbf{b} \cdot \mathbf{a} \quad \mathbf{a} \cdot \mathbf{c} \times \mathbf{b} \quad \mathbf{b} \cdot \mathbf{a} \times \mathbf{c} \quad \mathbf{c} \cdot \mathbf{b} \times \mathbf{a}.$$

We shall now prove the following two important results :

I. *A cyclic permutation of three vectors does not change the value of the scalar triple product and an anti-cyclic permutation changes the value in sign but not in magnitude.*

II. *The positions of dot and cross can be interchanged without any change in the value of the scalar triple product.* (*Patna 2003*)

Firstly suppose that **a, b, c** is a right-handed system so have

$$V = \mathbf{a} \times \mathbf{b} \cdot \mathbf{c}.$$

The vector triads **b, c, a** and **c, a, b** are also right-handed and the parallelopiped with *OA, OB, OC* as coterminous edges is the same as that with *OB, OC, OA* or with *OC, OA, OB* as coterminous edges. Thus,

$$V = \mathbf{b} \times \mathbf{c} \cdot \mathbf{a} \quad \text{and} \quad V = \mathbf{c} \times \mathbf{a} \cdot \mathbf{b}.$$

Also since scalar multiplication is commutative, we shall

$$\mathbf{a} \cdot \mathbf{b} \times \mathbf{c} = \mathbf{b} \times \mathbf{c} \cdot \mathbf{a} = V,$$

$$\mathbf{b} \cdot \mathbf{c} \times \mathbf{a} = \mathbf{c} \times \mathbf{a} \cdot \mathbf{b} = V,$$

$$\mathbf{c} \cdot \mathbf{a} \times \mathbf{b} = \mathbf{a} \times \mathbf{b} \cdot \mathbf{c} = V.$$

Thus, six out of the twelve scalar triple products are equal to *V* and in each of these cases, we have the same cyclic order of the vectors and interchange of dot and cross has not mattered.

Again **a, c, b; b, a, c; c, b, a** being left-handed systems, we have

$$\mathbf{a} \times \mathbf{c} \cdot \mathbf{b} = -V, \quad \mathbf{b} \times \mathbf{a} \cdot \mathbf{c} = -V, \quad \mathbf{c} \times \mathbf{b} \cdot \mathbf{a} = -V$$

and

$$\mathbf{a} \cdot \mathbf{c} \times \mathbf{b} = \mathbf{c} \times \mathbf{b} \cdot \mathbf{a} = -V,$$

$$\mathbf{b} \cdot \mathbf{a} \times \mathbf{c} = \mathbf{a} \times \mathbf{c} \cdot \mathbf{b} = -V,$$

$$\mathbf{c} \cdot \mathbf{b} \times \mathbf{a} = \mathbf{b} \times \mathbf{a} \cdot \mathbf{c} = -V.$$

Each of these latter six scalar products involves an anti-cyclic permutation of the vectors in $\mathbf{a} \times \mathbf{b} \cdot \mathbf{c}$.

Notation. In view of the properties of scalar triple product obtained above, we write

$$\mathbf{a} \times \mathbf{b} \cdot \mathbf{c} = [\mathbf{a}\ \mathbf{b}\ \mathbf{c}]$$

This notation takes note of the cyclic order of the vectors and disregards the positions of dot and cross which are really not important. We have

$$[\mathbf{a}\ \mathbf{b}\ \mathbf{c}] = [\mathbf{b}\ \mathbf{c}\ \mathbf{a}] = [\mathbf{c}\ \mathbf{a}\ \mathbf{b}]$$

$$= -[\mathbf{a}\ \mathbf{c}\ \mathbf{b}] = -[\mathbf{b}\ \mathbf{a}\ \mathbf{c}] = -[\mathbf{c}\ \mathbf{b}\ \mathbf{a}].$$

Note. It is very important to notice that the scalar triple product

$$[\mathbf{a}\ \mathbf{b}\ \mathbf{c}]$$

is zero if, and only if, the three vectors **a, b, c** *are coplanar.* In particular, the scalar triple product is zero if any two of the three vectors are the same.

Note. If **i, j, k** constitute an orthonormal right-handed triad, then

$$[\mathbf{i}\ \mathbf{j}\ \mathbf{k}] = \mathbf{i} \times \mathbf{j} \cdot \mathbf{k} = \mathbf{k} \cdot \mathbf{k} = 1.$$

5.5.2. Volume of a Tetrahedron — Theorem

The volume of a tetrahedron ABCD is

$$\frac{1}{6} | \vec{AB} \times \vec{AC} \cdot \vec{AD} |.$$

We know that the volume of a tetrahedron $ABCD$ is $\frac{1}{3} \times$ area of ΔABC $\times$ height of D from the plane ABC. Also the volume, V, of the parallelopiped with AB, AC, AD as its coterminous edges is equal to the area of the parallelogram with AB, AC as adjacent sides $\times$ height of D from the plane ABC.

$\therefore$ volume of the tetrahedron

$$= \frac{1}{6} | \vec{AB} \times \vec{AC} \cdot \vec{AD} |.$$

Cor. Volume of the tetrahedron, the position vectors of whose vertices A, B, C, D are **a, b, c, d** is

$$= \frac{1}{6} | \vec{AB} \times \vec{AC} \cdot \vec{AD} |$$

$$= \frac{1}{6} | (\mathbf{b} - \mathbf{a}) \times (\mathbf{c} - \mathbf{a}) \cdot (\mathbf{d} - \mathbf{a}) |.$$

Ex. 1. Show that each of the four faces of a tetrahedron subtends the same volume at the centroid.

Ex. 2. Compare the volume of a tetrahedron with that of the tetrahedron formed by the centroids of its faces.

Ex. 3. G_1, G_2, G_3 are the centroids of the triangular faces OBC, OCA, OAB of a tetrahedron $OABC$, compare the volume of the tetrahedron $OABC$ with that of the parallelopiped constructed with OG_1, OG_2, OG_3 as coterminous edges.

5.6. DISTRIBUTIVE LAW

Theorem. $\mathbf{a} \times (\mathbf{b} + \mathbf{c}) = \mathbf{a} \times \mathbf{b} + \mathbf{a} \times \mathbf{c},$
where **a, b, c,** *are any three vectors.*

We shall now prove that the vector product distributes the sum of vectors, *i.e.*,

$$\mathbf{a} \times (\mathbf{b} + \mathbf{c}) = \mathbf{a} \times \mathbf{b} + \mathbf{a} \times \mathbf{c}.$$

This result will be proved by showing that the scalar product of the vector $\mathbf{a} \times (\mathbf{b} + \mathbf{c}) - \mathbf{a} \times \mathbf{b} - \mathbf{a} \times \mathbf{c}$ with *every* vector is zero.

Let, **r**, be *any* vector whatsoever. We have

$$\mathbf{r} \,.\, [\mathbf{a} \times (\mathbf{b} + \mathbf{c}) - \mathbf{a} \times \mathbf{b} - \mathbf{a} \times \mathbf{c}]$$
$$= \mathbf{r} \,.\, \mathbf{a} \times (\mathbf{b} + \mathbf{c}) - \mathbf{r} \,.\, \mathbf{a} \times \mathbf{b} - \mathbf{r} \,.\, \mathbf{a} \times \mathbf{c},$$

(for scalar multiplication distributes vector addition)

$$\mathbf{r} \times \mathbf{a} \,.\, (\mathbf{b} + \mathbf{c}) - \mathbf{r} \times \mathbf{a} \,.\, \mathbf{b} - \mathbf{r} \times \mathbf{a} \,.\, \mathbf{c},$$

(for the positions of dot and cross are interchangeable)

$$= \mathbf{r} \times \mathbf{a} \,.\, [(\mathbf{b} + \mathbf{c}) - \mathbf{b} - \mathbf{c}]$$

(for scalar multiplication distributes vector addition)

$$= \mathbf{r} \times \mathbf{a} \,.\, \mathbf{0} = 0.$$

Thus,

$$\mathbf{r} \,.\, [\mathbf{a} \times (\mathbf{b} + \mathbf{c}) - \mathbf{a} \times \mathbf{b} - \mathbf{a} \times \mathbf{c}] = 0,$$

for *every* vector **r**. It follows that

$$\mathbf{a} \times (\mathbf{b} + \mathbf{c}) - \mathbf{a} \times \mathbf{b} - \mathbf{a} \times \mathbf{c} = \mathbf{0}$$
$$\Rightarrow \quad \mathbf{a} \times (\mathbf{b} + \mathbf{c}) = \mathbf{a} \times \mathbf{b} + \mathbf{a} \times \mathbf{c}$$

Cor. 1. $(\mathbf{b} + \mathbf{c}) \times \mathbf{a} = -\,\mathbf{a} \times (\mathbf{b} + \mathbf{c})$
$$= -[\mathbf{a} \times \mathbf{b} + \mathbf{a} \times \mathbf{c}]$$
$$= -\,\mathbf{a} \times \mathbf{b} - \mathbf{a} \times \mathbf{c} = \mathbf{b} \times \mathbf{a} + \mathbf{c} \times \mathbf{a}.$$

Cor. 2. $\mathbf{a} \times (\mathbf{b} - \mathbf{c}) = \mathbf{a} \times [\mathbf{b} + (-\,\mathbf{c})]$
$$= \mathbf{a} \times \mathbf{b} + \mathbf{a} \times (-\,\mathbf{c}) = \mathbf{a} \times \mathbf{b} - \mathbf{a} \times \mathbf{c}.$$

Similarly

$$(\mathbf{b} - \mathbf{c}) \times \mathbf{a} = \mathbf{b} \times \mathbf{a} - \mathbf{c} \times \mathbf{a}.$$

Cor. 3. Scalar product in terms of an orthonormal base. If

$$\mathbf{a} = a_1\mathbf{i} + a_2\,\mathbf{j} + a_3\mathbf{k}, \quad \mathbf{b} = b_1\mathbf{i} + b_2\,\mathbf{j} + b_3\mathbf{k}, \quad \mathbf{c} = c_1\mathbf{i} + c_2\,\mathbf{j} + c_3\mathbf{k},$$

we have

$$\mathbf{a} \times \mathbf{b} \,.\, \mathbf{c} = [(a_2b_3 - a_3b_2)\,\mathbf{i} + (a_3b_1 - a_1b_3)\,\mathbf{j} + (a_1b_2 - a_2b_1)\,\mathbf{k}].$$
$$(c_1\mathbf{i} + c_2\,\mathbf{j} + c_3\mathbf{k})$$

$$= \begin{vmatrix} a_1 & a_2 & a_3 \\ b_1 & b_2 & b_3 \\ c_1 & c_2 & c_3 \end{vmatrix}$$

5.7. SOME PROPERTIES OF SCALAR TRIPLE PRODUCT

5.7.1. *To express the scalar triple product* $[\mathbf{a}\ \mathbf{b}\ \mathbf{c}]$ *in terms of any three non-coplanar vectors* **l, m, n.**

Let $\mathbf{a} = a_1\mathbf{l} + a_2\mathbf{m} + a_3\mathbf{n};\ \mathbf{b} = b_1\mathbf{l} + b_2\mathbf{m} + b_3\mathbf{n};\ \mathbf{c} = c_1\mathbf{l} + c_2\mathbf{m} + c_3\mathbf{n}$

$$\therefore \quad \mathbf{a} \times \mathbf{b} = (a_1\mathbf{l} + a_2\mathbf{m} + a_3\mathbf{n}) \times (b_1 l + b_2\mathbf{m} + b_3\mathbf{n})$$

$$= (a_2b_3 - a_3b_2)\ (\mathbf{m} \times \mathbf{n}) + (a_3b_1 - a_1b_3)\ (\mathbf{n} \times \mathbf{l}) + (a_1b_2 - a_2b_1)\ (\mathbf{l} \times \mathbf{m})$$

$$\therefore \quad (\mathbf{a} \times \mathbf{b}) \cdot \mathbf{c} = [(a_2b_3 - a_3b_2)\ (\mathbf{m} \times \mathbf{n}) + (a_3b_1 - a_1b_3)\ (\mathbf{n} \times \mathbf{l}) + (a_1b_2 - a_2b_1)\ (\mathbf{l} \times \mathbf{m})] \cdot (c_1\mathbf{l} + c_2\mathbf{m} + c_3\mathbf{n})$$

$$= c_1\,(a_2b_3 - a_3b_2)\ [\mathbf{m}\ \mathbf{n}\ \mathbf{l}] + c_2\,(a_3b_1 - a_1b_3)\ [\mathbf{n}\ \mathbf{l}\ \mathbf{m}] + c_3\,(a_1b_2 - a_2b_1)\ [\mathbf{l}\ \mathbf{m}\ \mathbf{n}]$$

But, as $[\mathbf{m}\ \mathbf{n}\ \mathbf{l}] = [\mathbf{n}\ \mathbf{l}\ \mathbf{m}] = [\mathbf{l}\ \mathbf{m}\ \mathbf{n}]$

we have

$$(\mathbf{a} \times \mathbf{b}) \cdot \mathbf{c} = [a_1\,(b_2c_3 - b_2c_3) - a_2\,(b_1c_3 - b_3c_1) + a_3\,(b_1c_2 - b_2c_1)\ [\mathbf{l}\ \mathbf{m}\ \mathbf{n}]$$

$$\Rightarrow \quad [\mathbf{a}\ \mathbf{b}\ \mathbf{c}] = \begin{vmatrix} a_1 & a_2 & a_3 \\ b_1 & b_2 & b_3 \\ c_1 & c_2 & c_3 \end{vmatrix} = [\mathbf{l}\ \mathbf{m}\ \mathbf{n}]$$

5.7.2. *To express any vector* **r** *as a linear combination of three non-coplanar vectors* **a, b, c.** (Kolkata, 97)

Let x, y, z be three scalars. Then

$$\mathbf{r} = x\mathbf{a} + y\mathbf{b} + z\mathbf{c}$$

$$\therefore \quad \mathbf{r} \cdot (\mathbf{b} \times \mathbf{c}) = x\,[\mathbf{a}\ \mathbf{b}\ \mathbf{c}] + y\,[\mathbf{b}\ \mathbf{b}\ \mathbf{c}] + z\,[\mathbf{c}\ \mathbf{b}\ \mathbf{c}]$$

$$\Rightarrow \quad [\mathbf{r}\ \mathbf{b}\ \mathbf{c}] = x\,[\mathbf{a}\ \mathbf{b}\ \mathbf{c}]$$

Similarly $\quad [\mathbf{r}\ \mathbf{c}\ \mathbf{a}] = y\,[\mathbf{a}\ \mathbf{b}\ \mathbf{c}]$

and $\quad [\mathbf{r}\ \mathbf{a}\ \mathbf{b}] = z\,[\mathbf{a}\ \mathbf{b}\ \mathbf{c}]$

$$\Rightarrow \quad \mathbf{r} = \frac{[\mathbf{r}\,\mathbf{b}\,\mathbf{c}]}{[\mathbf{a}\,\mathbf{b}\,\mathbf{c}]}\mathbf{a} + \frac{[\mathbf{r}\,\mathbf{c}\,\mathbf{a}]}{[\mathbf{a}\,\mathbf{b}\,\mathbf{c}]}\mathbf{b} + \frac{[\mathbf{r}\,\mathbf{a}\,\mathbf{b}]}{[\mathbf{a}\,\mathbf{b}\,\mathbf{c}]}\mathbf{c}.$$

5.7.3. Distributive Law for Scalar Products

(*i*) $[\mathbf{a},\ \mathbf{b} + \mathbf{c},\ \mathbf{d}] = [\mathbf{a}\ \mathbf{b}\ \mathbf{d}] + [\mathbf{a}\ \mathbf{c}\ \mathbf{d}]$

(*ii*) $[\mathbf{a},\ \mathbf{b} + \mathbf{c},\ \mathbf{d} + \mathbf{e}] = [\mathbf{a}\ \mathbf{b}\ \mathbf{d}] + [\mathbf{a}\ \mathbf{b}\ \mathbf{e}] + [\mathbf{a}\ \mathbf{c}\ \mathbf{d}] + [\mathbf{a}\ \mathbf{c}\ \mathbf{e}]$

Proof : (*i*) $\quad [\mathbf{a},\ \mathbf{b} + \mathbf{c},\ \mathbf{d}] = [\mathbf{a}\ {}^{\prime}\ (\mathbf{b}\ {}^{\prime}\ \mathbf{c})] \cdot \mathbf{d}$

$$= [\mathbf{a} \times \mathbf{b} + \mathbf{a} \times \mathbf{c}] \cdot \mathbf{d}$$

$$= (\mathbf{a} \times \mathbf{b}) \cdot \mathbf{d} + (\mathbf{a} \times \mathbf{c}) \cdot \mathbf{d}$$

$$= [\mathbf{a}\ \mathbf{b}\ \mathbf{d}] + [\mathbf{a}\ \mathbf{c}\ \mathbf{d}]$$

Similarly part (*ii*) can be proved.

EXAMPLES

Example 1. *Find the volume of the parallelopiped whose coterminous edges are* $2\mathbf{i} - 3\mathbf{j} + \mathbf{k}$, $\mathbf{i} - \mathbf{j} + 2\mathbf{k}$ *and* $2\mathbf{i} + \mathbf{j} - \mathbf{k}$.

Solution. Volume of the parallelopiped

$$= \begin{vmatrix} 2 & -3 & 1 \\ 1 & -1 & 2 \\ 2 & 1 & -1 \end{vmatrix}$$

$$= 2\,(1 - 2) + 3\,(-1 - 4) + 1\,(1 + 2)$$

$$= -14 \text{ or to say } 14 \text{ units.}$$

Example 2. *Show that the vectors* $\mathbf{a} - 2\mathbf{b} + 3\mathbf{c}$, $-2\mathbf{a} + 3\mathbf{b} - 4\mathbf{c}$ *and* $\mathbf{a} - 3\mathbf{b} + 5\mathbf{c}$ *are coplanar.*

Solution. Let $\mathbf{p} = \mathbf{a} - 2\mathbf{b} + 3\mathbf{c}$, $\mathbf{q} = -2\mathbf{a} + 3\mathbf{b} - 4\mathbf{c}$, $\mathbf{r} = \mathbf{a} - 3\mathbf{b} + 5\mathbf{c}$. By § 5.7.1.

$$[\mathbf{p}\,\mathbf{q}\,\mathbf{r}] = \begin{vmatrix} 1 & -2 & 3 \\ -2 & 3 & -4 \\ 1 & -3 & 5 \end{vmatrix} = [\mathbf{a}\,\mathbf{b}\,\mathbf{c}]$$

$$= 0 \times [\mathbf{a}\,\mathbf{b}\,\mathbf{c}] = 0$$

showing that $\mathbf{p}, \mathbf{q}, \mathbf{r}$ are coplanar.

Example 3. *Prove that if* $\mathbf{l}, \mathbf{m}, \mathbf{n}$ *be three non-coplanar vectors, then*

$$[\mathbf{l}\,\mathbf{m}\,\mathbf{n}]\,[\mathbf{a}\times\mathbf{b}] = \begin{vmatrix} \mathbf{l}.\mathbf{a} & \mathbf{l}.\mathbf{b} & \mathbf{l} \\ \mathbf{m}.\mathbf{a} & \mathbf{m}.\mathbf{b} & \mathbf{m} \\ \mathbf{n}.\mathbf{a} & \mathbf{n}.\mathbf{b} & \mathbf{n} \end{vmatrix}$$

(Awadh 99, 2005, Rohilkhand 98, 2004, Garhwal 94, 99, 2006)

Solution. Let $\mathbf{a} = a_1\mathbf{i} + a_2\mathbf{j} + a_3\mathbf{k}$, $\mathbf{b} = b_1\mathbf{i} + b_2\mathbf{j} + b_3\mathbf{k}$,

$\mathbf{l} = l_1\mathbf{i} + l_2\mathbf{j} + l_3\mathbf{k}$, $\mathbf{m} = m_1\mathbf{i} + m_2\mathbf{j} + m_3\mathbf{k}$,

$\mathbf{n} = n_1\mathbf{i} + n_2\mathbf{j} + n_3\mathbf{k}$. Then

$$[\mathbf{l}\,\mathbf{m}\,\mathbf{n}]\,[\mathbf{a}\times\mathbf{b}] = \begin{vmatrix} l_1 & l_2 & l_3 \\ m_1 & m_2 & m_3 \\ n_1 & n_2 & n_3 \end{vmatrix} \begin{vmatrix} \mathbf{i} & \mathbf{j} & \mathbf{k} \\ a_1 & a_2 & a_3 \\ b_1 & b_2 & b_3 \end{vmatrix}$$

$$= \begin{vmatrix} l_1\mathbf{i}+l_2\mathbf{j}+l_3\mathbf{k} & l_1a_1+l_2a_2+l_3a_3 & l_1b_1+l_2b_2+l_3b_3 \\ m_1\mathbf{i}+m_2\mathbf{j}+m_3\mathbf{k} & m_1a_1+m_2a_2+m_3a_3 & m_1b_1+m_2b_2+m_3b_3 \\ n_1\mathbf{i}+n_2\mathbf{j}+n_3\mathbf{k} & n_1a_1+n_2a_2+n_3a_3 & n_1b_1+n_2b_2+n_3b_3 \end{vmatrix}$$

$$= \begin{vmatrix} \mathbf{l} & \mathbf{l}.\mathbf{a} & \mathbf{l}.\mathbf{b} \\ \mathbf{m} & \mathbf{m}.\mathbf{a} & \mathbf{m}.\mathbf{b} \\ \mathbf{n} & \mathbf{n}.\mathbf{a} & \mathbf{n}.\mathbf{b} \end{vmatrix} = \begin{vmatrix} \mathbf{l}.\mathbf{a} & \mathbf{l}.\mathbf{b} & \mathbf{l} \\ \mathbf{m}.\mathbf{a} & \mathbf{m}.\mathbf{b} & \mathbf{m} \\ \mathbf{n}.\mathbf{a} & \mathbf{n}.\mathbf{b} & \mathbf{n} \end{vmatrix}$$

Example 4. *Find the value of p so that the vectors* **2i − j + k**, **i** + 2**j** − 3**k** *and* 3**i** + *p***j** + 5**k** *are coplanar.* (*Avadh 2000*)

Solution. Given vectors will be coplanar if

$$\begin{vmatrix} 2 & -1 & 1 \\ 1 & 2 & -3 \\ 3 & p & 5 \end{vmatrix} = 0$$

$$\Rightarrow \quad 2\,(10 + 3p) + 1\,(5 + 9) + 1\,(p - 6) = 0$$

$$\Rightarrow \quad 7p + 28 = 0 \quad \Rightarrow \quad p = -\,4.$$

Example 5. *Prove that*

$$\mathbf{a} \times \mathbf{b} = [(\mathbf{i} \times \mathbf{a}) \,.\, \mathbf{b}]\,\mathbf{i} + [(\mathbf{j} \times \mathbf{a}) \,.\, \mathbf{b}]\,\mathbf{j} + [(\mathbf{k} \times \mathbf{a}) \,.\, \mathbf{k}]\,\mathbf{k}.$$

Solution. For any vector **r**, we have

$$\mathbf{r} = (\mathbf{i} \,.\, \mathbf{r})\,\mathbf{i} + (\mathbf{j} \,.\, \mathbf{r})\,\mathbf{j} + (\mathbf{k} \,.\, \mathbf{r})\,\mathbf{k}$$

Replacing **r** by (**a** × **b**), we get

$$(\mathbf{a} \times \mathbf{b}) = [\mathbf{i} \,.\, (\mathbf{a} \times \mathbf{b})]\,\mathbf{i} + [\mathbf{j} \,.\, (\mathbf{a} \times \mathbf{b})]\,\mathbf{j} + [\mathbf{k} \,.\, (\mathbf{a} \times \mathbf{b})]\,\mathbf{k}$$

$$= [(\mathbf{i} \times \mathbf{a}) \,.\, \mathbf{b}]\,\mathbf{i} + [(\mathbf{j} \times \mathbf{a}) \,.\, \mathbf{b}]\,\mathbf{j} + [(\mathbf{k} \times \mathbf{a}) \,.\, \mathbf{b}]\,\mathbf{k}.$$

since the position of the dot and cross can be interchanged in a scalar triple product.

EXERCISES

1. If **i**, **j**, **k** are three mutually perpendicular unit vectors, compute

   (*a*) (2**i** − **j** + **k**) × (3**i** + **k**), (*b*) (**i** − 2**j** + 3**k**) × (2**j** − 3**k**)

   (*c*) (3**i** + 4**j**) × (**i** − **j** + **k**),

   and verify in each case with the help of scalar product that the vector product is perpendicular to the given vectors.

   Find also the lengths of these vector products.

2. Compute the following scalar triple products :

   (*a*) (**i** − 2**j** + 3**k**) × (2**i** + **j** − **k**) . (**j** + **k**).

   (*b*) (2**i** − 3**j** + **k**) . (**i** − **j** + 2**k**) × (2**i** + **j** − **k**).

3. Compute the following vector products :

   (*a*) [(**i** − **j** + **k**) × (2**i** − 3**j** − **k**)] × [(− 3**i** + **j** + **k**) × (2**j** + **k**)]

   (*b*) [(3**i** − 2**j** − 2**k**) × (**i** − **k**)] × [(**i** + **j** + **k**) × (**i** − 2**j** + 3**k**)].

4. Show that the vectors

   (*i*) **j** − 2**k**, **i** − **j** + **k**, − 2**i** + 3**j** − 4**k**;

   (*ii*) 2**i** − **j** + 2**k**, 4**i** + **j**, 3**i** + **k**,

   are coplanar.

5. Show that the vectors **i + j**, **i + k**, **k + j** are not coplanar.

6. Find the volume of the parallelopiped whose coterminous edges are represented by **a** = 2**i** – 3**j** + 4**k**, **b** = **i** + 2**j** – **k**, **c** = 3**i** – **j** + 2**k**.

7. If **a**, **b**, **c** are three vectors such that **a** × **b** = **c** and **b** × **c** = **a**, show that three vectors **a**, **b**, **c** are orthogonal in pairs and | **b** | = 1, | **c** | = | **a** |.

8. Prove that

$$[\mathbf{l}\,\mathbf{m}\,\mathbf{n}]\;[\mathbf{a}\,\mathbf{b}\,\mathbf{c}] = \begin{vmatrix} \mathbf{l}.\mathbf{a} & \mathbf{l}.\mathbf{b} & \mathbf{l}.\mathbf{c} \\ \mathbf{m}.\mathbf{a} & \mathbf{m}.\mathbf{b} & \mathbf{m}.\mathbf{c} \\ \mathbf{n}.\mathbf{a} & \mathbf{n}.\mathbf{b} & \mathbf{n}.\mathbf{c} \end{vmatrix}$$

(Kumaon 2006, Nagpur 2005, Avadh, 2005)

Also deduce that

$$[\mathbf{a}\,\mathbf{b}\,\mathbf{c}]^2 = \begin{vmatrix} \mathbf{a}.\mathbf{a} & \mathbf{a}.\mathbf{b} & \mathbf{a}.\mathbf{c} \\ \mathbf{b}.\mathbf{a} & \mathbf{b}.\mathbf{b} & \mathbf{b}.\mathbf{c} \\ \mathbf{c}.\mathbf{a} & \mathbf{c}.\mathbf{b} & \mathbf{c}.\mathbf{c} \end{vmatrix}$$

9. Show that the four points **a**, **b**, **c**, **d** are coplanar if

$$[\mathbf{b}\ \mathbf{c}\ \mathbf{d}] + [\mathbf{c}\ \mathbf{a}\ \mathbf{d}] + [\mathbf{a}\ \mathbf{b}\ \mathbf{d}] = [\mathbf{a}\ \mathbf{b}\ \mathbf{c}].$$

EXAMPLES

Example 1. *If one diagonal of a quadrilateral bisects the other, then it also bisects the quadrilateral.*

Solution. Let $OABC$ be the given quadrilateral such that its diagonal $\overrightarrow{OB}$ bisects the diagonal $\overrightarrow{AC}$.

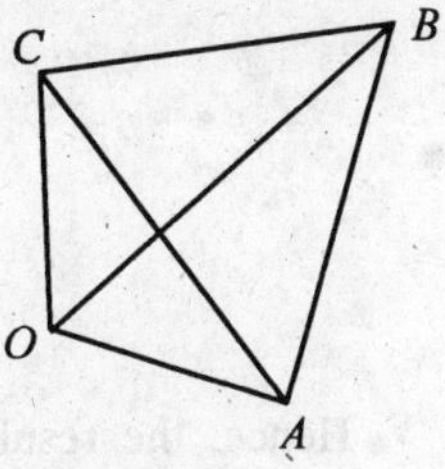

Fig. 5.8.

Let $\overrightarrow{OA} = \mathbf{a},\ \overrightarrow{OB} = \mathbf{b},\ \overrightarrow{OC} = \mathbf{c}.$

Since the mid-points $\dfrac{\mathbf{a}+\mathbf{c}}{2}$ of $\overrightarrow{AC}$ lies on $\overrightarrow{OB}$, there exists a scalar t such that

$$\frac{\mathbf{a}+\mathbf{c}}{2} = t\mathbf{b} \;\Rightarrow\; \mathbf{a}+\mathbf{c} = 2t\mathbf{b}.$$

Multiplying both sides vectorially with **b**, we have

$$(\mathbf{a} + \mathbf{c}) \times \mathbf{b} = 2t\, \mathbf{b} \times \mathbf{b} = \mathbf{0}$$

$$\Rightarrow \quad \mathbf{a} \times \mathbf{b} = \mathbf{b} \times \mathbf{c}$$

$$\Rightarrow \quad \frac{1}{2}\mathbf{a}\times\mathbf{b} = \frac{1}{2}\mathbf{b}\times\mathbf{c}$$

$$\Rightarrow \quad \frac{1}{2}\,|\,\mathbf{a}\times\mathbf{b}\,| = \frac{1}{2}\,|\,\mathbf{b}\times\mathbf{c}\,|$$

$\Rightarrow \quad \Delta\, OAB = \Delta\, OBC$

for $OABC$ is a plane quadrilateral.

Hence, the diagonal OB bisects the quadrilateral.

Example 2. *P, Q are the mid-points of the non-parallel sides BC and AD of a trapezium ABCD. Show that* $\Delta\, APD = \Delta\, CQB$.

Solution. Let $\overrightarrow{AB} = \mathbf{b}$ and $\overrightarrow{AD} = \mathbf{d}$.

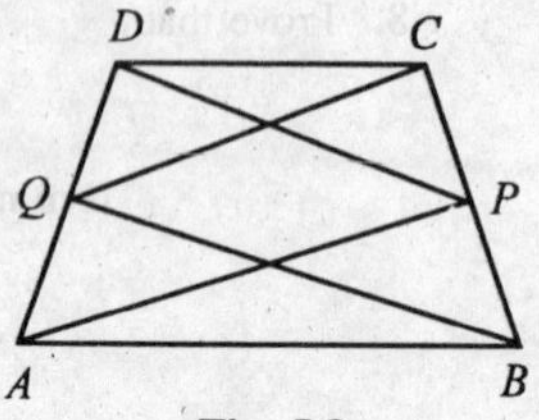

Fig. 5.9.

Now DC is parallel to AB

$\Rightarrow$ there exists a scalar t such that

$$\overrightarrow{DC} = t\, \overrightarrow{AB} = t\, \mathbf{b}.$$

$$\therefore \quad \overrightarrow{AC} = \overrightarrow{AD} + \overrightarrow{DC} = \mathbf{d} + t\, \mathbf{b}.$$

The position vectors of P and Q are $\frac{1}{2}(\mathbf{b} + \mathbf{d} + t\, \mathbf{b})$ and $\frac{1}{2}\mathbf{d}$ respectively.

Now
$$2\, \Delta\, APD = \overrightarrow{AP} \times \overrightarrow{AD}$$
$$= \frac{1}{2}(\mathbf{b} + \mathbf{d} + t\mathbf{b}) \times \mathbf{d} = \frac{1}{2}(1 + t)\,(\mathbf{b} \times \mathbf{d}).$$

Also
$$2\, \Delta\, CQB = \overrightarrow{CQ} \times \overrightarrow{CB} = \left[\frac{1}{2}\mathbf{d} - (\mathbf{d} + t\mathbf{b})\right] \times [\mathbf{b} - (\mathbf{d} + t\, \mathbf{b})]$$
$$= \left[-\frac{1}{2}\mathbf{d} - t\mathbf{b}\right] \times [-\mathbf{d} + (1 - t)\, \mathbf{b}]$$
$$= -\frac{1}{2}(1 - t)\, \mathbf{d} \times \mathbf{b} + t\mathbf{b} \times \mathbf{d}$$
$$= \frac{1}{2}(1 - t)\, \mathbf{b} \times \mathbf{d} + t\mathbf{b} \times \mathbf{d} = \frac{1}{2}(1 - t + 2t)\, \mathbf{b} \times \mathbf{d}$$
$$= \frac{1}{2}(1 + t)\, \mathbf{b} \times \mathbf{d} = 2\, \Delta\, \overrightarrow{APD}.$$

Hence, the result.

Example 3. *Given that* **a, b, c** *are the position vectors of the vertices of a* $\Delta\, ABC$*, find the vector area of the triangle.*

Solution. The required vector area

$$= \frac{1}{2}(\overrightarrow{AB} \times \overrightarrow{AC})$$
$$= \frac{1}{2}[(\mathbf{b} - \mathbf{a}) \times (\mathbf{c} - \mathbf{a})]$$
$$= \frac{1}{2}[\mathbf{a} \times \mathbf{b} + \mathbf{b} \times \mathbf{c} + \mathbf{c} \times \mathbf{a}].$$

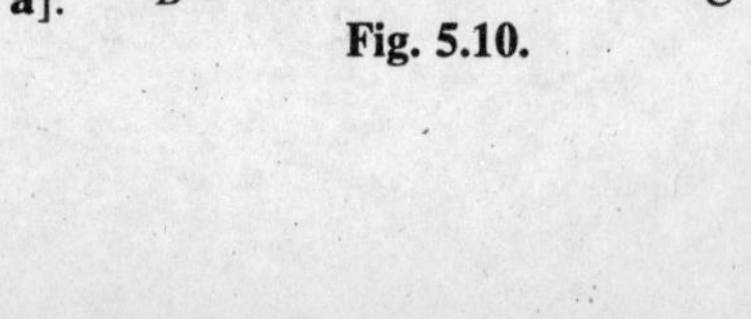

Fig. 5.10.

Example 4. *Find the area of the triangle whose vertices are the points with rectangular cartesian co-ordinates*

$$(1,\ 2,\ 3),\quad (-2,\ 1,\ -4),\quad (3,\ 4,\ -2).$$

Solution. Let **i, j, k** denote unit vectors along the three rectangular axes. If O be the origin and A, B, C denote the given vertices, we have

$$\vec{OA} = \mathbf{i}+2\mathbf{j}+3\mathbf{k},\ \vec{OB} = -2\mathbf{i}+\mathbf{j}-4\mathbf{k},\ \vec{OC} = 3\mathbf{i}+4\mathbf{j}-2\mathbf{k}.$$

$$\therefore\quad \vec{AB} = \vec{OB} - \vec{OA} = -3\mathbf{i}-\mathbf{j}-7\mathbf{k},$$

$$\vec{AC} = \vec{OC} - \vec{OA} = 2\mathbf{i}+2\mathbf{j}-5\mathbf{k}.$$

$$\vec{AB}\times\vec{AC} = (-3\mathbf{i}-\mathbf{j}-7\mathbf{k})\times(2\mathbf{i}+2\mathbf{j}-5\mathbf{k})$$

$$= 19\mathbf{i} - 29\mathbf{j} - 4\mathbf{k}.$$

$\therefore$ the required area $= \dfrac{1}{2}|\vec{AB}\times\vec{AC}| = \dfrac{1}{2}|19\mathbf{i}-29\mathbf{j}-4\mathbf{k}|$

$$= \frac{1}{2}\sqrt{(19^2+29^2+4^2)} = \frac{\sqrt{1218}}{2}.$$

Example 5. *Find the volume of the tetrahedron the rectangular cartesian co-ordinates of whose vertices are*

$$(0,\ 1,\ 2),\quad (3,\ 0,\ 1),\quad (4,\ 3,\ 6),\quad (2,\ 3,\ 2).$$

Solution. Let, as usual, **i, j, k** denote unit vectors along the three rectangular axes. If A, B, C, D denote the given vertices, we have

$$\vec{OA} = \mathbf{j}+2\mathbf{k},\qquad \vec{OB} = 3\mathbf{i}+\mathbf{k},$$

$$\vec{OC} = 4\mathbf{i}+3\mathbf{j}+6\mathbf{k},\qquad \vec{OD} = 2\mathbf{i}+3\mathbf{j}+2\mathbf{k}.$$

The required volume is

$$\frac{1}{6}|\vec{AB}\times\vec{AC}\cdot\vec{AD}|.$$

We have

$$\vec{AB} = \vec{OB} - \vec{OA} = 3\mathbf{i}-\mathbf{j}-\mathbf{k}.$$

$$\vec{AC} = \vec{OC} - \vec{OA} = 4\mathbf{i}+2\mathbf{j}+4\mathbf{k}.$$

$$\vec{AD} = \vec{OD} - \vec{OA} = 2\mathbf{i}+2\mathbf{j}.$$

$$\therefore\quad \vec{AB}\times\vec{AC} = (3\mathbf{i}-\mathbf{j}-\mathbf{k})\times(4\mathbf{i}+2\mathbf{j}+4\mathbf{k})$$

$$= -2\mathbf{i} - 16\mathbf{j} + 10\mathbf{k}.$$

$$\therefore \quad \overrightarrow{AB} \times \overrightarrow{AC} \cdot \overrightarrow{AD} = (-2\mathbf{i} - 16\mathbf{j} + 10\mathbf{k}) \cdot (2\mathbf{i} + 2\mathbf{j}) = -36.$$

Thus, the required volume = 36.

Example 6. *Show that the perpendicular distance of a point whose position vector is* **a** *from the plane through three points with position vectors* **b, c, d,** *is*

$$[\mathbf{b}\ \mathbf{c}\ \mathbf{d}] + [\mathbf{c}\ \mathbf{a}\ \mathbf{d}] + [\mathbf{a}\ \mathbf{b}\ \mathbf{d}] - [\mathbf{a}\ \mathbf{b}\ \mathbf{c}]\ /\ |\,\mathbf{b} \times \mathbf{c} + \mathbf{c} \times \mathbf{d} + \mathbf{d} \times \mathbf{b}\,|$$

Solution. Let *ABCD* be the tetrahedron whose vertices *A, B, C, D* have position vectors **a, b, c, d** respectively.

Volume of tetrahedron *ABCD*

$$= \frac{1}{3}(\text{area } \Delta\, BCD) \times h,$$

where *h* is the perpendicular distance of *A* from the plane *BCD*

$$\therefore \quad h = \frac{3 \times \text{volume of tetrahedron } ABCD}{\text{Area of } \Delta\, BCD}$$

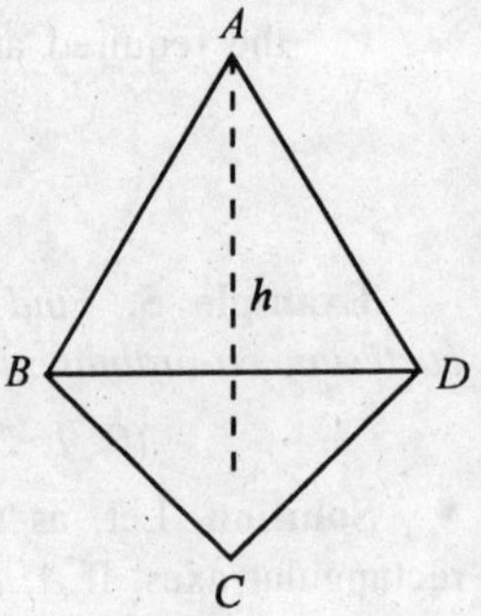

Fig. 5.11.

$$= 3 \cdot \frac{\frac{1}{6}[\overrightarrow{AB}, \overrightarrow{AC}, \overrightarrow{AD}]}{\left(\frac{1}{2}\right)|\,\overrightarrow{BC} \times \overrightarrow{BD}\,|}$$

$$= \frac{[\mathbf{b}-\mathbf{a},\ \mathbf{c}-\mathbf{a},\ \mathbf{d}-\mathbf{a}]}{|\,(\mathbf{c}-\mathbf{b}) \times (\mathbf{d}-\mathbf{b})\,|}$$

$$= \frac{(\mathbf{b}-\mathbf{a}) \cdot [(\mathbf{c}-\mathbf{a}) \times (\mathbf{d}-\mathbf{a})]}{|\,\mathbf{c} \times \mathbf{d} - \mathbf{c} \times \mathbf{b} - \mathbf{b} \times \mathbf{d}\,|}$$

$$= \frac{(\mathbf{b}-\mathbf{a}) \cdot [\mathbf{c} \times \mathbf{d} - \mathbf{c} \times \mathbf{a} - \mathbf{a} \times \mathbf{d}]}{|\,\mathbf{b} \times \mathbf{c} + \mathbf{c} \times \mathbf{d} + \mathbf{d} \times \mathbf{b}\,|}$$

$$= \frac{[\mathbf{b}\,\mathbf{c}\,\mathbf{d}] - [\mathbf{b}\,\mathbf{c}\,\mathbf{a}] - [\mathbf{b}\,\mathbf{a}\,\mathbf{d}] - [\mathbf{a}\,\mathbf{c}\,\mathbf{d}]}{|\,\mathbf{b} \times \mathbf{c} + \mathbf{c} \times \mathbf{d} + \mathbf{d} \times \mathbf{b}\,|}$$

$$= \frac{[\mathbf{b}\,\mathbf{c}\,\mathbf{d}] + [\mathbf{c}\,\mathbf{a}\,\mathbf{d}] + [\mathbf{a}\,\mathbf{b}\,\mathbf{d}] - [\mathbf{a}\,\mathbf{b}\,\mathbf{c}]}{|\,\mathbf{b} \times \mathbf{c} + \mathbf{c} \times \mathbf{d} + \mathbf{d} \times \mathbf{b}\,|}$$

Example 7. *Prove that the formula for the volume of a tetrahedron in terms of the lengths of three concurrent edges and their mutual inclinations is*

$$V^2 = \frac{a^2b^2c^2}{36}\begin{vmatrix} 1 & \cos\phi & \cos\psi \\ \cos\phi & 1 & \cos\theta \\ \cos\psi & \cos\theta & 1 \end{vmatrix}$$

Solution. Let $OABC$ be the tetrahedron with O as origin. Let **a, b, c** be the position vectors of A, B, C and **a, b, c** be their magnitudes.

Let $\mathbf{a} = a_1\mathbf{i} + a_2\mathbf{j} + a_3\mathbf{k}$, $\mathbf{b} = b_1\mathbf{i} + b_2\mathbf{j} + b_3\mathbf{k}$, $\mathbf{c} = c_1\mathbf{i} + c_2\mathbf{j} + c_3\mathbf{k}$

Then

$$V = \frac{1}{6}[\mathbf{a}\ \mathbf{b}\ \mathbf{c}] = \frac{1}{6}\begin{vmatrix} a_1 & a_2 & a_3 \\ b_1 & b_2 & b_3 \\ c_1 & c_2 & c_3 \end{vmatrix}$$

$$\therefore \quad V^2 = \frac{1}{36}\begin{vmatrix} a_1 & a_2 & a_3 \\ b_1 & b_2 & b_3 \\ c_1 & c_2 & c_3 \end{vmatrix} \times \begin{vmatrix} a_1 & a_2 & a_3 \\ b_1 & b_2 & b_3 \\ c_1 & c_2 & c_3 \end{vmatrix}$$

$$= \frac{1}{36}\begin{vmatrix} a_1^2 + a_2^2 + a_3^2 & a_1b_1 + a_2b_2 + a_3b_3 & a_1c_1 + a_2c_2 + a_3c_3 \\ a_1b_1 + a_2b_2 + a_3b_3 & b_1^2 + b_2^2 + b_3^2 & b_1c_1 + b_2c_2 + b_3c_3 \\ a_1c_1 + a_2c_2 + a_3c_3 & b_1c_1 + b_2c_2 + b_3c_3 & c_1^2 + c_2^2 + c_3^2 \end{vmatrix}$$

$$= \frac{1}{36}\begin{vmatrix} |\mathbf{a}|^2 & \mathbf{a}\cdot\mathbf{b} & \mathbf{a}\cdot\mathbf{c} \\ \mathbf{a}\cdot\mathbf{b} & |\mathbf{b}|^2 & \mathbf{b}\cdot\mathbf{c} \\ \mathbf{a}\cdot\mathbf{c} & \mathbf{b}\cdot\mathbf{c} & |\mathbf{c}|^2 \end{vmatrix}$$

$$= \frac{1}{36}\begin{vmatrix} a^2 & ab\cos\phi & ca\cos\psi \\ ab\cos\phi & b^2 & bc\cos\theta \\ ca\cos\psi & bc\cos\theta & c^2 \end{vmatrix}$$

$$= \frac{\mathrm{a}^2\mathrm{b}^2\mathrm{c}^2}{36}\begin{vmatrix} 1 & \cos\phi & \cos\psi \\ \cos\phi & 1 & \cos\theta \\ \cos\psi & \cos\theta & 1 \end{vmatrix}$$

Example 8. *Prove that each of the four faces of a tetrahedron subtends the same volume at the centroid.*

Solution. $ABCD$ be a tetrahedron with G as centroid. With G as origin, let the position vectors of A, B, C, D be **a, b, c** and **d** respectively. Then

$$\frac{\mathbf{a}+\mathbf{b}+\mathbf{c}+\mathbf{d}}{4} = 0 \Rightarrow \mathbf{a}+\mathbf{b}+\mathbf{c}+\mathbf{d} = 0.$$

Volume of tetrahedron $GABC = \frac{1}{6}[\mathbf{a}\ \mathbf{b}\ \mathbf{c}]$

Volume of tetrahedron $GBCD = \frac{1}{6}[\mathbf{b}\ \mathbf{c}\ \mathbf{d}]$

$$= \frac{1}{6}(\mathbf{b} \times \mathbf{c}) \cdot (-\mathbf{a} - \mathbf{b} - \mathbf{c})$$

$$= -\frac{1}{6}(\mathbf{b} \times \mathbf{c}) \cdot \mathbf{a} = \frac{1}{6}[\mathbf{a}\ \mathbf{b}\ \mathbf{c}]$$

Similarly other volumes can be shown.

Example 9. *Compare the volume of a tetrahedron with that of the tetrahedron formed by the centroids of its faces.*

Solution. Let **a, b, c, d** be the position vectors of the vertices *A, B, C, D* respectively of given tetrahedron. Then

$$V = \frac{1}{6}[\mathbf{b} - \mathbf{a},\ \mathbf{c} - \mathbf{a},\ \mathbf{d} - \mathbf{a}]$$

$$= \frac{1}{6}[\mathbf{a} - \mathbf{b},\ \mathbf{a} - \mathbf{c},\ \mathbf{a} - \mathbf{d}] \quad \text{(numerically)}$$

Now, the position vectors $\mathbf{g}_1$, $\mathbf{g}_2$, $\mathbf{g}_3$, $\mathbf{g}_4$ of the centroids G_1, G_2, G_3, G_4 of the four faces *BCD, ACD, ABD* are *ABC* respectively are given by

$$\mathbf{g}_1 = \frac{\mathbf{b} + \mathbf{c} + \mathbf{d}}{3}, \qquad \mathbf{g}_2 = \frac{\mathbf{a} + \mathbf{c} + \mathbf{d}}{3},$$

$$\mathbf{g}_3 = \frac{\mathbf{a} + \mathbf{b} + \mathbf{d}}{3}, \qquad \mathbf{g}_4 = \frac{\mathbf{a} + \mathbf{b} + \mathbf{c}}{3}.$$

$$\therefore \quad V' = \frac{1}{6}[\mathbf{g}_2 - \mathbf{g}_1,\ \mathbf{g}_3 - \mathbf{g}_1,\ \mathbf{g}_4 - \mathbf{g}_1]$$

$$= \frac{1}{6}\left[\frac{1}{3}(\mathbf{a} - \mathbf{b}),\ \frac{1}{3}(\mathbf{a} - \mathbf{c}),\ \frac{1}{3}(\mathbf{a} - \mathbf{d})\right]$$

$$= \frac{1}{27} \cdot \frac{1}{6}[(\mathbf{a} - \mathbf{b}),\ (\mathbf{a} - \mathbf{c}),\ (\mathbf{a} - \mathbf{d})]$$

$$= \frac{1}{27} V.$$

EXERCISES

1. *ABC* is a triangle, *E* and *F* are the mid-points of *AC* and *AB* respectively. *CP* is drawn parallel to *AB* to meet BE produced in *P*, show that

$$\Delta\ FEP = \Delta\ FCE = \frac{1}{4} \Delta\ ABC.$$

2. Show that $\vec{AC} \times \vec{BD}$ represents twice the vector area of a plane quadrilateral *ABCD*.

3. *ABC* is a triangle and *EF* is any straight line parallel to *BC* meeting *AC*, *AB* in *E*, *F* respectively. If *BR* and *CQ* be drawn parallel to *AC*, *AB* respectively to meet *EF* in *R* and *Q* respectively, prove that

$$\Delta\, ARB = \Delta\, ACQ.$$

4. Two triangles of equal area are on the opposite sides of the same base; prove that the straight line joining their vertices is bisected by the base.

5. *ABCD* is a quadrilateral such that

$$\vec{AB} = \mathbf{b},\ \vec{AD} = \mathbf{d},\ \vec{AC} = m\mathbf{b} + p\mathbf{d};$$

show that the area of the quadrilateral *ABCD* is

$$\frac{1}{2} \mid m + p \mid \mid \mathbf{b} \times \mathbf{d} \mid.$$

6. *ABCD* is a plane quadrilateral with $\vec{AB} = a,\ \vec{BC} = b,\ \vec{CD} = c$. Prove that the area of the quadrilateral is

$$\frac{1}{2} \mid \mathbf{a} \times \mathbf{b} + \mathbf{b} \times \mathbf{c} - \mathbf{c} \times \mathbf{a} \mid.$$

7. Find the areas of the triangles with the following vertices :

(*a*) (0, 0, 0), (1, 2, 3), (2, – 1, 4);

(*b*) (1, 0, 0), (0, 1, 0), (1, 1, 1);

(*c*) (– 1, 2, 3), (2, – 1, – 1), (1, 1, – 1);

(*d*) (*a*, 0, 0), (0, *b*, 0), (0, 0, *c*).

8. Find the volumes of the tetrahedron with the following vertices :

(*a*) (0, 0, 0), (1, 1, – 1), (1, – 1, 1), (– 1, 1, 1).

(*b*) (– 1, 0, 1), (2, – 1, 0), (3, 2, 5), (1, 2, 1).

9. Show that the volume of the tetrahedron, the co-ordinates of whose vertices are (x_1, y_1, z_1), (x_2, y_2, z_2), (x_3, y_3, z_3) and (x_4, y_4, z_4) is the absolute value of

$$\frac{1}{6}\begin{vmatrix} x_2 - x_1 & y_2 - y_1 & z_2 - z_1 \\ x_3 - x_1 & y_3 - y_1 & z_3 - z_1 \\ x_4 - x_1 & y_4 - y_1 & z_4 - z_1 \end{vmatrix}$$

10. Show that the volume of the tetrahedron, the position vectors of whose vertices are **a, b, c, d** is

$$\frac{1}{6}\left| [\mathbf{b}\ \mathbf{c}\ \mathbf{d}] + [\mathbf{c}\ \mathbf{a}\ \mathbf{d}] + [\mathbf{a}\ \mathbf{b}\ \mathbf{d}] - [\mathbf{a}\ \mathbf{b}\ \mathbf{c}] \right|$$

Deduce the condition for the points with position vectors **a, b, c, d** to be coplanar.

11. Show that the volume of a tetrahedron bounded by the four planes $\mathbf{r} \cdot (m\mathbf{j} + n\mathbf{k}) = 0$, $\mathbf{r} \cdot (n\mathbf{k} + l\mathbf{i}) = 0$, $\mathbf{r} \cdot (l\mathbf{i} + m\mathbf{j}) = 0$ and $\mathbf{r} \cdot (l\mathbf{i} + m\mathbf{j} + n\mathbf{k}) = p$ is $2p^3 / 3lmn$. (*Avadh, 1998*)

5.8. VECTOR TRIPLE PRODUCTS

Let **a, b, c** be any three vectors. Then an expression of the form

$$(\mathbf{a} \times \mathbf{b}) \times \mathbf{c}$$

which is the vector product of $\mathbf{a} \times \mathbf{b}$ with **c**, is known as vector triple product.

It will be shown that this product is a linear combination of **a** and **b** and precisely that

$$(\mathbf{a} \times \mathbf{b}) \times \mathbf{c} = (\mathbf{a} \cdot \mathbf{c})\, \mathbf{b} - (\mathbf{b} \cdot \mathbf{c})\, \mathbf{a}.$$

The vector $(\mathbf{a} \times \mathbf{b}) \times \mathbf{c}$ being perpendicular to the vector $\mathbf{a} \times \mathbf{b}$, is coplanar with **a** and **b** and, as such, it is a linear combination of **a** and **b.** Let

$$(\mathbf{a} \times \mathbf{b}) \times \mathbf{c} = l\mathbf{a} + m\mathbf{b}. \qquad \text{...}(i)$$

Also the vector on the left is perpendicular to the vector **c** so that its scalar product with, **c**, is zero.

Multiplying both sides of (*i*) scalarly with, **c**, we obtain

$$l\, \mathbf{a} \cdot \mathbf{c} + m\, \mathbf{b} \cdot \mathbf{c} = 0. \qquad \text{...}(ii)$$

From (*i*) and (*ii*), we obtain an equality of the form

$$(\mathbf{a} \times \mathbf{b}) \times \mathbf{c} = k\, [(\mathbf{a} \cdot \mathbf{c})\, \mathbf{b} - (\mathbf{b} \cdot \mathbf{c})\, \mathbf{a}] \qquad \text{...}(iii)$$

This equality has been obtained by using the fact that *the vector* $(\mathbf{a} \times \mathbf{b}) \times \mathbf{c}$ *is coplanar with* **a** *and* **b** *and is perpendicular to* **c.**

It will here below be shown that $k = 1$. The proof of this fact involves an examination of the *magnitudes* of the vectors whereas the equality (*ii*) has been obtained by considering *directions* only. A proof of the result will now be given.

Note. The vector multiplication is not associative, *i.e.*, $(\mathbf{a} \times \mathbf{b}) \times \mathbf{c}$ and $\mathbf{a} \times (\mathbf{b} \times \mathbf{c})$ are not equal for all vectors **a, b, c.** In fact, as seen above, $(\mathbf{a} \times \mathbf{b}) \times \mathbf{c}$ is a linear combination of **a** and **b** and $\mathbf{a} \times (\mathbf{b} \times \mathbf{c})$ is a linear combination of **b** and **c**.

The student may compute

$$\{(\mathbf{i} - \mathbf{j}) \times (\mathbf{i} + \mathbf{j} + \mathbf{k})\} \times (\mathbf{j} - \mathbf{k})$$

and $$(\mathbf{i} - \mathbf{j}) \times \{(\mathbf{i} + \mathbf{j} + \mathbf{k})\} \times (\mathbf{j} - \mathbf{k})\}$$

and see that they are *not* equal.

Theorem. *To prove that*

$$(\mathbf{a} \times \mathbf{b}) \times \mathbf{c} = (\mathbf{a} \cdot \mathbf{c})\, \mathbf{b} - (\mathbf{b} \cdot \mathbf{c})\, \mathbf{a}$$ (*Kolkata, 99, 2005 Garhwal 98*)

where **a, b, c** *are three vectors.*

The proof will be given by suitably selecting an orthonormal base consisting of vectors **i, j, k** related to **a, b, c.**

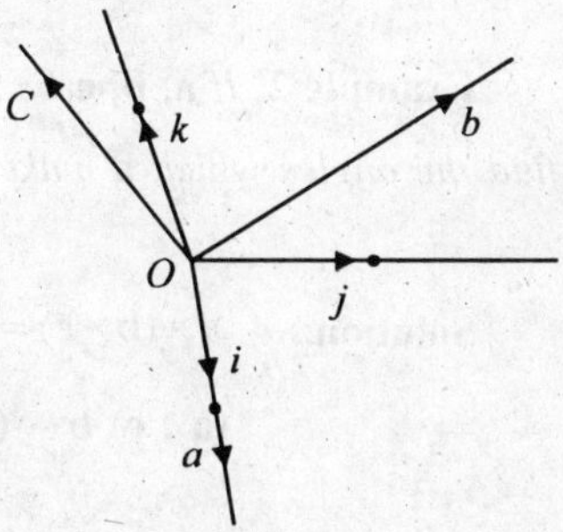

Fig. 5.11.

i, is a unit vector along **a**;

j, is a unit vector perpendicular to **a** and in the plane of **a** and **b**.

k, is a unit vector perpendicular to **i** and **j** such that **i, j, k** is a right-handed triad.

Thus, we have expressions of the form

$$\mathbf{a} = a_1\mathbf{i},$$

$$\mathbf{b} = b_1\mathbf{i} + b_2\mathbf{j},$$

$$\mathbf{c} = c_1\mathbf{i} + c_2\mathbf{j} + c_3\mathbf{k},$$

Now $\mathbf{a} \times \mathbf{b} = a_1\mathbf{i} \times (b_1\mathbf{i} + b_2\mathbf{j})$

$$= a_1 b_2 \mathbf{k}.$$

$$(\mathbf{a} \times \mathbf{b}) \times \mathbf{c} = a_1 b_2 \mathbf{k} \times (c_1\mathbf{i} + c_2\mathbf{j} + c_3\mathbf{k})$$

$$= -a_1 b_2 c_2 \mathbf{i} + a_1 b_2 c_1 \mathbf{j}.$$

Also $(\mathbf{a} \cdot \mathbf{c})\,\mathbf{b} - (\mathbf{b} \cdot \mathbf{c})\,\mathbf{a} = a_1 c_1 (b_1\mathbf{i} + b_2\mathbf{j}) - (b_1 c_1 + b_2 c_2)\, a_1\mathbf{i}$

$$= -a_1 b_2 c_2 \mathbf{i} + a_1 b_2 c_1 \mathbf{j}.$$

Hence $(\mathbf{a} \times \mathbf{b}) \times \mathbf{c} = (\mathbf{a} \cdot \mathbf{c})\,\mathbf{b} - (\mathbf{b} \cdot \mathbf{c})\,\mathbf{a}$ ...(*ii*)

Cor. $\mathbf{a} \times (\mathbf{b} \times \mathbf{c}) = -(\mathbf{b} \times \mathbf{c}) \times \mathbf{a}$

$$= -[(\mathbf{b} \cdot \mathbf{a})\,\mathbf{c} - (\mathbf{c} \cdot \mathbf{a})\,\mathbf{b}]$$

$$= (\mathbf{c} \cdot \mathbf{a})\,\mathbf{b} - (\mathbf{b} \cdot \mathbf{a})\,\mathbf{c} = (\mathbf{a} \cdot \mathbf{c})\,\mathbf{b} - (\mathbf{a} \cdot \mathbf{b})\,\mathbf{c}.$$

Rule to Remember. As a help to memory, we notice that each of the two scalar products on the right of (*ii*) involves the outer vector **c**, and of the two other vectors, **a, b** in the bracket; **a** is remote from **c** and **b** is adjacent **c**. Thus, we have the rule :

(*remote* × *adjacent*) × *outer* = (*outer* . *remote*) *adjacent* –
(*outer* . *adjacent*) *remote.*

The formula in the cor. is also written by the same rule.

EXAMPLES

Example 1. *Prove* $\mathbf{i} \times (\mathbf{a} \times \mathbf{i}) + \mathbf{j} \times (\mathbf{a} \times \mathbf{j}) + \mathbf{k} \times (\mathbf{a} \times \mathbf{k}) = 2\mathbf{a}$.

(*U.P.P.C.S. 2004, Kumaon 2002 Awadh 2000, 2001, Rohilkhand 2000*)

Solution. L.H.S. $= (\mathbf{i} \cdot \mathbf{i})\,\mathbf{a} - (\mathbf{i} \cdot \mathbf{a})\,\mathbf{i} + (\mathbf{j} \cdot \mathbf{j})\,\mathbf{a}$

$$- (\mathbf{j} \cdot \mathbf{a})\,\mathbf{j} + (\mathbf{k} \cdot \mathbf{k})\,\mathbf{a} - (\mathbf{k} \cdot \mathbf{a})\,\mathbf{k}$$

$$= 3\mathbf{a} \,.\, [(\mathbf{i} \cdot \mathbf{a})\,\mathbf{i} + (\mathbf{j} \cdot \mathbf{a})\,\mathbf{j} + (\mathbf{k} \cdot \mathbf{a})\,\mathbf{k}]$$

$$= 3\mathbf{a} - \mathbf{a} = 2\mathbf{a}.$$

Example 2. *If* **a, b, c** *be three unit vectors such that* $\mathbf{a}\times(\mathbf{b}\times\mathbf{c})=\frac{1}{2}\mathbf{b}$, *find the angles which* **a** *makes with* **b** *and* **c, b** *and* **c** *being non-parallel.*

(Awadh 99, 2001, 2003; Kumaon 98)

Solution. $\mathbf{a}\times(\mathbf{b}\times\mathbf{c})=\frac{1}{2}\mathbf{b}$

$\Rightarrow$ $(\mathbf{a}\,.\,\mathbf{c})\,\mathbf{b}-(\mathbf{a}\,.\,\mathbf{b})\,\mathbf{c}=\frac{1}{2}\mathbf{b}$

$\Rightarrow$ $\left(\mathbf{a}\,.\,\mathbf{c}-\frac{1}{2}\right)\mathbf{b}-(\mathbf{a}\,.\,\mathbf{b})\,\mathbf{c}=0$

Since **b** and **c** are non-parallel, hence coefficients of **b** and **c** vanish separately.

$\therefore$ $\mathbf{a}\,.\,\mathbf{c}-\frac{1}{2}=0$ and $\mathbf{a}\,.\,\mathbf{b}=0$

$\Rightarrow$ Angle between **a** and $\mathbf{c}=\cos^{-1}\left(\frac{1}{2}\right)=\frac{\pi}{3}$ and Angle between **a** and $\mathbf{b}=\pi/2$.

Example 3. *Show that* $\mathbf{a}\times(\mathbf{b}\times\mathbf{c})=(\mathbf{a}\times\mathbf{b})\times\mathbf{c}$ *if and only if either* $\mathbf{b}=0$, *or* **c** *is collinear with* **a**, *or* **b** *is perpendicular to both* **a** *and* **c.**

(Purvanchal 2004, Avadh 2002)

Solution. $\mathbf{a}\times(\mathbf{b}\times\mathbf{c})=(\mathbf{a}\times\mathbf{b})\times\mathbf{c}$

$\Leftrightarrow$ $(\mathbf{a}\,.\,\mathbf{c})\,\mathbf{b}-(\mathbf{a}\,.\,\mathbf{b})\,\mathbf{c}=(\mathbf{a}\,.\,\mathbf{c})\,\mathbf{b}-(\mathbf{b}\,.\,\mathbf{c})\,\mathbf{a}$

$\Leftrightarrow$ $(\mathbf{a}\,.\,\mathbf{b})\,\mathbf{c}=(\mathbf{b}\,.\,\mathbf{c})\,\mathbf{a}$ ...(*i*)

The condition is necessary. If **c** and **a** are collinear then theorem is proved. But if **c** and **a** are not collinear, then (1) implies that

$$\mathbf{a}\,.\,\mathbf{b}=0 \text{ and } \mathbf{b}\,.\,\mathbf{c}=0$$

It is true if $\mathbf{b}=0$ or **b** is perpendicular to both **a** and **c.**

The condition is sufficient. Let $\mathbf{b}=0$, or **b** be perpendicular to both **a** and **c.** Then both sides of (1) are equal and hence the given relation holds. Again, if **c** is collinear with **a**, then there exists a non-zero scalar x such that $\mathbf{c}=x\mathbf{a}$. Then, we have

$$(\mathbf{a}\,.\,\mathbf{b})\,\mathbf{c}=(\mathbf{a}\,.\,\mathbf{b})\,x\mathbf{a}=x\,(\mathbf{a}\,.\,\mathbf{b})\,\mathbf{a}$$

$$=(x\mathbf{a}\,.\,\mathbf{b})\,.\,\mathbf{a}=(\mathbf{c}\,.\,\mathbf{b})\,\mathbf{a}=(\mathbf{b}\,.\,\mathbf{c})\,\mathbf{a}.$$

Hence the proposition.

Example 4. *Prove that* $[\mathbf{a}\times\mathbf{b},\ \mathbf{b}\times\mathbf{c},\ \mathbf{c}\times\mathbf{a}]=[\mathbf{a}\ \mathbf{b}\ \mathbf{c}]^2$.

(Patna 2003, Avadh 2005, Kumaon 2002, 2005, Rohilkhand 2005, 2006)

Solution. Let $\mathbf{b}\times\mathbf{c}=\mathbf{d}$

Then $(\mathbf{b}\times\mathbf{c})\times(\mathbf{c}\times\mathbf{a})=\mathbf{d}\times(\mathbf{c}\times\mathbf{a})=(\mathbf{d}\,.\,\mathbf{a})\,\mathbf{c}-(\mathbf{d}\,.\,\mathbf{c})\,\mathbf{a}$

$=[\mathbf{b}\ \mathbf{c}\ \mathbf{a}]\,\mathbf{c}-[\mathbf{b}\ \mathbf{c}\ \mathbf{c}]\,\mathbf{a}$

$=[\mathbf{b}\ \mathbf{c}\ \mathbf{a}]\,\mathbf{c}$ $(\because\ [\mathbf{b}\,\mathbf{c}\,\mathbf{c}]=0)$

$$\therefore\ [\mathbf{a} \times \mathbf{b},\ \mathbf{b} \times \mathbf{c},\ \mathbf{c} \times \mathbf{a}] = (\mathbf{a} \times \mathbf{b}) \cdot [(\mathbf{b} \times \mathbf{c}) \times (\mathbf{c} \times \mathbf{a})]$$
$$= (\mathbf{a} \times \mathbf{b}) \cdot [\mathbf{b}\ \mathbf{c}\ \mathbf{a}]\ \mathbf{c}$$
$$= [\mathbf{b}\ \mathbf{c}\ \mathbf{a}]\ [\mathbf{a}\ \mathbf{b}\ \mathbf{c}] = [\mathbf{a}\ \mathbf{b}\ \mathbf{c}]^2.$$

EXERCISES

1. Show that

$$(\mathbf{a} \times \mathbf{b}) \times \mathbf{c} = \mathbf{a} \times (\mathbf{b} \times \mathbf{c})$$

if, and only if, the vectors **a** and **c** are collinear.

2. Show that

(*i*) $[\mathbf{a} + \mathbf{b}, \mathbf{b} + \mathbf{c}, \mathbf{c} + \mathbf{a}] = 2\ [\mathbf{a}\ \mathbf{b}\ \mathbf{c}]$.

(*Kumaon 2000, 2005, Avadh 97, Purvanchal 2004, Rohilkhand 99, Garhwal 2000*)

(*ii*) $(\mathbf{a} \times \mathbf{b}) \times (\mathbf{a} \times \mathbf{c}) \cdot \mathbf{d} = \mathbf{a} \cdot \mathbf{d}\ [\ \mathbf{a}\ \mathbf{b}\ \mathbf{c}]$.

3. Prove that

(*i*) $\mathbf{a} \times (\mathbf{b} + \mathbf{c}) + \mathbf{b} \times (\mathbf{c} + \mathbf{a}) + \mathbf{c} \times (\mathbf{a} + \mathbf{b}) = \mathbf{0}$.

(*Awadh 99, Rohilkhand 98*)

(*ii*) $\mathbf{a} \times (\mathbf{b} \times \mathbf{c}) + \mathbf{b} \times (\mathbf{c} \times \mathbf{a}) + \mathbf{c} \times (\mathbf{a} \times \mathbf{b}) = \mathbf{0}$.

(*Rohilkhand 98, U.P.P.C.S. 2000, Garhwal 99*)

4. Show that

$$(\mathbf{b} \times \mathbf{c}) \cdot (\mathbf{a} \times \mathbf{d}) + (\mathbf{c} \times \mathbf{a}) \cdot (\mathbf{b} \times \mathbf{d}) + (\mathbf{a} \times \mathbf{b}) \cdot (\mathbf{c} \times \mathbf{d}) = 0.$$

5. Prove that

$$\mathbf{d} \cdot [\mathbf{a} \times \{\mathbf{b} \times (\mathbf{c} \times \mathbf{d})\}] = (\mathbf{b} \cdot \mathbf{d})\ [\mathbf{a}\ \mathbf{c}\ \mathbf{d}].$$

6. **p**, **q**, **r** are three vectors defined by the relations

$$\mathbf{p} = \frac{\mathbf{b} \times \mathbf{c}}{[\mathbf{a}\ \mathbf{b}\ \mathbf{c}]},\ \mathbf{q} = \frac{\mathbf{c} \times \mathbf{a}}{[\mathbf{a}\ \mathbf{b}\ \mathbf{c}]},\ \mathbf{r} = \frac{\mathbf{a} \times \mathbf{b}}{[\mathbf{a}\ \mathbf{b}\ \mathbf{c}]},$$

and $[\mathbf{a}\ \mathbf{b}\ \mathbf{c}] \neq \mathbf{0}$.

Prove that

(*i*) $\mathbf{a} \times \mathbf{p} + \mathbf{b} \times \mathbf{q} + \mathbf{c} \times \mathbf{r} = 0$,

(*ii*) $\mathbf{a} \cdot \mathbf{p} + \mathbf{b},\ \mathbf{q} + \mathbf{c} \cdot \mathbf{r} = 3$,

(*iii*) $[\mathbf{a}\ \mathbf{b}\ \mathbf{c}]\ [\mathbf{p}\ \mathbf{q}\ \mathbf{r}] = 1$,

(*iv*) $\mathbf{a} = \frac{\mathbf{q} \times \mathbf{r}}{[\mathbf{p}\ \mathbf{q}\ \mathbf{r}]},\ \mathbf{b} = \frac{\mathbf{r} \times \mathbf{p}}{[\mathbf{p}\ \mathbf{q}\ \mathbf{r}]},\ \mathbf{a} = \frac{\mathbf{p}\mathbf{q}}{[\mathbf{p}\ \mathbf{q}\ \mathbf{r}]}$

7. **a**, **b**, **c**; **a′**, **b′**, **c′** are two systems of non-coplanar vectors and

$$\mathbf{p} = \mathbf{a} \times (\mathbf{b}' \times \mathbf{c}'),\ \ \mathbf{q} = \mathbf{b} \times (\mathbf{c}' \times \mathbf{a}'),\ \ \mathbf{r} = \mathbf{c} \times (\mathbf{a}' \times \mathbf{b}');$$
$$\mathbf{p}' = \mathbf{a}' \times (\mathbf{b} \times \mathbf{c}),\ \ \mathbf{q}' = \mathbf{b}' \times (\mathbf{c} \times \mathbf{a}),\ \ \mathbf{r}' = \mathbf{c}' \times (\mathbf{a} \times \mathbf{b}).$$

Show that

$$[\mathbf{p}\ \mathbf{q}\ \mathbf{r}] \neq 0, \quad \Leftrightarrow \quad [\mathbf{p}', \mathbf{q}', \mathbf{r}'] \neq 0.$$

8. **a, b, c** are three non-coplanar vectors; express $\mathbf{b} \times \mathbf{c}, \mathbf{c} \times \mathbf{a}, \mathbf{a} \times \mathbf{b}$ as linear combination of **a, b, c** and show that if

$$\mathbf{b} \times \mathbf{c} = l_{11}\mathbf{a} + l_{12}\mathbf{b} + l_{13}\mathbf{c}, \quad \mathbf{c} \times \mathbf{a} = l_{21}\mathbf{a} + l_{22}\mathbf{b} + l_{23}\mathbf{c},$$
$$\mathbf{a} \times \mathbf{b} = l_{31}\mathbf{a} + l_{32}\mathbf{b} + l_{33}\mathbf{c},$$

then

$$l_{ij} = A_{ij} / [\mathbf{a}\ \mathbf{b}\ \mathbf{c}],$$

where A_{ij} is the cofactor of the element in the ith row and jth column in the determinantal expression for $[\mathbf{a}\ \mathbf{b}\ \mathbf{c}]^2$.

SUMMARY

1. The vector multiplication is neither commutative nor associative.

2. If **i, j, k** are mutually perpendicular unit vectors, then

$$\mathbf{i} \times \mathbf{i} = 0, \quad \mathbf{j} \times \mathbf{j} = 0, \quad \mathbf{k} \times \mathbf{k} = 0,$$
$$\mathbf{i} \times \mathbf{j} = \mathbf{k}, \quad \mathbf{j} \times \mathbf{k} = \mathbf{i}, \quad \mathbf{k} \times \mathbf{i} = \mathbf{j},$$
$$\mathbf{j} \times \mathbf{i} = -\mathbf{k}, \ \mathbf{k} \times \mathbf{j} = -\mathbf{i}, \ \mathbf{i} \times \mathbf{k} = -\mathbf{j},$$

3. The area of a triangle OAB where $\vec{OA} = \mathbf{a}$ and $\vec{OB} = \mathbf{b}$ is

$$\frac{1}{2} |\mathbf{a} \times \mathbf{b}|.$$

4. The scalar triple product $\mathbf{a} \cdot \mathbf{b} \times \mathbf{c}$ is 0 if and only if the three vectors **a, b, c** are coplanar.

5. A cyclic permutation of the three vectors does not change the value of the scalar triple product and an anti-cyclic permutation changes the value in sign but not in magnitude.

6. The positions of dot and cross can be interchanged without any change in the value of a scalar triple product.

7. The expression $\frac{1}{6} \mathbf{a} \cdot \mathbf{b} \times \mathbf{c}$ denotes the volume of the tetrahedron with coterminous edges

$$\vec{OA} = \mathbf{a}, \ \vec{OB} = \mathbf{b}, \ \vec{OC} = \mathbf{c}$$

if the vectors **a, b, c** form a right-handed system.

8. Vector multiplication distributes the sum of vectors, *i.e.*,

$$\mathbf{a} \times (\mathbf{b} + \mathbf{c}) = \mathbf{a} \times \mathbf{b} + \mathbf{a} \times \mathbf{c}$$

for all vectors **a, b, c.**

9. $(a_1\mathbf{i} + a_2\mathbf{j} + a_3\mathbf{k}) \times (b_1\mathbf{i} + b_2\mathbf{j} + b_3\mathbf{k})$

$$= (a_2b_3 - a_3b_2)\,\mathbf{i} + (a_3b_1 - a_1b_3)\,\mathbf{j} + (a_1b_2 - a_2b_1)\,\mathbf{k}$$

which is usually denoted by

$$\begin{vmatrix} \mathbf{i} & \mathbf{j} & \mathbf{k} \\ a_1 & a_2 & a_3 \\ b_1 & b_2 & b_3 \end{vmatrix}$$

10. If $\mathbf{a} = a_1\mathbf{i} + a_2\mathbf{j} + a_3\mathbf{k}$, $\mathbf{b} = b_1\mathbf{i} + b_2\mathbf{j} + b_3\mathbf{k}$,

and $\mathbf{c} = c_1\mathbf{i} + c_2\mathbf{j} + c_3\mathbf{k}$,

then $$\mathbf{a} \,.\, \mathbf{b} \times \mathbf{c} = \begin{vmatrix} a_1 & a_2 & a_3 \\ b_1 & b_2 & b_3 \\ c_1 & c_2 & c_3 \end{vmatrix}.$$

11. $(\mathbf{a} \times \mathbf{b}) \times \mathbf{c} = (\mathbf{a} \,.\, \mathbf{c})\, \mathbf{b} - (\mathbf{b} \,.\, \mathbf{c})\, \mathbf{a}$.

APPLICATION OF SCALAR TRIPLE PRODUCTS TO GEOMETRY

5.9. EQUATIONS OF PLANES PARALLEL TO GIVEN VECTORS AND PASSING THROUGH GIVEN POINTS

5.9.1. *Equation of the plane through three points A, B, C with position vector* **a, b, c.**

We have

$$\vec{AB} = \mathbf{b} - \mathbf{a},$$

$$\vec{AC} = \mathbf{c} - \mathbf{a}.$$

The vector

$$\vec{AB} \times \vec{AC} = (\mathbf{b} - \mathbf{a}) \times (\mathbf{c} - \mathbf{a})$$

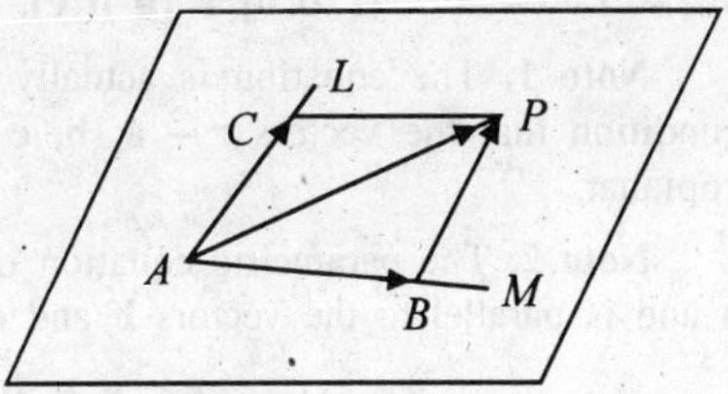

Fig. 5.12.

is normal to the plane. Thus, the equation of the plane is

$$(\mathbf{r} - \mathbf{a}) \,.\, (\mathbf{b} - \mathbf{a}) \times (\mathbf{c} - \mathbf{a}) = 0$$

$$\Rightarrow \quad \mathbf{r} \,.\, (\mathbf{b} \times \mathbf{c} + \mathbf{c} \times \mathbf{a} + \mathbf{a} \times \mathbf{b}) = [\mathbf{a}\ \mathbf{b}\ \mathbf{c}]$$

which is the required equation.

Note. The equation can also be deduced from the parametric equation $\mathbf{r} = \mathbf{a} + t(\mathbf{b} - \mathbf{a}) + p(\mathbf{c} - \mathbf{a})$ on multiplying both sides of

$$\mathbf{r} - \mathbf{a} = t(\mathbf{b} - \mathbf{a}) + p(\mathbf{c} - \mathbf{a}) \text{ scalarly with } (\mathbf{b} - \mathbf{a}) \times (\mathbf{c} - \mathbf{a}).$$

Cor. Condition for coplanarity of four points. Four points **a, b, c, d** will be coplanar if, and only if, the point **d**, lies on the plane through the three points **a, b, c,** *i.e.,* **d** satisfies the equation

$$\mathbf{r} \cdot (\mathbf{b} \times \mathbf{c} + \mathbf{c} \times \mathbf{a} + \mathbf{a} \times \mathbf{b}) = [\mathbf{a}\ \mathbf{b}\ \mathbf{c}].$$

Thus, the required condition is

$$[\mathbf{d}\ \mathbf{b}\ \mathbf{c}] + [\mathbf{d}\ \mathbf{c}\ \mathbf{a}] + [\mathbf{d}\ \mathbf{a}\ \mathbf{b}] = [\mathbf{a}\ \mathbf{b}\ \mathbf{c}].$$

Note. As seen in chapter 2 four points with position vectors **a, b, c, d** are coplanar, if and only if, there exist four scalars $x, y, z, t,$ not all zero, such that

$$x\mathbf{a} + y\mathbf{b} + z\mathbf{c} + t\mathbf{d} = 0, \qquad x + y + z + t = 0.$$

From these eliminating t, we obtain

$$x(\mathbf{a} - \mathbf{d}) + y(\mathbf{b} - \mathbf{d}) + z(\mathbf{c} - \mathbf{d}) = 0$$

Now supposing that $x \neq 0$ and multiplying scalarly with

$$(\mathbf{b} - \mathbf{d}) \times (\mathbf{c} - \mathbf{d}),$$

we obtain

$$x(\mathbf{a} - \mathbf{d}) \cdot (\mathbf{b} - \mathbf{d}) \times (\mathbf{c} - \mathbf{d}) = 0$$

$$\Rightarrow \quad [\mathbf{a}\ \mathbf{b}\ \mathbf{c}] = [\mathbf{d}\ \mathbf{b}\ \mathbf{c}] + [\mathbf{d}\ \mathbf{c}\ \mathbf{a}] + [\mathbf{d}\ \mathbf{a}\ \mathbf{b}],$$

which is the condition as obtained above.

5.9.2. *Equation of the plane which passes through a given point A with position vector* **a** *and is parallel to the given vectors* **b** *and* **c**.

The vector $\mathbf{b} \times \mathbf{c}$ is normal to the plane and the point with position vector **a** lies on the plane. The equation of the plane, therefore, is

$$(\mathbf{r} - \mathbf{a}) \cdot \mathbf{b} \times \mathbf{c} = 0$$

$$\Leftrightarrow \quad [\mathbf{r}\ \mathbf{b}\ \mathbf{c}] = [\mathbf{a}\ \mathbf{b}\ \mathbf{c}].$$

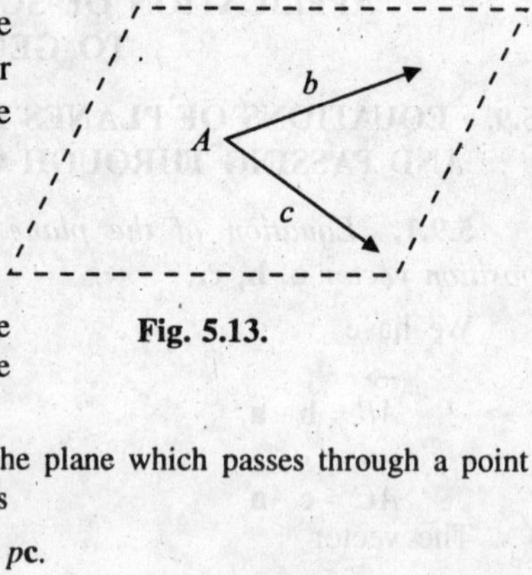

Fig. 5.13.

Note 1. The equation is actually the condition that the vectors $\mathbf{r} - \mathbf{a}, \mathbf{b}, \mathbf{c}$ are coplanar.

Note 2. The parametric equation of the plane which passes through a point **a** and is parallel to the vectors **b** and **c** is

$$\mathbf{r} = \mathbf{a} + t\mathbf{b} + p\mathbf{c}.$$

$$\Rightarrow \quad \mathbf{r} - \mathbf{a} = t\mathbf{b} + p\mathbf{c}.$$

and on multiplying both sides scalarly with $\mathbf{b} \times \mathbf{c}$, we get

$$(\mathbf{r} - \mathbf{a}) \cdot (\mathbf{b} \times \mathbf{c}) = t\mathbf{b}, \qquad \mathbf{b} \times \mathbf{c} + p\mathbf{c} \cdot \mathbf{b} \times \mathbf{c} = 0$$

5.9.3. *Equation of the plane through two given points A, B with position vectors* **a, b** *and parallel to a given vector* **c**.

The vector $(\mathbf{a} - \mathbf{b}) \times \mathbf{c}$ is normal to the plane. Thus, we require the equation of the plane which passes through a point **a** and which is normal to the vector

$$(\mathbf{a} - \mathbf{b}) \times \mathbf{c}.$$

The required equation therefore is

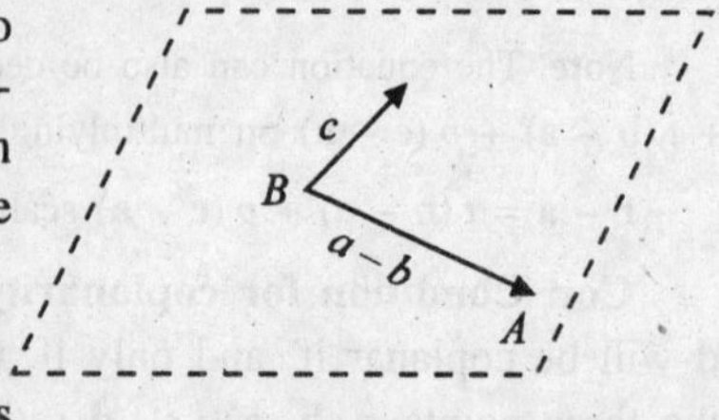

Fig. 5.14.

$$(\mathbf{r}-\mathbf{a})\,.\,(\mathbf{a}-\mathbf{b})\times\mathbf{c}=0$$

$$\Leftrightarrow \qquad \mathbf{r}\,.\,[\mathbf{b}\times\mathbf{c}+\mathbf{c}\times\mathbf{a}]=[\mathbf{a}\;\mathbf{b}\;\mathbf{c}]. \qquad ...(i)$$

Note 1. We may also obtain this result on taking the parametric equation

$$\mathbf{r}=\mathbf{a}+t\,(\mathbf{b}-\mathbf{a})+p\mathbf{c} \quad\Leftrightarrow\quad \mathbf{r}-\mathbf{a}=t\,(\mathbf{b}-\mathbf{a})+p\mathbf{c}$$

and multiplying both sides scalarly with $(\mathbf{b}-\mathbf{a})\times\mathbf{c}$.

Note 2. The vector $\mathbf{b}\times\mathbf{c}+\mathbf{c}\times\mathbf{a}$ is normal to the plane which passes through points with position vectors **a, b** and which is parallel to the vector **c.**

5.10. COPLANARITY OF TWO LINES

To find the condition for the lines

$$\mathbf{r}=\mathbf{a}+t\mathbf{b}, \qquad \mathbf{r}=\mathbf{c}+p\mathbf{d}$$

to be coplanar.

The equation of the plane which contains the line $\mathbf{r}=\mathbf{a}+t\mathbf{b}$ and which is parallel to the vector **d** is

$$(\mathbf{r}-\mathbf{a})\,.\,(\mathbf{b}\times\mathbf{d})=0$$

The two lines will be coplanar if this plane passes through the point with position vector **c**, *i.e.*, if $[\mathbf{c}\;\mathbf{b}\;\mathbf{d}]=[\mathbf{a}\;\mathbf{b}\;\mathbf{d}]$. ...(*i*)

Cor. 1. Plane through two coplanar lines. Assuming the condition for coplanarity to be satisfied, we now find the equation of the plane through the two lines.

The plane which passes through the point with position vector **a** and which is parallel to the vectors **b** and **d** is the plane through the two coplanar lines. The equation of the required plane therefore is

$$(\mathbf{r}-\mathbf{a})\,.\,(\mathbf{b}\times\mathbf{d})=0 \quad\Leftrightarrow\quad \mathbf{r}\,.\,\mathbf{b}\times\mathbf{d}=[\mathbf{a}\;\mathbf{b}\;\mathbf{d}] \qquad ...(ii)$$

The condition (*i*) of coplanarity is actually the condition for the plane (*ii*) to contain the point **c** of the line $\mathbf{r}=\mathbf{c}+p\mathbf{d}$.

5.11. SHORTEST DISTANCE BETWEEN TWO LINES

To find the shortest distance between the lines

$$\mathbf{r}=\mathbf{a}+t\mathbf{b}, \quad \mathbf{r}=\mathbf{c}+p\mathbf{d}. \qquad (Kolkata, 99)$$

The line of shortest distance between two given lines is the line which meets the two lines perpendicularly.

Being perpendicular to each of the two given lines, the line of shortest distance *LM* is parallel to the vector

$$\mathbf{b}\times\mathbf{d}.$$

Also the length *LM* is the projection of $\overrightarrow{AC}$ upon $\overrightarrow{LM}$.

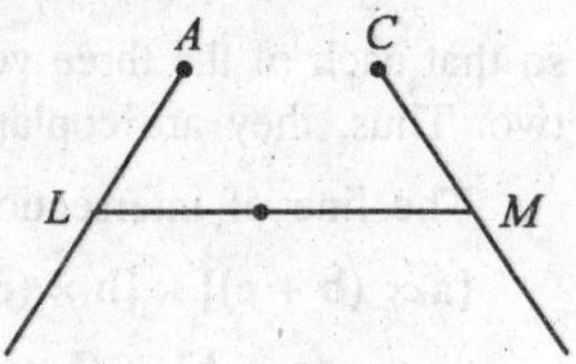

Fig. 5.15.

Thus

$$LM = \frac{|(\mathbf{c}-\mathbf{a}).(\mathbf{b}\times\mathbf{d})|}{|\mathbf{b}\times\mathbf{d}|} = \frac{|[\mathbf{c}\ \mathbf{b}\ \mathbf{d}]-[\mathbf{a}\ \mathbf{b}\ \mathbf{d}]|}{|\mathbf{b}\times\mathbf{d}|}.$$

Also the line of shortest distance is the line of intersection of the two planes through the given lines and the line LM. The equations of these planes are

$$(\mathbf{r}-\mathbf{a})\,.\,\mathbf{b}\times(\mathbf{b}\times\mathbf{d})=0, \quad (\mathbf{r}-\mathbf{c})\,.\,\mathbf{d}\times(\mathbf{b}\times\mathbf{d})=0.$$

The line of intersection of these planes is the required line of shortest distance.

EXAMPLES

Example 1. *Show that the planes through each of the three edges of a trihedral angle bisecting the internal angle between the other two are coaxal.*

Find also the common line in terms of the vectors along the three edges.

Solution. Suppose that OA, OB, OC are three edges of a trihedral angle at O. We suppose that

$$\overrightarrow{OA}=\mathbf{a},\ \overrightarrow{OB}=\mathbf{b},\ \overrightarrow{OC}=\mathbf{c}$$

are unit vectors.

Let OD, OE, OF bisect the internal angles between OA, OB and OC taken in pairs. Then the bisectors OD, OE and OF lie along the vectors

$$\mathbf{b}+\mathbf{c},\ \ \mathbf{c}+\mathbf{a},\ \ \mathbf{a}+\mathbf{b}$$

respectively. (Refer Ex. 5, Page 38-39). The equations of the planes

$$OAD,\quad OBE,\quad OCF$$

are

$$\mathbf{r}\,.\,\mathbf{a}\times(\mathbf{b}+\mathbf{c})=0,\quad \mathbf{r}\,.\,\mathbf{b}\times(\mathbf{c}+\mathbf{a})=0,\quad \mathbf{r}\,.\,\mathbf{c}\times(\mathbf{a}+\mathbf{b})=0.$$

These planes have a point, O, in common and will thus be coaxal if the normals to them are coplanar, *i.e.*, the vectors

$$\mathbf{a}\times(\mathbf{b}+\mathbf{c}),\ \ \mathbf{b}\times(\mathbf{c}+\mathbf{a}),\ \ \mathbf{c}\times(\mathbf{a}+\mathbf{b}) \qquad ...(i)$$

are coplanar. Now we have the identity

$$\mathbf{a}\times(\mathbf{b}+\mathbf{c})+\mathbf{b}\times(\mathbf{c}+\mathbf{a})+\mathbf{c}\times(\mathbf{a}+\mathbf{b})=0,$$

so that each of the three vectors in (*i*) is a linear combination of the other two. Thus, they are coplanar.

The line of intersection of these planes is parallel to the vector

$$[\mathbf{a}\times(\mathbf{b}+\mathbf{c})]\times[\mathbf{b}\times(\mathbf{c}+\mathbf{a})]=(\mathbf{a}\times\mathbf{b}+\mathbf{a}\times\mathbf{c})\times(\mathbf{b}\times\mathbf{c}+\mathbf{b}\times\mathbf{a})$$

$$=(\mathbf{a}\times\mathbf{b})\times(\mathbf{b}\times\mathbf{c})+(\mathbf{a}\times\mathbf{c})\times(\mathbf{b}\times\mathbf{c})+(\mathbf{a}\times\mathbf{c})\times(\mathbf{b}\times\mathbf{a})$$

$$=[\mathbf{a}\ \mathbf{b}\ \mathbf{c}]\,\mathbf{b}-[\mathbf{a}\ \mathbf{c}\ \mathbf{b}]\,\mathbf{c}-[\mathbf{a}\ \mathbf{c}\ \mathbf{b}]\,\mathbf{a}$$

$$= [\mathbf{a}\ \mathbf{b}\ \mathbf{c}]\ [\mathbf{a} + \mathbf{b} + \mathbf{c}].$$

Thus, the line of intersection is parallel to the vector

$$\mathbf{a} + \mathbf{b} + \mathbf{c}$$

where **a, b, c** are unit vectors along the three edges.

Example 2. *In each of the three planes determined by two of the lines*

$$OA,\ OB,\ OC$$

a straight line is drawn through O perpendicular to the third line, prove that the three lines so determined are coplanar.

Solution. Let $\overrightarrow{OA} = \mathbf{a},\ \overrightarrow{OB} = \mathbf{b},\ \overrightarrow{OC} = \mathbf{c}.$

The plane containing OA and OB is

$$\mathbf{r} \cdot \mathbf{a} \times \mathbf{b} = 0.$$

Any line in this plane is perpendicular to the vector

$$\mathbf{a} \times \mathbf{b}.$$

Thus, the line in this plane perpendicular to OC is parallel to the vector

$$(\mathbf{a} \times \mathbf{b}) \times \mathbf{c}$$

so that the three lines in question are

$$\mathbf{r} = t\,(\mathbf{a} \times \mathbf{b}) \times \mathbf{c},\quad \mathbf{r} = p\,(\mathbf{b} \times \mathbf{c}) \times \mathbf{a},\quad \mathbf{r} = k\,(\mathbf{c} \times \mathbf{a}) \times \mathbf{b}.$$

We now have the identity

$$(\mathbf{a} \times \mathbf{b}) \times \mathbf{c} + (\mathbf{b} \times \mathbf{c}) \times \mathbf{a} + (\mathbf{c} \times \mathbf{a}) \times \mathbf{b} = 0.$$

so that the three vectors

$$(\mathbf{a} \times \mathbf{b}) \times \mathbf{c},\quad (\mathbf{b} \times \mathbf{c}) \times \mathbf{a},\quad (\mathbf{c} \times \mathbf{a}) \times \mathbf{b}$$

are coplanar.

Thus, the three lines are coplanar.

Example 3. *l_1, l_2, l_3 and m_1, m_2, m_3 are two sets of lines such that the lines of shortest distances between l_1, l_2, l_3 taken in pairs are separately parallel to m_1, m_2, m_3; show that the lines of shortest distances between m_1, m_2, m_3 taken in pairs are also, separately parallel to l_1, l_2, l_3.*

Solution. Let the given lines be parallel to the vectors

$$\mathbf{a}_1, \mathbf{a}_2, \mathbf{a}_3 \text{ and } \mathbf{b}_1, \mathbf{b}_2, \mathbf{b}_3.$$

The given conditions are equivalent to

$$k_1\,\mathbf{b}_1 = \mathbf{a}_2 \times \mathbf{a}_3,\quad k_2\,\mathbf{b}_2 = \mathbf{a}_3 \times \mathbf{a}_1,\quad k_3\,\mathbf{b}_3 = \mathbf{a}_1 \times \mathbf{a}_2,$$

where k_1, k_2, k_3 are some scalars.

$$\therefore \quad k_2\,\mathbf{b}_2 \times k_3\,\mathbf{b}_3 = (\mathbf{a}_3 \times \mathbf{a}_1) \times (\mathbf{a}_1 \times \mathbf{a}_2)$$

$$= [\mathbf{a}_3\ \mathbf{a}_1\ \mathbf{a}_2]\ \mathbf{a}_1 = [\mathbf{a}_1\ \mathbf{a}_2\ \mathbf{a}_3]\ \mathbf{a}_1$$

so that the vector $\mathbf{a}_1$ is parallel to the line of shortest distance between the lines m_2 and m_3.

Hence the result.

Example 4. *A square PQRS is folded along the diagonal PR so that the planes PRQ and PRS are perpendicular to one another. Determine the shortest distance between PQ and RS in terms of p, the length of the side of the square.*

Solution. Let the new position of Q be Q'. Let **i, j, k** denote unit vector along PQ', $Q'R$, PS so that they form a right-handed orthonormal system.

Taking P as the origin of reference, the position vectors of Q', R and S are

$$p\mathbf{i},\quad p\mathbf{i} + p\mathbf{j},\quad p\mathbf{k}$$

respectively.

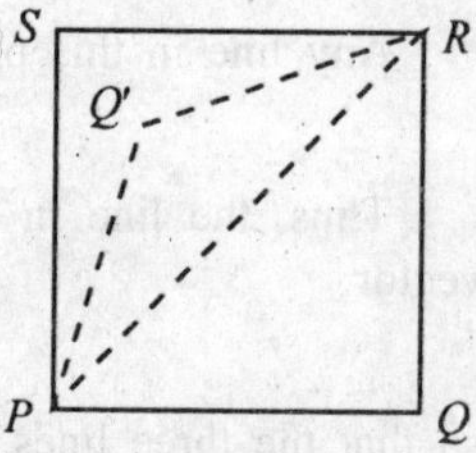

Fig. 5.16.

The equations of PQ' and RS are

$$\mathbf{r} = tp\mathbf{i},$$

$$\mathbf{r} = p\mathbf{i} + p\mathbf{j} + t\,(p\mathbf{i} + p\mathbf{j} - p\mathbf{k}).$$

$\therefore$ the required shortest distance

$$= \frac{|(p\mathbf{i} + p\mathbf{j})\,.\,p\mathbf{i} \times (p\mathbf{i} + p\mathbf{j} - p\mathbf{k})|}{|p\mathbf{i} \times (p\mathbf{i} + p\mathbf{j} - p\mathbf{k})|}$$

$$= \frac{|(p\mathbf{i} + p\mathbf{j})\,.\,(p^2\mathbf{k} + p^2\mathbf{j})|}{|p^2\mathbf{k} + p^2\mathbf{j}|} = \frac{p^3}{\sqrt{(p^4 + p^4)}} = \frac{p}{\sqrt{2}}.$$

Example 5. *OA, OB, OC are three coterminous edges of a rectangular parallelopiped and OA = x, OB = y, OC = z. Show that the shortest distance between OA and either diagonal skew to it is*

$$yz / \sqrt{(y^2 + z^2)}$$

and the shortest distances divide OA in the ratios

$$z^2 / y^2 \quad \text{and} \quad y^2 / z^2.$$

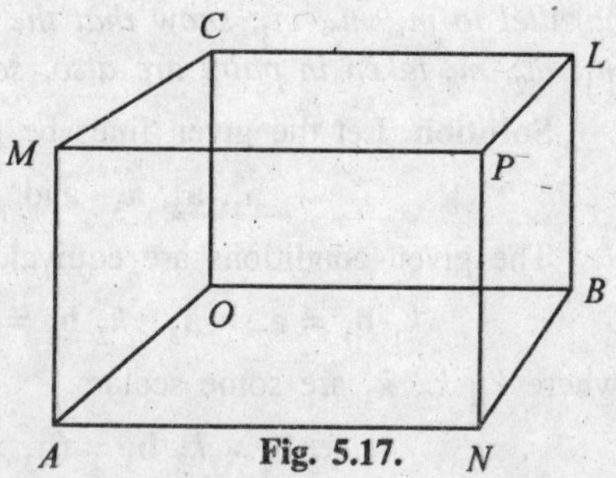

Fig. 5.17.

Solution. Let **i, j, k** denote unit vectors OA, OB, OC so that they form an orthonormal system.

The diagonals CN and BM are both skew to the edge OA.

We shall determine the shortest distance between OA and the diagonal CN.

The position vectors of A, C, N are

$$x\mathbf{i},\ z\mathbf{k},\ x\mathbf{i} + y\mathbf{j}$$

so that the equations of OA and CN are

$$\mathbf{r} = t\mathbf{i},\quad \mathbf{r} = z\mathbf{k} + p\,(x\mathbf{i} + y\mathbf{j} - z\mathbf{k}).$$

The length of the shortest distance between these lines is

$$= \frac{|(z\mathbf{k} - 0)\,.\,\mathbf{i} \times (x\mathbf{i} + y\mathbf{j} - z\mathbf{k})|}{|\mathbf{i} \times (x\mathbf{i} + y\mathbf{j} - z\mathbf{k})|}$$

$$= \frac{|z\mathbf{k}\,.\,(y\mathbf{k} + z\mathbf{j})|}{|y\mathbf{k} + z\mathbf{j}|} = \frac{yz}{\sqrt{(y^2 + z^2)}}.$$

Suppose now that the line of shortest distance meets the two lines in the points

$$t\mathbf{i},\quad z\mathbf{k} + p\,(x\mathbf{i} + y\mathbf{j} - z\mathbf{k})$$

so that their join is perpendicular to both the lines. Thus, we have

$$\begin{cases} [z\mathbf{k} + p(x\mathbf{i} + y\mathbf{j} - z\mathbf{k} - t\mathbf{i}]\,.\,\mathbf{i} = 0 \\ [z\mathbf{k} + p(x\mathbf{i} + y\mathbf{j} - z\mathbf{k}) - t\mathbf{i}]\,.\,(x\mathbf{i} + y\mathbf{j} - z\mathbf{k}) = 0 \end{cases}$$

$$\Rightarrow \quad \begin{cases} px - t = 0 \\ (px - t)\,x + py^2 - z\,(z - pz) = 0 \end{cases}$$

$$\Rightarrow \quad p = \frac{z^2}{y^2 + z^2},\quad t = \frac{xz^2}{y^2 + z^2}.$$

If the line of shortest distance meets OA in L', the position vector $\overrightarrow{OL'}$ of L' is

$$\frac{xz^2}{y^2 + z^2}\,\mathbf{i},$$

$$\Rightarrow \quad \overrightarrow{OL'} = xz^2 / (y^2 + z^2).$$

$$\therefore \quad LA = x - \frac{xz^2}{y^2 + z^2} = \frac{xy^2}{y^2 + z^2}$$

$$\Rightarrow \quad \frac{OL'}{L'A} = \frac{z^2}{y^2}.$$

We may also now consider the other diagonal BM and prove the corresponding result.

EXERCISES

1. Find the line through the point **a** parallel to the line of intersection of the planes
$$\mathbf{r} \cdot \mathbf{n}_1 = 1, \quad \mathbf{r} \cdot \mathbf{n}_2 = 1.$$
2. Find the line through the point **a** parallel to the planes
$$\mathbf{r} \cdot \mathbf{n}_1 = q_1, \quad \mathbf{r} \cdot \mathbf{n}_2 = q_2.$$
3. Find the line through the point **c** parallel to the plane $\mathbf{r} \cdot \mathbf{n} = 1$ and perpendicular to the line, $\mathbf{r} = \mathbf{a} + t\mathbf{b}$.
4. What is the equation of the line which passes through the point **a**, meets the line $\mathbf{r} = \mathbf{b} + t\mathbf{c}$ and is parallel to the plane $\mathbf{r} \cdot \mathbf{n} = 1$.
5. Find the plane which passes through a points **a** and is perpendicular to the two planes
$$\mathbf{r} \cdot \mathbf{n}_1 = q_1, \quad \mathbf{r} \cdot \mathbf{n}_2 = q_2.$$
6. Find the plane which passes through a point, **a**, is perpendicular to the plane $\mathbf{r} \cdot \mathbf{n} = q$ and is parallel to the line $\mathbf{r} = \mathbf{b} + t\mathbf{c}$.
7. Find the plane which passes through the two points **a**, **b** and is perpendicular to the plane $\mathbf{r} \cdot \mathbf{n} = q$.
8. Find the plane containing the line $\mathbf{r} = \mathbf{a} + t\mathbf{b}$ and perpendicular to the plane $\mathbf{r} \cdot \mathbf{n} = q$.
9. What is the equation of the plane which passes through the line of intersection of the planes
$$\mathbf{r} \cdot \mathbf{n}_1 = q_1, \quad \mathbf{r} \cdot \mathbf{n}_2 = q_2,$$
and is parallel to the line of intersection of the planes
$$\mathbf{r} \cdot \mathbf{n}_3 = q_3, \quad \mathbf{r} \cdot \mathbf{n}_4 = q_4.$$
10. What is the equation of the plane containing the parallel lines
$$\mathbf{r} = \mathbf{a} + t\mathbf{c}, \qquad \mathbf{r} = \mathbf{b} + p\mathbf{c}.$$
11. What is the condition for the lines
$$\mathbf{r} = \mathbf{a} + t\,(\mathbf{b} \times \mathbf{c}), \qquad \mathbf{r} = \mathbf{b} + p\,(\mathbf{c} \times \mathbf{a}).$$
to intersect. Assuming the condition satisfied, express the position vector of the point of intersection in terms of the non-coplanar vectors **a, b, c.**
12. Show that the lines
$$\mathbf{r} \times \mathbf{a} = \mathbf{b} \times \mathbf{a}, \quad \mathbf{r} \times \mathbf{b} = \mathbf{a} \times \mathbf{b}$$
intersect and find the point of intersection.
13. Find the equation of the line which passes through the point **a** and intersect the lines
$$\mathbf{r} = \mathbf{b} + t\mathbf{c}, \qquad \mathbf{r} = \mathbf{d} + p\mathbf{c}.$$
14. Find the equation of the plane through the line
$$\mathbf{r} = \mathbf{a} + t\mathbf{b},$$
and parallel to the line
$$\mathbf{r} = \mathbf{c} + p\mathbf{d},$$
and hence obtain the shortest distance between the two lines.
15. Show that the line of shortest distance between the lines
$$\mathbf{r} = \mathbf{a} + t\mathbf{b}, \qquad \mathbf{r} = \mathbf{c} + p\mathbf{d}$$
meet the lines at points given by
$$t = \frac{[\mathbf{d} \times (\mathbf{a} - \mathbf{c})] \cdot (\mathbf{b} \times \mathbf{d})}{(\mathbf{b} \times \mathbf{d})^2}, \quad p = \frac{[\mathbf{b} \times (\mathbf{a} - \mathbf{c})] \cdot (\mathbf{b} \times \mathbf{d})}{(\mathbf{b} \times \mathbf{d})^2}$$
16. The position vector of four points *A, B, C, D* relative to any origin *O*, are

denoted by a, b, c, d. Interpret geometrically the equations

(*i*) $(\mathbf{a} - \mathbf{b}) \times (\mathbf{c} \times \mathbf{d}) = 0$, (*ii*) $(\mathbf{a} - \mathbf{b}) \cdot (\mathbf{c} \times \mathbf{d}) = 0$.

17. OA, OB and OC are three concurrent lines and the lines OA_1, OB_1, OC_1 are normal to the planes BOC, COA and AOB respectively; show that the lines OA, OB, OC are also normal to the planes B_1OC_1, C_1OA_1 and A_1OB_1 respectively.

18. AA' is the common perpendicular to two skew lines PQA and $P'Q'A'$; P, Q being any two points on the first line and P', Q', any two points on the second. Prove that the common perpendicular of AA' and the line joining the mid-points of PP', QQ' bisects AA'.

19. If a, b, c, d are four lines in space and (ad) represents any plane parallel to a and d and so on and if (ad) is perpendicular to (bc) and (bd) is perpendicular to (ca), prove that (cd) is perpendicular to (ab).

20. OA, OB, OC are coterminous edges of a parallelopiped and P is the vertex opposite to O; show that the distance of O from the plane ABC is twice that of P from the same plane.

21. OA, OB and OC are three coterminous vectors. T_1 and T_2 are two parallelopipeds such that OA, OB, OC are coterminous edges of T_1 and are the altitudes through O of T_2; show that the product of the volumes of the two tetrahedron is

$$\alpha^2 \beta^2 \gamma^2$$

where α, β, γ are the lengths of OA, OB, OC.

22. OA, OB and OC are three mutually perpendicular straight lines and p is the length of the perpendicular from O to the plane ABC; show that

$$p^{-2} = a^{-2} + b^{-2} + c^{-2}$$

and the area of the $\Delta\, ABC$ is

$$\frac{1}{2}\sqrt{(b^2c^2 + c^2a^2 + a^2b^2)},$$

a, b, c being the lengths of OA, OB and OC respectively.

23. Two skew lines AP, BQ are met by the shortest distance between them at A, B and P, Q are points on them such that $AP = p$, $BQ = q$. If the planes APQ and BPQ are perpendicular, show that pq is constant.

24. Two skew lines AP, BQ inclined to one another at an angle of 60° are intersected by the shortest distance between them at A, B respectively and P, Q are points on the lines such that AQ is perpendicular to BP, prove that $AP \cdot BQ = 2AB^2$.

25. Points P and Q are taken on two skew lines so that PQ is always parallel to a given plane, show that the locus of a point R which divides PQ in a given ratio is a straight line.

26. A straight line intersects two skew lines. Show that the locus of a point which divides the intercept in a given ratio is a plane perpendicular to the line of shortest distance between the given lines. Find also the condition for the plane to bisect the line of shortest distance.

27. $ABCD$ is a skew quadrilateral. Points E, F, G and H are taken on AB, BC, CD and DA respectively such that

$AE : EB :: DG : GC$ and $BF : FC :: AH : HD$.

Show that the points EF and GH are coplanar.

28. l_1, l_2, l_3 are three given mutually skew lines m_1, m_2, m_3 are the lines of shortest distance between $l_2, l_3 : l_3, l_1$, and l_1, l_2 respectively. Show that the planes through any point P parallel to the pairs of lines l_1, m_1; l_2, m_2; l_3, m_3 are coaxal.

OBJECTIVE QUESTIONS

For each of the following questions, four alternatives are given for the answer. Only one of them is correct. Choose the correct alternative.

1. A vector that is perpendicular to both the vectors $\mathbf{a} = \mathbf{i} - 2\mathbf{j} + \mathbf{k}$ and $\mathbf{b} = \mathbf{i} - \mathbf{j} + \mathbf{k}$ is

(*a*) $-\mathbf{i} + \mathbf{k}$ (*b*) $-\mathbf{i} - 2\mathbf{j} + \mathbf{k}$

(*c*) $\mathbf{i} - 2\mathbf{j} + \mathbf{k}$ (*d*) $\mathbf{i} + \mathbf{k}$

2. Given $\mathbf{a} = \mathbf{i} + \mathbf{j} - \mathbf{k}$, $\mathbf{b} = -\mathbf{i} + 2\mathbf{j} + \mathbf{k}$ and $\mathbf{c} = -\mathbf{i} + 2\mathbf{j} - \mathbf{k}$, a unit vector perpendicular to both $\mathbf{a} + \mathbf{b}$ and $\mathbf{b} + \mathbf{c}$ is

(*a*) $\mathbf{i}$ (*b*) $\mathbf{j}$ (*c*) $\mathbf{k}$ (*d*) $(\mathbf{i} + \mathbf{j} + \mathbf{k})/\sqrt{3}$

3. If $\mathbf{a}$ is perpendicular to $\mathbf{b}$ and $\mathbf{c}$, then

(*a*) $\mathbf{a} \times (\mathbf{b} \times \mathbf{c}) = 1$ (*b*) $\mathbf{a} \times (\mathbf{b} \times \mathbf{c}) = \mathbf{0}$

(*c*) $\mathbf{a} \times (\mathbf{b} \times \mathbf{c}) = -1$ (*d*) None of these

4. $\mathbf{a} \times (\mathbf{b} + \mathbf{c}) + \mathbf{b} \times (\mathbf{c} + \mathbf{a}) + \mathbf{c} \times (\mathbf{a} + \mathbf{b})$ is equal to

(*a*) $2\,[\mathbf{a}\ \mathbf{b}\ \mathbf{c}]$ (*b*) $\mathbf{0}$

(*c*) 3 (*d*) None of these

5. If $\mathbf{a} \times \mathbf{b} = \mathbf{a} \times \mathbf{c}$, $\mathbf{a} \neq 0$, then

(*a*) $\mathbf{b} = \mathbf{c} + \lambda\mathbf{a}$ (*b*) $\mathbf{c} = \mathbf{a} + \lambda\mathbf{b}$

(*c*) $\mathbf{a} = \mathbf{b} + \lambda\mathbf{c}$ (*d*) None of these

6. If $\mathbf{a} \cdot \mathbf{b} = \mathbf{a} \cdot \mathbf{c}$ and $\mathbf{a} \times \mathbf{b} = \mathbf{a} \times \mathbf{c}$, then

(*a*) either $\mathbf{a} = 0$ or $\mathbf{b} = \mathbf{c}$ (*b*) $\mathbf{a}$ is parallel to $(\mathbf{b} - \mathbf{c})$

(*c*) $\mathbf{a}$ is perpendicular to $(\mathbf{b} - \mathbf{c})$ (*d*) None of these

7. If $\mathbf{a} \times \mathbf{b} = \mathbf{c}$ and $\mathbf{b} \times \mathbf{c} = \mathbf{a}$, then

(*a*) $\mathbf{a}, \mathbf{b}, \mathbf{c}$ are orthogonal in pairs and $|\mathbf{a}| = |\mathbf{c}|$ and $|\mathbf{b}| = 1$

(*b*) $\mathbf{a}, \mathbf{b}, \mathbf{c}$ are orthogonal to each other

(*c*) $\mathbf{a}, \mathbf{b}, \mathbf{c}$ are orthogonal in pairs but $|\mathbf{a}| \neq |\mathbf{c}|$

(*d*) $\mathbf{a}, \mathbf{b}, \mathbf{c}$ are orthogonal but $|\mathbf{b}| \neq 1$

8. The vector $\mathbf{a} \times (\mathbf{b} \times \mathbf{a})$ is :

(*a*) perpendicular to $\mathbf{a}$ (*b*) perpendicular to $\mathbf{b}$

(*c*) null vector (*d*) perpendicular to both $\mathbf{a}$ and $\mathbf{b}$

9. The triple product $(\mathbf{a} + \mathbf{c}) \cdot (\mathbf{b} + \mathbf{c}) \times (\mathbf{a} + \mathbf{b} + \mathbf{c})$ equals to

(*a*) $[\mathbf{a}\ \mathbf{b}\ \mathbf{c}]$ (*b*) $2\,[\mathbf{a}\ \mathbf{b}\ \mathbf{c}]$

(*c*) 0 (*d*) None of these

10. $(\mathbf{a} + 2\mathbf{b} - \mathbf{c}) \cdot (\mathbf{a} - \mathbf{b}) \times (\mathbf{a} - \mathbf{b} - \mathbf{c})$ equals to

(*a*) $[\mathbf{a}\ \mathbf{b}\ \mathbf{c}]$ (*b*) $-[\mathbf{a}\ \mathbf{b}\ \mathbf{c}]$

(*c*) $3\,[\mathbf{a}\ \mathbf{b}\ \mathbf{c}]$ (*d*) $-3\,[\mathbf{a}\ \mathbf{b}\ \mathbf{c}]$

11. **a . (b + c) × (a + b + c) =**

(*a*) 0 (*b*) 2 [**a b c**]

(*c*) [**a b c**] (*d*) None of these

12. Let **a** and 2**b** denote the diagonals of a parallelogram. Then area of the parallelogram is given by

(*a*) | **a** × **b** | (*b*) $\frac{1}{2}$ | **a** × **b** |

(*c*) 2 | **a** × **b** | (*d*) None of these

13. If **a, b, c** are three unit vectors, **b** × **c** and **a** × (**b** × **c**) = $\frac{1}{2}$ **b**, then angle between **a** and **c** is

(*a*) π / 6 (*b*) π / 2

(*c*) π / 3 (*d*) None of these

14. The value of $\left\{|\mathbf{a}\times\mathbf{b}|^2+(\mathbf{a}\cdot\mathbf{b})^2\right\}\div\left(\mathbf{a}^2\mathbf{b}^2\right)$ is

(*a*) unity (*b*) zero

(*c*) 2 (*d*) None of these

15. **(a – b) × (a + b) is equal to**

(*a*) 0 (*b*) **a** × **b**

(*c*) 2 (**a** × **b**) (*d*) | **a** $|^2$ + | **b** $|^2$

16. The unit vector perpendicular to each of the vectors 2**i** – **j** + **k** and 3**i** + 4**j** – **k** is

(*a*) – 3**i** + 5**j** + 11**k** (*b*) (– 3**i** + 5**j** + 11**k**) / 155

(*c*) (– 3**i** + 5**j** + 11**k**) / $\sqrt{(155)}$ (*d*) None of these

17. If θ is the angle between vectors **a** and **b**, and | **a** × **b** | = | **a . b** |, then θ is equal to

(*a*) 0° (*b*) 180° (*c*) 135° (*d*) 45°

18. If the vector **c, a** = *x***i** + *y***j** + *z***k** and **b** = **j** are such that **a, c** and **b** form a right handed system then **c** is

(*a*) *z***i** – *x***k** (*b*) **0**

(*c*) *y***j** (*d*) – *z***i** + *x***k**

19. The value of *x* such that the vectors **a** = 2**i** – **j** + **k**, **b** = *x***i** + 2**j** – 3**k** and **c** = 3**i** – 4**j** + 5**k** are coplanar is

(*a*) 1 (*b*) 3 (*c*) – 2 (*d*) 0

20. The value of *p* such that the vectors **i** + 3**j** – 2**k**, 2**i** – **j** + 4**k** and 3**i** + 2**j** + *p***k** are coplanar is

(*a*) 4 (*b*) 2 (*c*) 8 (*d*) 10

21. If **r** satisfies the equation **r** × (**i** + 2**j** + **k**) = **i** – **k**, then for any scalar *t*, **r** is equal to

(*a*) **i** + *t* (**i** + 2**j** + **k**) (*b*) **j** + *t* (**i** + 2**j** + **k**)

(*c*) **k** + *t* (**i** + 2**j** + **k**) (*d*) None of these

22. The area of the triangle whose two sides are given by $4\mathbf{i} - \mathbf{j} + \mathbf{k}$ and $3\mathbf{i} + \mathbf{j} - \mathbf{k}$ is

(*a*) $7\sqrt{2}$ (*b*) $14\sqrt{2}$ (*c*) $14/\sqrt{2}$ (*d*) $7/\sqrt{2}$

23. The area of the parallelogram whose adjacent sides are $2\mathbf{i} - 3\mathbf{k}$ and $4\mathbf{j} + 2\mathbf{k}$ is

(*a*) $2\sqrt{(14)}$ (*b*) $4\sqrt{(14)}$ (*c*) $16\sqrt{(14)}$ (*d*) $\sqrt{(14)}$

24. The area of the parallelogram having diagonals $\mathbf{a} = 3\mathbf{i} + \mathbf{j} - 2\mathbf{k}$ and $\mathbf{b} = \mathbf{i} - 3\mathbf{j} + 4\mathbf{k}$ is

(*a*) $10\sqrt{3}$ (*b*) $5\sqrt{3}$ (*c*) 8 (*d*) 4

25. If the non-zero vectors **a** and **b** are perpendicular to each other, then the solution of the equation $\mathbf{r} \times \mathbf{a} = \mathbf{b}$, is given by

(*a*) $\mathbf{r} = x\mathbf{a} + \dfrac{1}{\mathbf{a}\,.\,\mathbf{a}}(\mathbf{a} \times \mathbf{b});\ x \in R$

(*b*) $\mathbf{r} = x\mathbf{b} - \dfrac{1}{\mathbf{b}\,.\,\mathbf{b}}(\mathbf{a} \times \mathbf{b});\ x \in R$

(*c*) $\mathbf{r} = x\mathbf{a} \times \mathbf{b},\ x \in R$ (*d*) None of these

26. $[\mathbf{a} + \mathbf{b}, \mathbf{b} + \mathbf{c}, \mathbf{c} + \mathbf{a}] =$

(*a*) $[\mathbf{a}\ \mathbf{b}\ \mathbf{c}]$ (*b*) $\Sigma\,(\mathbf{a}\,.\,\mathbf{b})\,\mathbf{c}$

(*c*) $2\,[\mathbf{a}\ \mathbf{b}\ \mathbf{c}]$ (*d*) $|\mathbf{a}|\,|\mathbf{b}|\,|\mathbf{c}|$

27. The value of $[\mathbf{a} - \mathbf{b}, \mathbf{b} - \mathbf{c}, \mathbf{c} - \mathbf{a}]$ where $|\mathbf{a}| = 1$, $|\mathbf{b}| = 5$, $|\mathbf{c}| = 3$ is

(*a*) 0 (*b*) 1 (*c*) 6 (*d*) None of these

28. $[\mathbf{a}, \mathbf{b}, \mathbf{a} \times \mathbf{b}]$ is equal to

(*a*) $|\mathbf{a} \times \mathbf{b}|$ (*b*) $|\mathbf{a} \times \mathbf{b}|^2$

(*c*) $|\mathbf{a}\,.\,\mathbf{b}|$ (*d*) $|\mathbf{a}|\,|\mathbf{b}|$

29. $\mathbf{i} \times (\mathbf{a} \times \mathbf{i}) + \mathbf{j} \times (\mathbf{a} \times \mathbf{j}) + \mathbf{k} \times (\mathbf{a} \times \mathbf{k}) =$

(*a*) **a** (*b*) 2**a** (*c*) 3**a** (*d*) **0** (*U.P.P.C.S.* 2

30. If $\mathbf{u} = \mathbf{i} \times (\mathbf{a} \times \mathbf{i}) + \mathbf{j} \times (\mathbf{a} \times \mathbf{j}) + \mathbf{k} \times (\mathbf{a} \times \mathbf{k})$, then

(*a*) **u** is a unit vector (*b*) $\mathbf{u} = \mathbf{a} + \mathbf{i} + \mathbf{j} + \mathbf{k}$

(*c*) $\mathbf{u} = 2\mathbf{a}$ (*d*) $\mathbf{u} = 8\,(\mathbf{i} + \mathbf{j} + \mathbf{k})$

31. If $\mathbf{u} = \mathbf{a} - \mathbf{b}$, $\mathbf{v} = \mathbf{a} + \mathbf{b}$ and $|\mathbf{a}| = |\mathbf{b}| = 2$, then $|\mathbf{u} \times \mathbf{v}|$ is equal to

(*a*) $2\sqrt{\left[16 - (\mathbf{a} \cdot \mathbf{b})^2\right]}$ (*b*) $\sqrt{\left[16 - (\mathbf{a} \cdot \mathbf{b})^2\right]}$

(*c*) $2\sqrt{\left[4 - (\mathbf{a} \cdot \mathbf{b})^2\right]}$ (*d*) $\left[4 - (\mathbf{a} \cdot \mathbf{b})^2\right]$

32. For non-zero vectors **a, b, c**, $|(\mathbf{a} \times \mathbf{b})\,.\,\mathbf{c}| = |\mathbf{a}|\,|\mathbf{b}|\,|\mathbf{c}|$ holds if and only if (*a*) **a** . **b** = 0, **b** . **c** = 0 (*b*) **b** . **c** = 0, **c** . **a** = 0

(*c*) **c** . **a** = 0, **a** . **b** = 0 (*d*) **a** . **b** = **b** . **c** = **c** . **a** = 0

33. If **u** = **a** × (**b** × **c**) + **b** × (**c** × **a**) + **c** × (**a** × **b**) then

(*a*) **u** is a unit vector (*b*) **u** = **a** + **b** + **c**

(*c*) **u** = **0** (*d*) **u** ≠ **0**

34. The vector **a** lies in the plane of vectors **b** and **c**, which of the following is correct :

(*a*) **a** . (**b** × **c**) = 0 (*b*) **a** . (**b** × **c**) = 1

(*c*) **a** . (**b** × **c**) = – 1 (*d*) **a** . (**b** × **c**) = 3

35. Let *a, b, c* be distinct non-negative numbers. If the vectors $a\mathbf{i} + a\mathbf{j} + c\mathbf{k}$, **i** + **k** and $c\mathbf{i} + c\mathbf{j} + b\mathbf{k}$ lie in a plane, then **c** is

(*a*) the A.M. of *a* and *b* (*b*) the G.M. of *a* and *b*

(*c*) the H.M. of *a* and *b* (*d*) equal to zero

36. Let **a** = **i** + **j** and **b** = 2**i** – **k**, the point of intersection of the lines **r** × **a** = **b** × **a** and **r** × **b** = **a** × **b** is

(*a*) – **i** + **j** + **k** (*b*) 3**i** – **j** + **k**

(*c*) 3**i** + **j** + **k** (*d*) **i** – **j** – **k**

37. If the vectors $(-bc, b^2 + bc, c^2 + bc)$, $(a^2 + ac, -ac, c^2 + ac)$ and $(a^2 + ab, b^2 + ab, -ab)$ are coplanar, where none of *a, b* or *c* is zero, then

(*a*) $a^2 + b^2 + c^2 = 1$ (*b*) $bc + ca + ab = 0$

(*c*) $a + b + c = 0$ (*d*) $a^2 + b^2 + c^2 = bc + ca + ab$

ANSWERS

1. (*a*)	**2.** (*c*)	**3.** (*b*)	**4.** (*b*)	**5.** (*a*)
6. (*a*)	**7.** (*a*)	**8.** (*a*)	**9.** (*a*)	**10.** (*c*)
11. (*a*)	**12.** (*a*)	**13.** (*c*)	**14.** (*a*)	**15.** (*c*)
16. (*c*)	**17.** (*d*)	**18.** (*a*)	**19.** (*a*)	**20.** (*b*)
21. (*b*)	**22.** (*d*)	**23.** (*b*)	**24.** (*b*)	**25.** (*a*)
26. (*c*)	**27.** (*a*)	**28.** (*b*)	**29.** (*b*)	**30.** (*c*)
31. (*a*)	**32.** (*d*)	**33.** (*c*)	**34.** (*a*)	**35.** (*b*)
36. (*c*)	**37.** (*b*)			

6

Geometry with Cartesian Co-ordinates

Cartesian Equations. We shall now see how the properties of scalar products enable us to obtain various results in Cartesian Geometry, where points, lines and planes are given in terms of cartesian co-ordinates determined with reference to three mutually perpendicular concurrent lines taken as cartesian axes.

6.1. NORMAL FORM OF THE CARTESIAN EQUATION OF A PLANE

We have already shown in § 4.1, page 100 that if **n** be a unit vector normal to a given plane and p be the length of the perpendicular from the origin to the plane, then the vector equation of the plane is

$$\mathbf{r} \cdot \mathbf{n} = p.$$

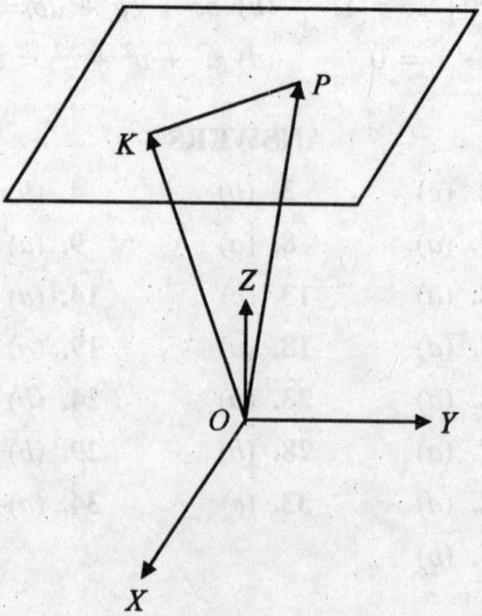

Fig. 6.1.

Substituting

$$\mathbf{r} = x\mathbf{i} + y\mathbf{j} + z\mathbf{k}; \quad \mathbf{n} = a\mathbf{i} + b\mathbf{j} + c\mathbf{k}$$

in the equation, we see that $ax + by + cz = p$ is the cartesian equation of the plane such that a, b, c are the direction cosines of the normal to the plane and p is the length of the perpendicular from the origin to the plane.

Note 1. We may see that $ax + by + cz = d$ is also the equation of a plane. Here, however, a, b, c are the direction ratios of the normal to the plane.

The coefficients a, b, c are proportional to the direction cosines of the normal to the plane whose equation is

$$ax + by + cz + d = 0.$$

The actual direction cosines are

$$a \Big/ \sqrt{\Sigma a^2},\ b \Big/ \sqrt{\Sigma a^2},\ c \Big/ \sqrt{\Sigma a^2}.$$

Note 2. Equation of a plane is of the first degree.

6.1.1. *To find the equation of the plane passing through the three points*

$$P_1(x_1, y_1, z_1),\quad P_2(x_2, y_2, z_2),\quad P_3(x_3, y_3, z_3).$$

We have

$$\vec{OP_1} = x_1\mathbf{i} + y_1\mathbf{j} + z_1\mathbf{k}, \text{ etc.,}$$

where O is the origin and **i, j, k** are unit vectors along the coordinate axes. If $P(x, y, z)$ be any point of the plane, we have

$$\vec{P_1P} \times \vec{P_1P_2} \cdot \vec{P_1P_3} = 0. \qquad \text{...}(i)$$

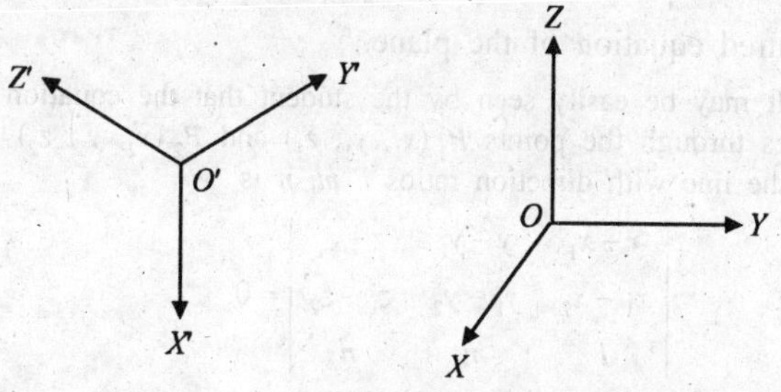

Fig 6.2.

Now

$$\vec{P_1P} = (x - x_1)\mathbf{i} + (y - y_1)\mathbf{j} + (z - z_1)\mathbf{k},$$

$$\vec{P_1P_2} = (x_2 - x_1)\mathbf{i} + (y_2 - y_1)\mathbf{j} + (z_2 - z_1)\mathbf{k},$$

$$\vec{P_1P_3} = (x_3 - x_1)\mathbf{i} + (y_3 - y_1)\mathbf{j} + (z_3 - z_1)\mathbf{k}.$$

By Cor. 3, Page 130 (*i*) gives the following required equation :

$$\begin{vmatrix} x - x_1 & y - y_1 & z - z_1 \\ x_2 - x_1 & y_2 - y_1 & z_2 - z_1 \\ x_3 - x_1 & y_3 - y_1 & z_3 - z_1 \end{vmatrix} = 0$$

6.1.2. *To find the equation of the plane which passes through P_1 (x_1, y_1, z_1) and which is parallel to two lines with direction ratios l_1, m_1, n_1 and l_2, m_2, n_2.*

Let P (x, y, z) be any point of the plane so that

$$\vec{OP} = x\mathbf{i} + y\mathbf{j} + z\mathbf{k}.$$

We have

$$\vec{OP_1} = x_1\mathbf{i} + y_1\mathbf{j} + z_1\mathbf{k}$$

so that

$$\vec{P_1P} = (x - x_1)\,\mathbf{i} + (y - y_1)\,\mathbf{j} + (z - z_1)\,\mathbf{k}.$$

Also we have

$$\mathbf{a} = l_1\mathbf{i} + m_1\mathbf{j} + n_1\mathbf{k}, \quad \mathbf{b} = l_2\mathbf{i} + m_2\mathbf{j} + n_2\mathbf{k}.$$

The vectors $\vec{P_1P}$, **a** and **b** being coplanar, the scalar product

$$\vec{P_1P} \times \mathbf{a} \,.\, \mathbf{b} = 0$$

so that

$$\begin{vmatrix} x - x_1 & y - y_1 & z - z_1 \\ l_1 & m_1 & n_1 \\ l_2 & m_2 & n_2 \end{vmatrix} = 0$$

is the required equation of the plane.

Note. It may be easily seen by the student that the equation of the plane which passes through the points P_1 (x_1, y_1, z_1) and P_2 (x_2, y_2, z_2) and which is parallel to the line with direction ratios l, m, n is

$$\begin{vmatrix} x - x_1 & y - y_1 & z - z_1 \\ x_1 - x_2 & y_1 - y_2 & z_1 - z_2 \\ l & m & n \end{vmatrix} = 0.$$

6.2. CARTESIAN EQUATIONS OF A LINE

To find the line which passes through a point A (x_1, y_1, z_1) having direction ratios $p, q. r.$

The position vector of the point A is $x_1\mathbf{i} + y_1\mathbf{j} + z_1\mathbf{k}$ and the vector $p\mathbf{i} + q\mathbf{j} + r\mathbf{k}$ is parallel to the given line. Replacing the vectors **a** and **b** by $x_1\mathbf{i} + y_1\mathbf{j} + z_1\mathbf{k}$ and $p\mathbf{i} + q\mathbf{j} + r\mathbf{k}$ respectively in the parametric equation

$$\mathbf{r} = \mathbf{a} + t\mathbf{b}$$

we obtain

$$x\mathbf{i} + y\mathbf{j} + z\mathbf{k} = (x_1\mathbf{i} + x_2\mathbf{j} + x_3\mathbf{k}) + t\,(p\mathbf{i} + q\mathbf{j} + r\mathbf{k})$$

$$\Rightarrow (x - x_1 - tp)\,\mathbf{i} + (y - y_1 - tq)\,\mathbf{j} + (z - z_1 - tr)\,\mathbf{k} = 0$$

$$\Rightarrow x - x_1 - tp = 0,\quad y - y_1 - tq = 0,\quad z - z_1 - tr = 0$$

$$\Rightarrow x = x_1 - tp,\quad y = y_1 - tq,\quad z = z_1 + tr$$

$$\Rightarrow \frac{x - x_1}{p} = \frac{y - y_1}{q} = \frac{z - z_1}{r}$$

which are the required equations of the line.

Cor. *To find the line through two points* $P_1(x_1, y_1, z_1)$ *and* $P_2(x_2, y_2, z_2)$.

We write

$$\overrightarrow{OP_1} = \mathbf{a} = x_1\mathbf{i} + y_1\mathbf{j} + z_1\mathbf{k}$$

$$\mathbf{b} = (x_2 - x_1)\,\mathbf{i} + (y_2 - y_1)\,\mathbf{j} + (z_2 - z_1)\,\mathbf{k}$$

so that **a** is the position vector of a point on the line and **b** is a vector parallel to the line. The required equations therefore are

$$\frac{x - x_1}{x_2 - x_1} = \frac{y - y_1}{y_2 - y_1} = \frac{z - z_1}{z_2 - z_1},$$

6.3. COPLANARITY OF LINES

To find the condition for the lines

$$\frac{x - x_1}{l_1} = \frac{y - y_1}{m_1} = \frac{z - z_1}{n_1},$$

$$\frac{x - x_2}{l_2} = \frac{y - y_2}{m_2} = \frac{z - z_2}{n_2},$$

to be coplanar.

In vector notation, we can rewrite these equations as

$$\mathbf{r} = \mathbf{a} + t\mathbf{b},\quad \mathbf{r} = \mathbf{c} + t\mathbf{d}$$

where

$$\mathbf{a} = x_1\mathbf{i} + y_1\mathbf{j} + z_1\mathbf{k},\quad \mathbf{b} = l_1\mathbf{i} + m_1\mathbf{j} + n_1\mathbf{k}.$$

$$\mathbf{c} = x_2\mathbf{i} + y_2\mathbf{j} + z_2\mathbf{k},\quad \mathbf{d} = l_2\mathbf{i} + m_2\mathbf{j} + n_2\mathbf{k}.$$

The condition for coplanarity is

$$(\mathbf{a} - \mathbf{c}) \cdot \mathbf{b} \times \mathbf{d} = 0$$

$$\begin{vmatrix} x_2 - x_1 & y_2 - y_1 & z_2 - z_1 \\ l_1 & m_1 & n_1 \\ l_2 & m_2 & n_2 \end{vmatrix} = 0.$$

Assuming the condition to be satisfied, the equation of the plane containing the two lines is

$$(\mathbf{a} - \mathbf{a}) \cdot \mathbf{b} \times \mathbf{d} = 0$$

$$\Leftrightarrow \begin{vmatrix} x-x_1 & y-y_1 & z-z_1 \\ l_1 & m_1 & n_1 \\ l_2 & m_2 & n_2 \end{vmatrix} = 0.$$

6.4. SHORTEST DISTANCE

Adopting the preceding notation, the length of shortest distance is

$$\frac{(\mathbf{c} - \mathbf{a}) \cdot \mathbf{b} \times \mathbf{d}}{|\mathbf{b} \times \mathbf{d}|}$$

where $$(\mathbf{c} - \mathbf{a}) \cdot \mathbf{b} \times \mathbf{d} = \begin{vmatrix} x_2-x_1 & y_2-y_1 & z_2-z_1 \\ l_1 & m_1 & n_1 \\ l_2 & m_2 & n_2 \end{vmatrix}$$

and $$\mathbf{b} \times \mathbf{d} = \Sigma\,(m_1 n_2 - m_2 n_1)\,\mathbf{i}$$

$$\Rightarrow \quad |\mathbf{b} \times \mathbf{d}| = \sqrt{\Sigma (m_1 n_2 - m_2 n_1)^2}.$$

Of course the required shortest distance is the absolute value of

$$\frac{(\mathbf{c} - \mathbf{a}) \cdot \mathbf{b} \times \mathbf{d}}{|\mathbf{b} \times \mathbf{d}|}.$$

6.5. CHANGE OF AXES

Let

$$OX,\ OY,\ OZ \text{ and } O'X',\ O'Y',\ O'Z'$$

be two sets of rectangular co-ordinate axes.

Let (f, g, h) be the co-ordinates of O' and let

$$l_1, m_1, n_1;\quad l_2, m_2, n_2;\quad l_3, m_3, n_3$$

be the direction ratios of OX', OY', OZ', with respect to OX, OY, OZ.

Let (x, y, z), (x', y', z') be the co-ordinates of a point P with respect to the two sets of axes.

If, $\mathbf{i}, \mathbf{j}, \mathbf{k} : \mathbf{i}', \mathbf{j}', \mathbf{k}'$ be unit vectors along the two sets of axes, we have

$$\overrightarrow{OP} = x\mathbf{i} + y\mathbf{j} + z\mathbf{k}.$$

$$\overrightarrow{O'P} = x'\mathbf{i}' + y'\mathbf{j}' + z'\mathbf{k}', \quad \overrightarrow{OO'} = f\,\mathbf{i} + g\mathbf{j} + h\mathbf{k},$$

$$\mathbf{i}' = l_1 \mathbf{i} + m_1 \mathbf{j} + n_1 \mathbf{k},$$

$$\mathbf{j}' = l_2\mathbf{i} + m_2\,\mathbf{j} + n_2\mathbf{k},$$
$$\mathbf{k}' = l_3\mathbf{i} + m_3\,\mathbf{j} + n_3\mathbf{k}.$$

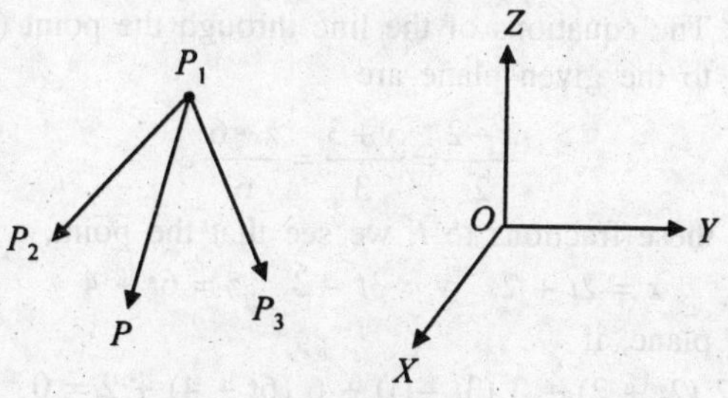

Fig. 6.3.

Now $\qquad \vec{OP} = \vec{OO'} + \vec{O'P}$

$\Rightarrow \qquad x\mathbf{i} + y\mathbf{j} + z\mathbf{k} = f\,\mathbf{i} + g\mathbf{j} + h\mathbf{k} + x'\mathbf{i}' + y'\mathbf{j}' + z'\mathbf{k}'$

$\Rightarrow \qquad (x - f)\,\mathbf{i} + (y - g)\,\mathbf{j} + (z - h)\,\mathbf{k} = x'\,(l_1\mathbf{i} + m_1\,\mathbf{j} + n_1\mathbf{k}) + y'\,(l_2\mathbf{i} + m_2\,\mathbf{j} + n_2\mathbf{k}) + z'\,(l_3\mathbf{i} + m_3\,\mathbf{j} + n_3\mathbf{k})$

$$\Rightarrow \qquad \begin{cases} x = f + x'\,l_1 + y'\,l_2 + z'\,l_3, \\ y = g + x'\,m_1 + y'\,m_2 + z'\,m_3, \\ z = h + x'\,n_1 + y'\,n_2 + z'\,n_3, \end{cases}$$

which are the required formulae for the transformation of co-ordinates from one set of axes to another.

EXAMPLES

Example 1. *If OX, OY, OZ and O′ X′, O′ Y′, O′ Z′ are two sets of rectangular co-ordinate axes and*

$$l_1, m_1, n_1; \quad l_2, m_2, n_2; \quad l_3, m_3, n_3$$

denote the direction cosines of the members of either set with respect to other, then

$$\begin{vmatrix} l_1 & m_1 & n_1 \\ l_2 & m_2 & n_2 \\ l_3 & m_3 & n_3 \end{vmatrix} = 1.$$

Solution. If **i, j, k**; **i′, j′, k′** denote unit vectors along the axes, we have

$$\mathbf{i}' = l_1\mathbf{i} + m_1\,\mathbf{j} + n_1\mathbf{k},$$
$$\mathbf{j}' = l_2\mathbf{i} + m_2\,\mathbf{j} + n_2\mathbf{k},$$
$$\mathbf{k}' = l_3\mathbf{i} + m_3\,\mathbf{j} + n_3\mathbf{k}.$$

Also we have

$$[\mathbf{i}'\,\mathbf{j}'\,\mathbf{k}'] = \begin{vmatrix} l_1 & m_1 & n_1 \\ l_2 & m_2 & n_2 \\ l_3 & m_3 & n_3 \end{vmatrix} = [\mathbf{i}\,\mathbf{j}\,\mathbf{k}]$$

As [**i j k**] = [**i**′, **j**′, **k**′], we have the required result.

Example 2. *Find the reflection of the point (2, – 3, 4) in the plane $2x + 3y + 5z + 2 = 0$.*

Solution. The equations of the line through the point (2, – 3, 4) and perpendicular to the given plane are

$$\frac{x-2}{2}=\frac{y+3}{3}=\frac{z-6}{6}$$

Equating these fractions to t, we see that the point,

$$x = 2t + 2, \quad y = 3t - 3, \quad z = 6t + 4 \qquad ...(i)$$

lies on given plane, if

$$2\,(2t + 2) + 3\,(3t - 3) + 6\,(6t + 4) + 2 = 0$$

$$4t + 4 + 9t - 9 + 36t + 24 + 2 = 0$$

$$\Rightarrow \qquad 49t = -21$$

$$t = \frac{-3}{7}.$$

Putting this value of t in (i) we see that the foot of the perpendicular is

$$2\left(\frac{-3}{7}\right)+2,\ 3\left(\frac{-3}{7}\right)-3,\ 6\left(\frac{-3}{7}\right)+4, \text{ i.e., } \left(\frac{8}{7}, \frac{-30}{7}, \frac{10}{7}\right)$$

i.e., $$\left(\frac{8}{7}, \frac{-30}{7}, \frac{10}{7}\right).$$

If (a, b, c) be the reflection P of the given point in the given plane, then the mid-point of AP, *viz.*,

$$\frac{a+2}{2}, \frac{b-3}{2}, \frac{c+4}{2}$$

must be the foot of the perpendicular.

$$\therefore \qquad \frac{a+2}{2}=\frac{8}{7}, \frac{b-3}{2}=\frac{-30}{7}, \frac{c+4}{2}=\frac{10}{7}$$

$$\Rightarrow \qquad a=\frac{16}{7}-2,\ b=-\frac{60}{7}+3,\ c=\frac{20}{7}-4$$

$$\Rightarrow \qquad a=\frac{2}{7},\ b=\frac{-39}{7},\ c=\frac{-8}{7}.$$

Thus $$\left(\frac{2}{7}, \frac{-39}{7}, \frac{-8}{7}\right)$$

is the required reflection of the given point in the given plane.

Example 3. *Find the reflection of the point A (1, 0, 0) in the line*

$$\frac{x-1}{2}=\frac{y+1}{-3}=\frac{z+10}{8}.$$

Solution. Any point P on the line is

$$2t+1, \quad -3t-1, \quad 8t-10. \qquad ...(i)$$

The line AP with direction ratios $2t$, $-3t-1$, $8t-10$ will be perpendicular to the given line, if

$$2(2t) - 3(-3t-1) + 8(8t-10) = 0$$

$$\Rightarrow \quad 4t + 9t + 3 + 64t - 80 = 0$$

$$\Rightarrow \quad 77t - 77 = 0 \Rightarrow t = 1.$$

Putting this value of t in the co-ordinates (i) r of the point P we see that the co-ordinates of the foot of the perpendicular, P are $(3, -4, -2)$.

If $B(a, b, c)$ be the reflection of the point $A(1, 0, 0)$, then the mid-point $\left(\frac{a+1}{2}, \frac{b}{2}, \frac{c}{2}\right)$ of AB must be the foot P of the perpendicular from the point on A the given line so that we have

$$\frac{a+1}{2} = 3, \quad \frac{b}{2} = -4, \quad \frac{c}{2} = -2$$

$$\Rightarrow \quad a = 5, \quad b = -8, \quad c = -4.$$

Thus, $(5, -8, -4)$ is the reflection of the given point in the given line.

Example 4. *Find the equation of the plane through the points A, B, C whose rectangular cartesian co-ordinates are*

$$(1, 1, 1), \quad (1, -1, 1), \quad (-1, -3, -5).$$

Solution. The position vectors of these points are

$$\mathbf{i}+\mathbf{j}+\mathbf{k}, \quad \mathbf{i}-\mathbf{j}+\mathbf{k}, \quad -\mathbf{i}-3\mathbf{j}+5\mathbf{k}.$$

respectively.

Let P

$$x\mathbf{i} + y\mathbf{j} + z\mathbf{k}$$

be any point on the plane. We have

$$(\vec{AB} \times \vec{AC}) \cdot \vec{AP} = 0$$

$$\Leftrightarrow \quad [(-2\mathbf{j} \times (-2\mathbf{i} - 4\mathbf{j} - 6\mathbf{k})] \cdot [(x-1)\,\mathbf{i} + (y-1)\,\mathbf{j} + (z-1)\,\mathbf{k}] = 0$$

$$\Leftrightarrow \quad (12\mathbf{i} - 4\mathbf{k}) \cdot [(x-1)\,\mathbf{i} + (y-1)\,\mathbf{j} + (z-1)\,\mathbf{k}] = 0$$

$$\Leftrightarrow \quad 12(x-1) - 4(z-1) = 0$$

$$\Leftrightarrow \quad 3x - z - 2 = 0.$$

which is the required equation.

Example 5. *Show that the points whose rectangular cartesian co-ordinates are*

$$(-6, 3, 2), \quad (3, -2, 4), \quad (5, 7, 3), \quad (-13, 17, -1)$$

are coplanar.

Solution. In the usual notation we have

$$\vec{OA} = -6\mathbf{i} + 3\mathbf{j} + 2\mathbf{k}, \quad \vec{OB} = 3\mathbf{i} - 2\mathbf{j} + 4\mathbf{k},$$

$$\vec{OC} = 5\mathbf{i}+7\mathbf{j}+3\mathbf{k},\ \vec{OD} = -13\mathbf{i}+17\mathbf{j}-\mathbf{k},$$

Thus,

$$\vec{AB} = 9\mathbf{i}-5\mathbf{j}+2\mathbf{k},\ \vec{AC} = 11\mathbf{i}+4\mathbf{j}+\mathbf{k},\ \vec{AD} = -7\mathbf{i}+14\mathbf{j}-3\mathbf{k}.$$

Finally we have

$$\vec{AB} \times \vec{AC} \cdot \vec{AD} = (9\mathbf{i}-5\mathbf{j}+2\mathbf{k}) \times (11\mathbf{i}+4\mathbf{j}+\mathbf{k}) \cdot (-7\mathbf{i}+14\mathbf{j}-3\mathbf{k}) = 0$$

$\Rightarrow$ $\vec{AB}, \vec{AC}, \vec{AD}$ are coplanar vectors.

$\Rightarrow$ A, B, C, D are coplanar points.

Example 6. *Find the length of the shortest distance between the lines*

$$\frac{x+3}{-4} = \frac{y-6}{3} = \frac{z}{2};\ \frac{x+2}{-4} = \frac{y}{1} = \frac{z-7}{1}.$$

Solution. The line $\dfrac{x+3}{-4} = \dfrac{y-6}{3} = \dfrac{z}{2}$ passes through the point (– 3, 6, 0), *i.e.*, through the point with position vector – **3i** + **6j** and is parallel to the line with direction ratios – 4, 3, 2, *i.e.*, parallel to the vector – **4i** + **3j** + **2k**.

The second line $\dfrac{x+2}{-4} = \dfrac{y}{1} = \dfrac{z-7}{1}$ passes through the point (– 2, 0, 7), *i.e.*, through the point that position vector – **2i** + **7k** and is parallel to the line with direction ratios – 4, 1, 1, *i.e.*, to the vector – **4i** + **j** + **k**.

The line of shortest distance is parallel to the line perpendicular to the given lines, *i.e.*, it is parallel to the vector

$$(-4\mathbf{i} + 3\mathbf{j} + 2\mathbf{k}) \times (-4\mathbf{i} + \mathbf{j} + \mathbf{k})$$

$$= \begin{vmatrix} \mathbf{i} & \mathbf{j} & \mathbf{k} \\ -4 & 3 & 2 \\ -4 & 1 & 1 \end{vmatrix} = \mathbf{i}-4\mathbf{j}+8\mathbf{k}.$$

The length of the shortest distance is the projection on this line of the join of the points on the two lines, *i.e.*, of the vector

$$= (-3\mathbf{i} + 6\mathbf{j}) - (-2\mathbf{i} + 7\mathbf{k}) = -\mathbf{i} + 6\mathbf{j} - 7\mathbf{k}.$$

Thus, the required length of the shortest distance is

$$\frac{(-\mathbf{i}+6\mathbf{j}-7\mathbf{k})\cdot(\mathbf{i}-4\mathbf{j}+8\mathbf{k})}{|\mathbf{i}-4\mathbf{j}+8\mathbf{k}|} = \frac{-81}{9} = -9.$$

Thus, the shortest distance between the given two lines is 9.

EXERCISES

1. Find the reflections of the points

(*i*) (– 5, 11, 6) in the plane $5x - 6y - 2z - 7 = 0$;

(*ii*) (3, 5, 8) in the plane $3x - y - z + 26 = 0$;

(*iii*) (5, – 9, – 6) in the plane $2x - y + 3z - 8 = 0$.

2. Find the feet of the perpendiculars from the origin to the lines

(*i*) $$\frac{x-3}{3} = \frac{y-8}{-4} = \frac{z-3}{-2}$$

(*ii*) $$\frac{x+1}{2} = \frac{y}{1} = \frac{z-4}{1}$$

(*iii*) $$\frac{x-3}{1} = \frac{1-y}{2} = \frac{z-5}{1}.$$

Find also the reflections in each case.

3. Find the reflection of the point (7, – 1, 2) in the line

$$\frac{x-9}{1} = \frac{y-5}{3} = \frac{z-5}{5}.$$

4. Given a triangle with vertices (1, 2, 3), (2, 1, 4), (5, 1, 3). Show that the orthocentre of this triangle is

$$\frac{1}{13}(23\mathbf{i} + \mathbf{j} + 9\mathbf{k}).$$

5. Examine if the following sets of points are coplanar :

(*a*) (3, 2, – 5), (– 3, 8, – 5), (– 3, 2, 1), (– 1, 4, – 3).

(*b*) (6, – 4, 4), (0, 0, – 4), (– 1, – 2, – 3), (1, 2, – 5).

(*c*) (– 2, – 1, 0), (1, – 2, – 1), (2, 1, 4), (0, 1, 0).

6. Find the equations of the planes through the points :

(*i*) (– 2, – 2, 2), (1, 1, 1), (1, – 1, 2)

(*ii*) (– 6, 3, 2), (3, – 2, 4), (5, 7, 3).

7. By vector method prove that the shortest distance between the straight line

$$y + z = 0, \quad z + x = 0$$

and the straight line

$$x + y = 0, \quad x + y + z - a = 0$$

is $\sqrt{6a/3}$.

8. A tetrahedron has vertices

$$A\ (0, 0, 0),\ B\ (1, 1, 0),\ C\ (0, 1, -1),\ D\ (1, 0, -1)$$

Show that the line joining the origin to the point (1, 1, 1) is perpendicular to the face *BCD*.

9. $A = (0, 1, 2)$, $B = (3, 0, 1)$, $C = (4, 3, 6)$, $D = (2, 3, 2)$ are the rectangular cartesian co-ordinates of four points. Find

(*i*) the area of the triangle *ABC*.

(*ii*) the perpendicular distance from *A* to the line *BC*.

(*iii*) the volume of the tetrahedron *ABCD*.

(*iv*) the perpendicular distance from *D* to the plane *ABC*.

(*v*) the shortest distance between the lines *AB* and *CD*.

10. Find the volume of the tetrahedron formed by the planes

$$my + nz = 0, \quad nz + lx = 0, \quad lx + my = 0,$$
$$lx + my + nz = p.$$

11. Find the equation of the plane which contains the line

$$x = 2, \quad y - z = 0$$

and is perpendicular to the plane $x + z = 3$.

Find the point where this plane meets the line

$$\frac{x}{2} = \frac{y}{3} = \frac{z}{1}.$$

12. Prove that a plane normally bisecting a diagonal of a cube meets the edges of the cube in points which are the vertices of a regular hexagon.

13. Prove that the lines

$$\frac{x+1}{3} = \frac{y+3}{5} = \frac{z+5}{7}; \quad \frac{x-2}{1} = \frac{y-4}{3} = \frac{z-6}{5}.$$

intersect. Find their point of intersection and the plane in which they lie.

14. Prove that the lines

$$\frac{x-a}{a'} = \frac{y-b}{b'} = \frac{z-c}{c'} \text{ and } \frac{x-a'}{a} = \frac{y-b'}{b} = \frac{z-c'}{c}$$

intersect and find the co-ordinates of the point of intersection and the equation of the plane in which they lie.

15. Prove that the lines

$$\frac{x-1}{2} = \frac{y+1}{-3} = \frac{z+10}{8}; \quad \frac{x-4}{1} = \frac{y+3}{-4} = \frac{z+1}{7}$$

intersect. Find also their point of intersection and the plane through them.

16. Find the magnitude and the equations of the line of shortest distance between the two lines :

(*i*) $\dfrac{x-3}{2} = \dfrac{y+15}{-7} = \dfrac{z-9}{5}; \quad \dfrac{x+1}{2} = \dfrac{y-1}{1} = \dfrac{z-9}{-3}.$

(*ii*) $\dfrac{x-3}{-1} = \dfrac{y-4}{2} = \dfrac{z+2}{1}; \quad \dfrac{x-1}{1} = \dfrac{y+7}{3} = \dfrac{z+2}{2}.$

17. Obtain the co-ordinates of the points where the shortest distance between the lines

$$\frac{x-23}{-6} = \frac{y-19}{-4} = \frac{z-25}{3}; \quad \frac{x-12}{-9} = \frac{y-1}{4} = \frac{z-5}{2}.$$

meets them.

18. Find the shortest distance between the lines

$$\frac{x-1}{2} = \frac{y-2}{3} = \frac{z-3}{4}; \quad \frac{x-2}{3} = \frac{y-3}{4} = \frac{z-4}{5}.$$

Show also that the lines are coplanar.

MISCELLANEOUS EXERCISES II

1. Show that the perpendicular distance of any point **a** from the line $\mathbf{r} = \mathbf{b} + t\mathbf{c}$ is
$$|(\mathbf{b} - \mathbf{a}) \times \mathbf{c}| / |\mathbf{c}|.$$

2. Two triangles of equal area are on the opposite sides of the same base. Prove that the straight line joining their vertices is bisected by the base.

3. If **a, b, c** are three vectors such that $\mathbf{a} + \mathbf{b} + \mathbf{c} = 0$, prove that
$$\mathbf{a} \times \mathbf{b} = \mathbf{b} \times \mathbf{c} = \mathbf{c} \times \mathbf{a}$$
and interpret the result geometrically.

4. A', B', C', D' are the projections of the four coplanar points A, B, C, D on any given plane; show that the volumes of the tetrahedron $AB'C'D'$ and $A'BCD$ are equal.

5. Lines are drawn in a given direction through the vertices O, A, B, C of a tetrahedron $OABC$ to meet the opposite faces in O', A', B', C'. Prove that
$$\text{Volume } O', A', B', C' = 3 \text{ Volume } OABC.$$

6. Show that the volume of a tetrahedron $ABCD$ is
$$\frac{1}{6} |\vec{AB} \times \vec{CD} \cdot \vec{AC}|$$
and deduce that if the lengths of two opposite edges of a tetrahedron are a, b and the angle between them is α and the shortest distance between the same is $2c$, then the volume of the tetrahedron is $\frac{1}{3} abc \sin \alpha$.

7. Two lines AB and CD of fixed lengths move along given straight lines; show that the volume of the tetrahedron remains constant.

Deduce that the volume of a tetrahedron $ABCD$ is
$$\frac{1}{6} |\vec{AB} \times \vec{CD} \cdot \vec{LM}|,$$
where LM is the line of shortest distance between the edges AB and CD.

8. A, B, C, D are any four points, show that
$$\vec{AB} \cdot \vec{CD} + \vec{BC} \cdot \vec{AD} + \vec{CA} \cdot \vec{BD} = 0.$$

9. $ABCD$ is a quadrilateral, show that
$$AC^2 + BD^2 = AB^2 + CD^2 + 2\,\vec{BC} \cdot \vec{AD}.$$

10. $OABC$ is a tetrahedron and the position vectors of A, B, C with respect to O as origin are **a, b, c** respectively.

Write down the position vectors of the mid-points P, Q, R, L, M, N of OA, OB, OC, BC, CA, AB respectively.

Determine the vector equations of the lines PL and QM and show that these lines intersect and find the position vector of the point of intersection.

Deduce that the lines PL, QM and RN are concurrent.

If a tetrahedron OABC is regular, that is all six edges have the same lengths, show that

(*i*) $\mathbf{a} \cdot \mathbf{b} = \mathbf{b} \cdot \mathbf{c} = \mathbf{c} \cdot \mathbf{a}$.

(*ii*) *PL* and *BC* are perpendicular.

11. Given that $\mathbf{p} = \mathbf{a} + \mathbf{b}$ and $\mathbf{q} = \mathbf{a} - \mathbf{b}$ and $|\mathbf{a}| = |\mathbf{b}| = k$ show that $\mathbf{p} \cdot \mathbf{q} = 0$ and that

$$|\mathbf{p} \times \mathbf{q}| = 2\,[k^4 - (\mathbf{a} \cdot \mathbf{b})^2]^{1/2}.$$

12. The point *A* is (2, 1, 6) and has position vector **a** relative to the origin *O*; *B* is (1, 3, 5) and has position vector **b**. The line *l* is given by $\mathbf{r} = \mathbf{b} + s\,(\mathbf{i} + 2\mathbf{j} + 3\mathbf{k})$ and the line *m* is given by $\mathbf{r} = \mathbf{a} + t\,(\mathbf{i} + 6\mathbf{j} + 5\mathbf{k})$. Show that the lines *l* and *m* intersect, and state the co-ordinates of the common point. Prove that *AB* is perpendicular to *l*, and hence write down the co-ordinates of the reflection of *A* in *l*.

13. Show that there is one and only one value of *p* for which the vectors

$$(p + 1)\,\mathbf{i} - 3\mathbf{j} + p\mathbf{k}, \quad p\mathbf{i} + (p + 1)\,\mathbf{j} - 3\mathbf{k}, \quad -3\mathbf{i} + p\mathbf{j} + (p + 1)\,\mathbf{k}$$

are linearly dependent and find this value of *p*.

14. Consider these two lines

$$\mathbf{r} = \mathbf{a} + \lambda\mathbf{l}, \qquad \mathbf{r} = \mathbf{b} + \mu\mathbf{m},$$

where

$$\mathbf{r} = 5\mathbf{i} + \mathbf{j} + 2\mathbf{k}, \qquad \mathbf{b} = -\mathbf{i} + 7\mathbf{j} + 8\mathbf{k},$$
$$\mathbf{l} = -4\mathbf{i} + \mathbf{j} - \mathbf{k}, \qquad \mathbf{m} = 2\mathbf{i} - 5\mathbf{j} - 7\mathbf{k}.$$

Show that the two lines intersect and find the position vector of the point *P* of their intersection. If the position vector of a point *Q* is $3\mathbf{i} + 7\mathbf{j} - 2\mathbf{k}$, show that *PQ* is perpendicular to *AB* where *A* and *B* are the points with position vectors **a** and **b**.

15. If $\mathbf{a} = \frac{1}{\sqrt{3}}(\mathbf{i} + \mathbf{j} + \mathbf{k})$, $\mathbf{b} = \sqrt{\left(\frac{2}{3}\right)}\left(\mathbf{i} - \frac{1}{2}\mathbf{j} - \frac{1}{2}\mathbf{k}\right)$ and

$$\mathbf{c} = \frac{1}{\sqrt{2}}(\mathbf{j} - \mathbf{k}).$$

show that **a**, **b** and **c** are mutually perpendicular unit vectors and express **i** in the form $p\mathbf{a} + q\mathbf{b} + r\mathbf{c}$.

16. Find the unit vectors which are perpendicular to both the vectors $\mathbf{i} + 4\mathbf{j}$ and $2\mathbf{i} + 4\mathbf{j} + 3\mathbf{k}$. Find also the angles between these unit vectors and the vector $\frac{1}{2}(\mathbf{i} + \mathbf{j} - \mathbf{k})$.

17. *ABCD* is a tetrahedron; *X*, *Y*, *Z* are the mid-points of *AB*, *AC*, *AD* and *P*, *Q*, *R*, are the mid-points of *CD*, *BD*, *BC*.

Prove that

$$AB^2 + BC^2 + 2PX^2 = AC^2 + BD^2 + 2QY^2 = AD^2 + BC^2 + 2RZ^2.$$

18. Find all the sets of vectors of **a**, **b** and **c** that satisfy the following conditions :

a is parallel to $\mathbf{i} + \mathbf{j}$ and $|\mathbf{a}| = 3$;

b is parallel to $\mathbf{i} - \mathbf{j}$ and $|\mathbf{b}| = 1$;

c satisfies the equation $\mathbf{a} = \mathbf{b} \times \mathbf{c}$ and $|\mathbf{c}| = 9$.

(Parallelism includes the case of opposite directions also.)

19. Find the general form of the vector x which satisfies the equation

$$x \times (\mathbf{i} + 2\mathbf{j} + \mathbf{k}) = \mathbf{i} - \mathbf{k}.$$

20. Points A, B, C have position vectors $\mathbf{a}$, $\mathbf{b}$, $\mathbf{c}$ and p, q, r are variable parameters subject to the condition $p + q + r = 1$. If the points are not collinear, prove that the plane ABC is represented by the equation $\mathbf{r} = p\mathbf{a} + q\mathbf{b} + r\mathbf{c}$.

Prove that the equation of the line of intersection of the two planes

$$\mathbf{r} = p_1\mathbf{i} + 2q_1\mathbf{j} + 3r_1\mathbf{k}, \quad p_1 + q_1 + r_1 = 1$$

and

$$\mathbf{r} = 2p_2\mathbf{i} + q_2\mathbf{j} + 2r_2\mathbf{k}, \quad p_2 + q_2 + r_2 = 1$$

can be written in terms of a single parametric t as

$$6\mathbf{r} = (3 + t)\,\mathbf{i} + 4t\mathbf{j} + 9\,(1 - t)\,\mathbf{k}.$$

21. The position vectors of the vertices A, B, C of a tetrahedron $OABC$ with respect to O as origin are

$$\vec{OA} = 2\mathbf{i} + \mathbf{j},\ \vec{OB} = \mathbf{j} + \mathbf{k},\ \vec{OC} = \mathbf{i} + 3\mathbf{j} - \mathbf{k}.$$

Find the angle between (*a*) the edges AB, CD (*b*) the faces OAB, OAC.

Prove that BC is perpendicular to the plane OAB and hence prove that the volume of the tetrahedron $OABC$ is 3/2.

22. Prove that the lines with vector equations

$$\mathbf{r} = (-\mathbf{i} + 2\mathbf{j}) + a\,(-\mathbf{i} + 2\mathbf{j} - \mathbf{k}).$$
$$\mathbf{r} = \mathbf{k} + b\,(\mathbf{i} + \mathbf{j} - \mathbf{k}),$$
$$\mathbf{r} = \mathbf{i} + \mathbf{j} - \mathbf{k} + c\,(\mathbf{i} + \mathbf{k})$$
$$\mathbf{r} = 2\mathbf{i} + \mathbf{j} + d\,(3\mathbf{i} - \mathbf{j}).$$

in the given order form a (skew) quadrilateral.

Prove that the ratio of the shortest distances between the two pairs of opposite sides of this quadrilateral is $\sqrt{7:1}$.

23. Referred to cartesian axes, A is the point (1, 2). If $\vec{AB} = 3\mathbf{i} - \mathbf{j}$ and $\vec{AC} = 3\mathbf{i} + 4\mathbf{j}$, write down the co-ordinates of the points B and C.

E and F are points such that $\vec{BE} = \frac{1}{2}\vec{AC}$ and $\vec{BF} = \vec{AB}$. Show that $\vec{CE} = \vec{EF}$ and state what can be deduced about the points C, E, F.

24. The line $\mathbf{r} = \mathbf{p} + t\mathbf{u}$ lies in the plane $\mathbf{a} \cdot \mathbf{r} = d$, where $\mathbf{a}$, $\mathbf{p}$, $\mathbf{u}$ and d are constants prove that $\mathbf{a} \cdot \mathbf{u} = 0$ and $\mathbf{a} \cdot \mathbf{p} = d$. Conversely, if $\mathbf{a} \cdot \mathbf{p} = d$ and $\mathbf{a} \cdot \mathbf{u} = 0$, show that the line lies in the plane.

25. Find the equation of the plane to which the vector $\mathbf{i} + \mathbf{j}$ is normal, and which contains the line l with equation

$$\mathbf{r} = (\mathbf{i} + \mathbf{j} + \mathbf{k}) + t\,(\mathbf{i} - \mathbf{j} - \mathbf{k}).$$

Find also the equation of the plane containing l and the point $\mathbf{j}$. Show that $\pi/6$ is an angle between the normals of these planes.

26. Unit vectors $\mathbf{i}$ and $\mathbf{j}$ are at right angles to each other. Also

$$\mathbf{p} = 3\mathbf{i} + 4\mathbf{j}, \quad \mathbf{q} = 5\mathbf{i}, \quad 4\mathbf{r} = \mathbf{p} + \mathbf{q}, \quad 2\mathbf{s} = \mathbf{p} - \mathbf{q}.$$

Show that

(*a*) $|\mathbf{p}| = |\mathbf{q}|$; (*b*) $|\mathbf{r}| = |\mathbf{s}|$;

(*c*) $\mathbf{r}$ is perpendicular to $\mathbf{s}$; (*d*) $|\mathbf{r} + \mathbf{s}| = |\mathbf{r} - \mathbf{s}|$;

(*e*) $\mathbf{r} + \mathbf{a}$ is perpendicular to $\mathbf{r} - \mathbf{a}$.

Show also that, for all real k,

$$|\mathbf{r} + k\mathbf{s}| = |\mathbf{r} - k\mathbf{s}|.$$

27. Show that the lines

$$l_1 : \mathbf{r} = \mathbf{j} + s(\mathbf{i} - \mathbf{j} + \mathbf{k}), \quad l_2 : \mathbf{r} = \mathbf{i} - \mathbf{j} + t(\mathbf{i} + \mathbf{j} + \mathbf{k})$$

have no point in common. If A lies on l_1; B lies on l_2 and AB is perpendicular to both l_1 and l_2, find the length of AB and the equation of AB. Find also the equation of

(*i*) the plane π_1 perpendicular to AB and containing l_1.

(*ii*) the plane π_2 perpendicular to AB and containing l_2.

Find also the cartesian equations of the lines l_1 l_2 and the planes π_1 π_2.

28. The six faces of a rectangular cuboid have vector equations

$$\mathbf{r}\,.\,\mathbf{k} = 0, \quad \mathbf{r}\,.\,(\mathbf{i} + \mathbf{j}) = 2, \quad \mathbf{r}\,.\,(\mathbf{j} - \mathbf{i}) = 2,$$
$$\mathbf{r}\,.\,\mathbf{k} = 2, \quad \mathbf{r}\,.\,(\mathbf{i} + \mathbf{j}) = 4, \quad \mathbf{r}\,.\,(\mathbf{j} - \mathbf{i}) = -2.$$

where **i, j, k** represent unit vectors along perpendicular axes Ox, Oy, Oz. Prove that the acute angle between the diagonals through the points $2\mathbf{i}$ and $2\mathbf{j}$ is $\cos^{-1}(1/7)$.

Find the vector equations of the planes parallel to Oz through these diagonals.

29. From the point $\mathbf{s}_1 = \mathbf{i} + \mathbf{j} + \mathbf{k}$ the perpendicular drawn to a plane π_1 is $\mathbf{i} - \mathbf{j} + \mathbf{k}$. From the point $\mathbf{s}_2 = 3\mathbf{i} - \mathbf{j} - \mathbf{k}$ the perpendicular drawn to a plane π_2 is $\mathbf{i} + 2\mathbf{j} + \mathbf{k}$. Find the equation of the plane π_3 containing the line of intersection of the planes π_1 and π_2 and passing through the mid-point of the line joining $\mathbf{s}_1$ and $\mathbf{s}_2$. Find also the length of the perpendicular from the origin to the plane π_3.

30. Three vectors **a, b** and **c** are such that $\mathbf{a} \neq 0$ and $\mathbf{a} \times \mathbf{b} = 2\mathbf{a} \times \mathbf{c}$. Show that $\mathbf{b} - 2\mathbf{c} = \lambda\mathbf{a}$ where λ is a scalar.

Given that $|\mathbf{a}| = |\mathbf{c}| = 1$, $|\mathbf{b}| = 4$ and the angle between **b** and **c** is $\cos^{-1}\left(\frac{1}{4}\right)$, show that $\lambda = +4$ or -4. For each of these cases find the cosine of the angle between **a** and **c**.

31. Three non-collinear points A, B and C have position vectors **a, b** and **c** respectively, relative to an origin O, not necessarily in the plane ABC. Prove that the area of the triangle ABC is equal to the magnitude of the vector

$$\frac{1}{2}(\mathbf{b} \times \mathbf{c} + \mathbf{c} \times \mathbf{a} + \mathbf{a} \times \mathbf{b}).$$

When O does lie in the plane ABC interpret this result in terms of the areas of the triangles OAB, OBC and OCA, considering the cases (*i*) O is inside the triangle ABC (*ii*) O is in the region bounded by AC and BC produced.

32. Prove that the line $\mathbf{r} = \mathbf{a} + t\mathbf{b}$ and the plane $\mathbf{r}\,.\,\mathbf{n} = p$ intersect in a point whose position vector is $\mathbf{a} + (p - \mathbf{a}\,.\,\mathbf{n})\,\mathbf{b}/(\mathbf{b}\,.\,\mathbf{n})$ given that $\mathbf{b}\,.\,\mathbf{n} \neq 0$.

33. The non-zero vectors **p, q** are such that their vector product is the zero vector. State what is implied by this relation about the directions of **p** and **q**.

The three non-zero vectors **a, b** and **c** satisfy the equation $\mathbf{a} \times \mathbf{b} = \mathbf{c} \times \mathbf{a}$. Deduce that $\mathbf{b} \times \mathbf{c} = k\mathbf{a}$ where k is a scalar. If also $\mathbf{a} \times \mathbf{b} = \mathbf{b} \times \mathbf{c} \neq 0$, show that $\mathbf{a} + \mathbf{b} + \mathbf{c} = 0$.

34. Given that $\mathbf{x} + \frac{1}{\mathbf{p}^2}(\mathbf{p}\,.\,\mathbf{x})\,\mathbf{p} = \mathbf{q}$, show that $\mathbf{p}\,.\,\mathbf{x} = \frac{1}{2}\mathbf{p}\,.\,\mathbf{q}$ and find **x** in terms of **p** and **q**.

35. In the plane of the triangle ABC, squares $ACXY$, $BCWZ$ are described, in the order given, externally to the triangle on AC and BC respectively. Taking

$$\vec{CX} = \mathbf{b},\ \vec{CA} = \mathbf{a},\ \vec{CW} = \mathbf{q},\ \vec{CB} = \mathbf{p}$$

prove that $\mathbf{a} \cdot \mathbf{p} + \mathbf{q} \cdot \mathbf{b} = 0$. Deduce that $\vec{AW} \cdot \vec{BX} = 0$.

36. In the parallelogram $OABC$, the angle AOC is acute and D is a point on BC such that $\dfrac{CD}{DB} = \dfrac{1}{2}$, $\vec{OA} = \mathbf{a}$ and $\vec{OC} = \mathbf{c}$. Prove that

$$9(AC^2 - OD^2) = 8(\mathbf{a}^2 - 3\mathbf{a} \cdot \mathbf{c}).$$

By considering the value of AD^2, prove that if $AC = OD$, then $AD = OC$.

37. Let $\mathbf{a}$ and $\mathbf{b}$ be non-zero vectors and d be a scalar.

(*i*) The equation $\mathbf{a} \cdot \mathbf{x} = d$ is to be solved for $\mathbf{x}$. Writing $\mathbf{x} = \lambda\mathbf{a} + \mathbf{e}$ where $\mathbf{e}$ is an arbitrary vector perpendicular to $\mathbf{a}$, show that the equation is satisfied if and only if λ has a particular value to be determined in terms of $\mathbf{a}$ and $\mathbf{d}$.

(*ii*) What condition must be satisfied by the vectors $\mathbf{a}$ and $\mathbf{b}$ if the equation $\mathbf{a} \times \mathbf{x} = \mathbf{b}$ has a solution ? Given that this condition is satisfied and writing

$$\mathbf{x} = \lambda\mathbf{a} + \mu\mathbf{b} + \gamma\mathbf{c}$$

where λ, μ, γ are scalars and $\mathbf{c}$ is a unit vector, perpendicular to $\mathbf{a}$ and $\mathbf{b}$ with $\mathbf{a}$, $\mathbf{b}$, $\mathbf{c}$ as right handed, show that the equation is satisfied if and only if μ, γ have particular values to be determined.

38. In the trapezium $OABC$, OA and CB are the parallel sides, and $CB = \dfrac{2}{3} OA$. M is the mid-point of the side OC, and the angle OMA is a right angle, $\vec{OA} = \mathbf{a}$ and $\vec{OC} = \mathbf{c}$.

(*i*) Express the vectors $\vec{OB}$, $\vec{AB}$ and $\vec{AM}$ in terms of $\mathbf{a}$ and $\mathbf{c}$. Prove that

(*ii*) $\mathbf{a} \cdot \mathbf{c} = \dfrac{1}{2}\mathbf{c}^2$.

(*iii*) $\vec{OB} \cdot \vec{AB} = \dfrac{1}{18}(21\, OC^2 - 4\, OA^2)$.

(*iv*) Given that the angle OAB is a right angle and that the length OC is 2 cm, calculate the length OA.

39. In the triangle OAB, $\vec{OA} = \mathbf{a}$ and $\vec{OB} = \mathbf{b}$. A point P is taken on OA such that $OP/PA = 3$, and a point Q is taken on OB such that $\dfrac{OQ}{QB} = \dfrac{1}{2}$. Express each of the vectors

$$\vec{AQ} \text{ and } \vec{BP}$$

in terms of $\mathbf{a}$ and $\mathbf{b}$.

Given that the lines AQ and BP are perpendicular, prove that

$$9\mathbf{a}^2 + 4\mathbf{b}^2 = 15\, \mathbf{a} \cdot \mathbf{b}.$$

Given also that $OA = 2$ and $OB = 3$; prove that the cosine of the angle AOB is 4/5, and calculate the length of PQ.

40. The points O, A, B, C, D are such that

$$\vec{OA} = \mathbf{a},\ \vec{OB} = \mathbf{b},\ \vec{OC} = 2\mathbf{a} + 3\mathbf{b},\ \vec{OD} = \mathbf{a} - 2\mathbf{b}.$$

Given that the length of $\vec{OA}$ is three times the length of $\vec{OB}$ show that $\vec{BD}$ and $\vec{AC}$ are perpendicular.

41. $ABCD$ is a plane quadrilateral whose sides CD, BA intersect at O. If P, Q are the mid-points of the diagonals AC and BD, prove that the area of $\Delta\, OPQ$ is one quarter of the area of the quadrilateral $ABCD$.

42. If $\mathbf{p}$ and $\mathbf{q}$ are the position vectors of two fixed points and if $\mathbf{r}$ is the position vector of a variable point, describe geometrically the loci given by the equations

$$(\mathbf{r} - \mathbf{p}) \cdot (\mathbf{r} - \mathbf{q}) = 0 \quad \text{and} \quad (\mathbf{r} - \mathbf{p}) \times (\mathbf{r} - \mathbf{q}) = 0.$$

43. AA', BB', CC' and DD' are edges of a cube perpendicular to the face $ABCD$. Find

(*i*) the perpendicular distance of D' from the diagonal AC',

(*ii*) the shortest distance between the edges DD' and the diagonal AC'.

44. In the triangle ABC, L is the mid-point of AB and M is a point in AC such that the ratio of the length of AM to the length of MC is 2 : 1. BM and LC intersect in N.

Find the position vector of N.

(*i*) Calculate the ratio of the area of triangle ANC to the area of triangle ANM.

(*ii*) Prove that the area of triangle BNC = area of triangle ANC.

(*iii*) Calculate the ratio of the length of BN to the length of NM.

(*iv*) Calculate the ratio of the area of triangle BNC to that of the triangle ABC.

45. If two opposite edges of tetrahedron are at right angles, show that every plane section of the tetrahedron parallel to this pair is a rectangle.

46. If P be any point on the circumference of the in-circle of an equilateral triangle ABC, show that $PA^2 + PB^2 + PC^2$ is constant.

47. OA, OB, OC are three mutually perpendicular straight lines through O and A, B, C also denote the angles of the triangle ABC; show that

(*i*) $\Delta\, ABC$ is acute angled.

(*ii*) $OA^2 \tan A = OB^2 \tan B = OC^2 \tan C$.

48. Show that the mid-points of the six edges of a cube which do not meet a particular diagonal are the six vertices of a regular hexagon. Also show that the plane of the hexagon bisects the diagonal normally.

49. The line AB is the common perpendicular to two skew lines AP and BQ, and C and R are the mid-points of AB and PQ respectively. Prove that CR and AB are perpendicular.

50. Let $\mathbf{a}$ be a given non-zero vector. For what vectors $\mathbf{u}$ is $\mathbf{a} \times \mathbf{u} = 0$?

Let $\mathbf{a} = \mathbf{i} + 2\mathbf{j} - 3\mathbf{k}$, $\mathbf{b} = 2\mathbf{i} + \mathbf{j} - \mathbf{k}$, where $\mathbf{i}, \mathbf{j}, \mathbf{k}$ denote the usual base vectors. Determine the set of vectors $\mathbf{v}$ for which

$$\mathbf{a} \times \mathbf{v} = \mathbf{a} \times \mathbf{b}.$$

Hence, find a vector $\mathbf{v}$ satisfying

$$\mathbf{a} \times \mathbf{v} = \mathbf{a} \times \mathbf{b}, \quad \mathbf{a} \cdot \mathbf{v} = 0.$$

7

Some Miscellaneous Topics

(Products of Four Vectors)

In this chapter, we shall consider products of four vectors and the solution of some vector equations.

7.1. SCALAR PRODUCT $(\mathbf{a}\times\mathbf{b}).(\mathbf{c}\times\mathbf{d})$ OF FOUR VECTORS

We shall prove that

$$(\mathbf{a}\times\mathbf{b}).(\mathbf{c}\times\mathbf{d})=(\mathbf{a}.\mathbf{c})(\mathbf{b}.\mathbf{d})-(\mathbf{a}.\mathbf{d})(\mathbf{b}.\mathbf{c})=\begin{vmatrix}\mathbf{a}.\mathbf{c} & \mathbf{a}.\mathbf{d}\\ \mathbf{b}.\mathbf{c} & \mathbf{b}.\mathbf{d}\end{vmatrix}.$$

Looking upon $(\mathbf{a}\times\mathbf{b}).(\mathbf{c}\times\mathbf{d})$ as a scalar triple product of the three vectors

$$\mathbf{a}\times\mathbf{b},\ \mathbf{c},\ \mathbf{d}$$

and interchanging the dot and cross, we have

$$(\mathbf{a}\times\mathbf{b}).(\mathbf{c}\times\mathbf{d}) = [(\mathbf{a}\times\mathbf{b})\times\mathbf{c}].\mathbf{d}$$
$$= [(\mathbf{a}.\mathbf{c})\,\mathbf{b}-(\mathbf{b}.\mathbf{c})\,\mathbf{a}].\mathbf{d}$$
$$= (\mathbf{a}.\mathbf{c})(\mathbf{b}.\mathbf{d})-(\mathbf{b}.\mathbf{c})(\mathbf{a}.\mathbf{d}).$$

7.2. VECTOR PRODUCT $(\mathbf{a}\times\mathbf{b})\times(\mathbf{c}\times\mathbf{d})$ OF FOUR VECTORS

We shall prove that (*Patna 2003*)

$$(\mathbf{a}\times\mathbf{b}).(\mathbf{c}\times\mathbf{d}) = [\mathbf{a}\ \mathbf{b}\ \mathbf{d}]\,\mathbf{c}-[\mathbf{a}\ \mathbf{b}\ \mathbf{c}]\,\mathbf{d}=[\mathbf{a}\ \mathbf{c}\ \mathbf{d}]\,\mathbf{b}-[\mathbf{b}\ \mathbf{c}\ \mathbf{d}]\,\mathbf{a}$$

We can look upon $(\mathbf{a}\times\mathbf{b})\times(\mathbf{c}\times\mathbf{d})$ as a vector triple product in two ways according as we put $\mathbf{a}\times\mathbf{b}=\mathbf{p}$ or $\mathbf{c}\times\mathbf{d}=\mathbf{q}$.

Thus, putting $\mathbf{a}\times\mathbf{b}=\mathbf{p}$, we have

$$(\mathbf{a}\times\mathbf{b}).(\mathbf{c}\times\mathbf{d}) = \mathbf{p}\times(\mathbf{c}\times\mathbf{d})$$
$$= (\mathbf{p}.\mathbf{d})\,\mathbf{c}-(\mathbf{p}.\mathbf{c})\,\mathbf{d}=[\mathbf{a}\ \mathbf{b}\ \mathbf{d}]\,\mathbf{c}-[\mathbf{a}\ \mathbf{b}\ \mathbf{c}]\,\mathbf{d}.$$

so that the vector product now appears as a linear combination of the vectors **c** and **d**.

Again, we have

$$(\mathbf{a}\times\mathbf{b})\times(\mathbf{c}\times\mathbf{d}) = (\mathbf{a}\times\mathbf{b})\times\mathbf{q}$$
$$= (\mathbf{a}.\mathbf{q})\,\mathbf{b}-(\mathbf{b}.\mathbf{q})\,\mathbf{a}=[\mathbf{a}\ \mathbf{c}\ \mathbf{d}]\,\mathbf{b}-[\mathbf{b}\ \mathbf{c}\ \mathbf{d}]\,\mathbf{a},$$

so that the vector product now appears as a linear combination of the vectors **a** and **b**.

Thus, we have

$$(\mathbf{a} \times \mathbf{b}) \times (\mathbf{c} \times \mathbf{d}) = [\mathbf{a}\ \mathbf{b}\ \mathbf{d}]\ \mathbf{c} - [\mathbf{a}\ \mathbf{b}\ \mathbf{c}]\ \mathbf{d} = [\mathbf{a}\ \mathbf{c}\ \mathbf{d}]\ \mathbf{b} - [\mathbf{b}\ \mathbf{c}\ \mathbf{d}]\ \mathbf{a}.$$

7.3. LINEAR RELATION CONNECTING FOUR VECTORS

To determine the Linear Relation connecting four given vectors.

Let **a, b, c, d** be four given vectors. Considering

$$(\mathbf{a} \times \mathbf{b}) \times (\mathbf{c} \times \mathbf{d})$$

as a vector triple product in two ways as in the preceding section, we obtain

$$[\mathbf{a}\ \mathbf{b}\ \mathbf{d}]\ \mathbf{c} - [\mathbf{a}\ \mathbf{b}\ \mathbf{c}]\ \mathbf{d} = [\mathbf{a}\ \mathbf{c}\ \mathbf{d}]\ \mathbf{b} - [\mathbf{b}\ \mathbf{c}\ \mathbf{d}]\ \mathbf{a}.$$

$$\Rightarrow\ [\mathbf{a}\ \mathbf{b}\ \mathbf{d}]\ \mathbf{c} - [\mathbf{a}\ \mathbf{b}\ \mathbf{c}]\ \mathbf{d} - [\mathbf{a}\ \mathbf{c}\ \mathbf{d}]\ \mathbf{b} + [\mathbf{b}\ \mathbf{c}\ \mathbf{d}]\ \mathbf{a} = \mathbf{0}.$$

$$\Rightarrow\ [\mathbf{b}\ \mathbf{c}\ \mathbf{d}]\ \mathbf{a} - [\mathbf{a}\ \mathbf{c}\ \mathbf{d}]\ \mathbf{b} + [\mathbf{a}\ \mathbf{b}\ \mathbf{d}]\ \mathbf{c} - [\mathbf{a}\ \mathbf{b}\ \mathbf{c}]\ \mathbf{d} = \mathbf{0} \qquad ...(i)$$

which is the required relation connecting the four vectors

$$\mathbf{a}, \mathbf{b}, \mathbf{c}, \mathbf{d}$$

If [**a b, c**] ≠ 0, *i.e.*, **a, b, c** are not coplanar vectors, we can rewrite (*i*) as

$$\mathbf{d} = \frac{[\mathbf{b}\,\mathbf{c}\,\mathbf{d}]\,\mathbf{a} - [\mathbf{a}\,\mathbf{c}\,\mathbf{d}]\,\mathbf{b} + [\mathbf{a}\,\mathbf{b}\,\mathbf{d}]\,\mathbf{c}}{[\mathbf{a}\,\mathbf{b}\,\mathbf{c}]}.$$

Thus, we have expressed any given vector **d** as a linear combination of three non-coplanar vectors **a, b,** and **c**.

Another Method. If **a, b, c** are not coplanar, then we have a relation of the form

$$\mathbf{d} = \lambda\mathbf{a} + \mu\mathbf{b} + \nu\mathbf{c}. \qquad ...(i)$$

Multiplying scalarly with **b** × **c**, we obtain

$$[\mathbf{d}\ \mathbf{b}\ \mathbf{c}] = \lambda\ [\mathbf{a}\ \mathbf{b}\ \mathbf{c}]. \qquad ...(ii)$$

Similarly on multiplying (*i*) scalarly with **c** × **a** and **a** × **b**, successively, we obtain

$$[\mathbf{d}\ \mathbf{c}\ \mathbf{a}] = \mu\ [\mathbf{b}\ \mathbf{c}\ \mathbf{a}] = \mu\ [\mathbf{a}\ \mathbf{b}\ \mathbf{c}] \qquad ...(iii)$$

$$[\mathbf{d}\ \mathbf{a}\ \mathbf{b}] = \nu\ [\mathbf{c}\ \mathbf{a}\ \mathbf{b}] = \nu\ [\mathbf{a}\ \mathbf{b}\ \mathbf{c}] \qquad ...(iv)$$

Substituting in (*i*) the values of λ, μ, ν, as obtained from (*ii*), (*iii*), (*iv*), we obtain the required relation.

7.4. PRODUCT OF TWO SCALAR TRIPLE PRODUCTS

To prove that

$$[\mathbf{p}\,\mathbf{q}\,\mathbf{r}]\,[\mathbf{p}'\,\mathbf{q}'\,\mathbf{r}'] = \begin{vmatrix} \mathbf{p}\cdot\mathbf{p}' & \mathbf{p}\cdot\mathbf{q}' & \mathbf{p}\cdot\mathbf{r}' \\ \mathbf{q}\cdot\mathbf{p}' & \mathbf{q}\cdot\mathbf{q}' & \mathbf{q}\cdot\mathbf{r}' \\ \mathbf{r}\cdot\mathbf{p}' & \mathbf{r}\cdot\mathbf{q}' & \mathbf{r}\cdot\mathbf{r}' \end{vmatrix}$$

where **p, q, r; p′, q′, r′** *are any vectors.*

Writing **p′** × **q′** = **a** and considering the four vectors **p, q, r, a,** we obtain the linear relation connecting these four vectors. As in § 7.1, above we have

$$[\mathbf{q\ r\ a}]\,\mathbf{p} - [\mathbf{p\ r\ a}]\,\mathbf{q} + [\mathbf{p\ q\ a}]\,\mathbf{r} - [\mathbf{p\ q\ r}]\,\mathbf{a} = \mathbf{0} \qquad ...(i)$$

Again

$$[\mathbf{q\ r\ a}] = \mathbf{q}\times\mathbf{r}\,.\,\mathbf{a} = (\mathbf{q}\times\mathbf{r})\cdot(\mathbf{p}'\times\mathbf{q}') = \begin{vmatrix} \mathbf{q}\cdot\mathbf{p}' & \mathbf{q}\cdot\mathbf{q}' \\ \mathbf{r}\cdot\mathbf{p}' & \mathbf{r}\cdot\mathbf{q}' \end{vmatrix}$$

$$[\mathbf{p\ r\ a}] = \mathbf{p}\times\mathbf{r}\,.\,\mathbf{a} = (\mathbf{p}\times\mathbf{r})\cdot(\mathbf{p}'\times\mathbf{q}') = \begin{vmatrix} \mathbf{p}\cdot\mathbf{p}' & \mathbf{p}\cdot\mathbf{q}' \\ \mathbf{r}\cdot\mathbf{p}' & \mathbf{r}\cdot\mathbf{q}' \end{vmatrix}$$

$$[\mathbf{p\ q\ a}] = \mathbf{p}\times\mathbf{q}\,.\,\mathbf{a} = (\mathbf{p}\times\mathbf{q})\cdot(\mathbf{p}'\times\mathbf{q}') = \begin{vmatrix} \mathbf{p}\cdot\mathbf{p}' & \mathbf{p}\cdot\mathbf{q}' \\ \mathbf{q}\cdot\mathbf{p}' & \mathbf{q}\cdot\mathbf{q}' \end{vmatrix}$$

We rewrite (*i*) as

$$[\mathbf{p\ q\ r}]\,\mathbf{p}'\times\mathbf{q}' = [\mathbf{q\ r\ a}]\,\mathbf{p} - [\mathbf{p\ r\ a}]\,\mathbf{q} + [\mathbf{p\ q\ a}]\,\mathbf{r}.$$

Multiplying scalarly with $\mathbf{r}'$, we obtain

$$[\mathbf{p\ q\ r}][\mathbf{p}'\ \mathbf{q}'\ \mathbf{r}'] = \begin{vmatrix} \mathbf{q}\cdot\mathbf{p}' & \mathbf{q}\cdot\mathbf{q}' \\ \mathbf{r}\cdot\mathbf{p}' & \mathbf{r}\cdot\mathbf{q}' \end{vmatrix}(\mathbf{p}\,.\,\mathbf{r}')$$

$$-\begin{vmatrix} \mathbf{p}\cdot\mathbf{p}' & \mathbf{p}\cdot\mathbf{q}' \\ \mathbf{r}\cdot\mathbf{p}' & \mathbf{r}\cdot\mathbf{q}' \end{vmatrix}(\mathbf{q}\,.\,\mathbf{r}') + \begin{vmatrix} \mathbf{p}\cdot\mathbf{p}' & \mathbf{p}\cdot\mathbf{q}' \\ \mathbf{q}\cdot\mathbf{q}' & \mathbf{p}\cdot\mathbf{p}' \end{vmatrix}(\mathbf{r}\,.\,\mathbf{r}')$$

$$= \begin{vmatrix} \mathbf{p}\cdot\mathbf{p}' & \mathbf{p}\cdot\mathbf{q}' & \mathbf{p}\cdot\mathbf{r}' \\ \mathbf{q}\cdot\mathbf{p}' & \mathbf{q}\cdot\mathbf{q}' & \mathbf{q}\cdot\mathbf{r}' \\ \mathbf{r}\cdot\mathbf{p}' & \mathbf{r}\cdot\mathbf{q}' & \mathbf{r}\cdot\mathbf{r}' \end{vmatrix}.$$

Cor.

$$[\mathbf{a\ b\ c}]^2 = \begin{vmatrix} \mathbf{a}\,.\,\mathbf{a} & \mathbf{a}\,.\,\mathbf{b} & \mathbf{a}\,.\,\mathbf{c} \\ \mathbf{b}\,.\,\mathbf{a} & \mathbf{b}\,.\,\mathbf{b} & \mathbf{b}\,.\,\mathbf{c} \\ \mathbf{c}\,.\,\mathbf{a} & \mathbf{c}\,.\,\mathbf{b} & \mathbf{c}\,.\,\mathbf{c} \end{vmatrix}$$

7.5. TWO USEFUL DECOMPOSITIONS

7.5.1. *If* **a, b, c** *are three non-coplanar vectors, then to prove that*

$$\mathbf{b}\times\mathbf{c},\quad \mathbf{c}\times\mathbf{a},\quad \mathbf{a}\times\mathbf{b}$$

are also non-coplanar and to express **a, b, c** *in terms of*

$$\mathbf{b}\times\mathbf{c},\quad \mathbf{c}\times\mathbf{a},\quad \mathbf{a}\times\mathbf{b}.$$

We are given that $[\mathbf{a\ b\ c}] \neq 0$ and we shall prove that the scalar triple product

$$[\mathbf{b}\times\mathbf{c},\quad \mathbf{c}\times\mathbf{a},\quad \mathbf{a}\times\mathbf{b}] \neq 0$$

We have

$$(\mathbf{b}\times\mathbf{c})\times(\mathbf{c}\times\mathbf{a})\,.\,(\mathbf{a}\times\mathbf{b}) = [(\mathbf{b}\,.\,\mathbf{c}\times\mathbf{a})\,\mathbf{c} - (\mathbf{c}\,.\,\mathbf{c}\times\mathbf{a})\,\mathbf{b}]\,.\,(\mathbf{a}\times\mathbf{b})$$

$$= [\mathbf{a\ b\ c}]\,\mathbf{c}\,.\,\mathbf{a}\times\mathbf{b}, \quad \text{for } \mathbf{c}\,.\,\mathbf{c}\times\mathbf{a} = \mathbf{a}$$

$$= [\mathbf{a\ b\ c}]\,[\mathbf{a\ b\ c}] \neq 0.$$

Hence, $\mathbf{b} \times \mathbf{c}$, $\mathbf{c} \times \mathbf{a}$, $\mathbf{a} \times \mathbf{b}$ are non-coplanar vectors. The vectors $\mathbf{b} \times \mathbf{c}$, $\mathbf{c} \times \mathbf{a}$, $\mathbf{a} \times \mathbf{b}$ being non-coplanar every vector and, in particular, **a, b, c** is expressible as linear combinations of the same.

Let

$$\mathbf{a} = l\mathbf{b} \times \mathbf{c} + m\mathbf{c} \times \mathbf{a} + n\mathbf{a} \times \mathbf{b}.$$

Multiplying both sides scalarly with **a, b, c,** successively, we get

$$\mathbf{a} \cdot \mathbf{a} = l\mathbf{b} \times \mathbf{c} \cdot \mathbf{a} = l\,[\mathbf{a}\ \mathbf{b}\ \mathbf{c}],\ \mathbf{a} \cdot \mathbf{b} = m\,[\mathbf{a}\ \mathbf{b}\ \mathbf{c}],\ \mathbf{a} \cdot \mathbf{c} = n\,[\mathbf{a}\ \mathbf{b}\ \mathbf{c}]$$

$$\Rightarrow \qquad l = \frac{\mathbf{a} \cdot \mathbf{a}}{[\mathbf{a}\ \mathbf{b}\ \mathbf{c}]},\ m = \frac{\mathbf{a} \cdot \mathbf{b}}{[\mathbf{a}\ \mathbf{b}\ \mathbf{c}]},\ n = \frac{\mathbf{a} \cdot \mathbf{c}}{[\mathbf{a}\ \mathbf{b}\ \mathbf{c}]},$$

$$\Rightarrow \qquad \mathbf{a} = \frac{1}{[\mathbf{a}\ \mathbf{b}\ \mathbf{c}]}[(\mathbf{a} \cdot \mathbf{a})\,\mathbf{b} \times \mathbf{c} + (\mathbf{a} \cdot \mathbf{b})\,\mathbf{c} \times \mathbf{a} + (\mathbf{a} \cdot \mathbf{c})\,\mathbf{a} \times \mathbf{b}].$$

We may similarly express **b** and **c** in terms of

$$\mathbf{b} \times \mathbf{c}, \quad \mathbf{c} \times \mathbf{a}, \quad \mathbf{a} \times \mathbf{b}.$$

7.5.2. *If* **a, b, c** *are three non-coplanar vectors, then to express* $\mathbf{b} \times \mathbf{c}$, $\mathbf{c} \times \mathbf{a}$, $\mathbf{a} \times \mathbf{b}$ *in terms of* **a, b, c.**

It has been already proved that the vectors $\mathbf{b} \times \mathbf{c}$, $\mathbf{c} \times \mathbf{a}$, and $\mathbf{a} \times \mathbf{b}$ are non-coplanar.

Let

$$\mathbf{b} \times \mathbf{c} = l\mathbf{a} + m\mathbf{b} + n\mathbf{c}.$$

Multiplying both sides scalarly with

$$\mathbf{b} \times \mathbf{c}, \quad \mathbf{c} \times \mathbf{a} \quad \text{and} \quad \mathbf{a} \times \mathbf{b}$$

successively, we get

$$(\mathbf{b} \times \mathbf{c}) \cdot (\mathbf{b} \times \mathbf{c}) = l\,[\mathbf{a}\ \mathbf{b}\ \mathbf{c}],$$
$$(\mathbf{b} \times \mathbf{c}) \cdot (\mathbf{c} \times \mathbf{a}) = m\,[\mathbf{a}\ \mathbf{b}\ \mathbf{c}],$$
$$(\mathbf{b} \times \mathbf{c}) \cdot (\mathbf{a} \times \mathbf{b}) = n\,[\mathbf{a}\ \mathbf{b}\ \mathbf{c}],$$

$$\Rightarrow \qquad l = \frac{(\mathbf{b} \times \mathbf{c}) \cdot (\mathbf{b} \times \mathbf{c})}{[\mathbf{a}\ \mathbf{b}\ \mathbf{c}]},$$

$$m = \frac{(\mathbf{b} \times \mathbf{c}) \cdot (\mathbf{c} \times \mathbf{a})}{[\mathbf{a}\ \mathbf{b}\ \mathbf{c}]},$$

$$n = \frac{(\mathbf{b} \times \mathbf{c}) \cdot (\mathbf{a} \times \mathbf{b})}{[\mathbf{a}\ \mathbf{b}\ \mathbf{c}]},$$

where the numerators can be transformed as in § 7.1 page 177.

Thus, we have expressed $\mathbf{b} \times \mathbf{c}$ in terms of **a, b, c.**

We may similarly express $\mathbf{c} \times \mathbf{a}$ and $\mathbf{a} \times \mathbf{b}$ in terms of **a, b, c.**

7.6. RECIPROCAL SYSTEM OF VECTORS (*Avadh 2005*)

If **a, b, c,** be three non-coplanar vectors so that $[\mathbf{a}\ \mathbf{b}\ \mathbf{c}] \neq 0$ and if $\mathbf{a}'$, $\mathbf{b}'$, $\mathbf{c}'$ be three other vectors such that

$$\mathbf{a}' = \frac{\mathbf{b} \times \mathbf{c}}{[\mathbf{a}\ \mathbf{b}\ \mathbf{c}]},\ \mathbf{b}' = \frac{\mathbf{c} \times \mathbf{a}}{[\mathbf{a}\ \mathbf{b}\ \mathbf{c}]},\ \mathbf{c}' = \frac{\mathbf{a} \times \mathbf{b}}{[\mathbf{a}\ \mathbf{b}\ \mathbf{c}]},$$

then $\mathbf{a}'$, $\mathbf{b}'$, $\mathbf{c}'$ are called reciprocal system to the vectors **a, b, c.**

Properties.

7.6.1. (I) *If* **a, b, c** *and* **a′, b′, c′** *be reciprocal systems of vectors, then*

$$\mathbf{a} \,.\, \mathbf{a}' = \mathbf{a} \cdot \frac{\mathbf{b} \times \mathbf{c}}{[\mathbf{a}\ \mathbf{b}\ \mathbf{c}]} = \frac{[\mathbf{a}\ \mathbf{b}\ \mathbf{c}]}{[\mathbf{a}\ \mathbf{b}\ \mathbf{c}]} = 1 \qquad (Kumaon\ 2005)$$

Similarly **b . b′ = c . c′** = 1.

Note. Due to this property, the two systems of vectors are called reciprocal systems.

(II) *To show that* **a . b′ = a . c′ = b . a′ = b . c′ = c . a′ = c . b′** = *0.*

$$\mathbf{a} \,.\, \mathbf{b}' = \mathbf{a} \cdot \frac{\mathbf{c} \times \mathbf{a}}{[\mathbf{a}\ \mathbf{b}\ \mathbf{c}]} = \frac{[\mathbf{a}\ \mathbf{c}\ \mathbf{a}]}{[\mathbf{a}\ \mathbf{b}\ \mathbf{c}]} = 0$$

Similarly other results follow.

(III) *To show that* [**a b c**] [**a′, b′, c′**] = *1.* (*Awadh 98, 2000, Kumaon 2006*)

We have $$[\mathbf{a}', \mathbf{b}', \mathbf{c}'] = \left[\frac{\mathbf{b} \times \mathbf{c}}{[\mathbf{a}\ \mathbf{b}\ \mathbf{c}]}, \frac{\mathbf{c} \times \mathbf{a}}{[\mathbf{a}\ \mathbf{b}\ \mathbf{c}]}, \frac{\mathbf{a} \times \mathbf{b}}{[\mathbf{a}\ \mathbf{b}\ \mathbf{c}]}\right]$$

$$= \frac{1}{[\mathbf{a}\ \mathbf{b}\ \mathbf{c}]^3} [\mathbf{b} \times \mathbf{c},\ \mathbf{c} \times \mathbf{a},\ \mathbf{a} \times \mathbf{b}]$$

$$= \frac{1}{[\mathbf{a}\ \mathbf{b}\ \mathbf{c}]^3} [\mathbf{a}\ \mathbf{b}\ \mathbf{c}]^2$$

$$= \frac{1}{[\mathbf{a}\ \mathbf{b}\ \mathbf{c}]}$$

∴ [**a′, b′, c′**] [**a b c**] = 1.

Since [**a b c**] ≠ 0, hence [**a′, b′, c′**] ≠ 0 ⇒ **a′, b′, c′** are also non-coplanar.

(IV) *The orthogonal triad of vectors* **i, j, k** *is self-reciprocal.*

Let **i′, j′, k′** be the system of vectors reciprocal to the system **i, j, k.** Then, we have

$$\mathbf{i}' = \frac{\mathbf{j} \times \mathbf{k}}{[\mathbf{i}\ \mathbf{j}\ \mathbf{k}]} = \mathbf{i}$$

Similarly, **j′ = j** and **k′ = k.**

7.6.2. Theorem

If **a, b, c** *be three non-coplanar vectors for which* [**a b c**] ≠ 0 *and* **a′, b′, c′** *constitute the reciprocal system of vectors, then any vector* **r** *can be expressed as*

$$\mathbf{r} = (\mathbf{r} \,.\, \mathbf{a}')\,\mathbf{a} + (\mathbf{r} \,.\, \mathbf{b}')\,\mathbf{b} + (\mathbf{r} \,.\, \mathbf{c}')\,\mathbf{c}. \qquad (Avadh\ 2005)$$

Since **a, b, c** are non-coplanar vectors, **r** can be expressed as a linear combination in the form

$$\mathbf{r} = x\mathbf{a} + y\mathbf{b} + z\mathbf{c} \qquad ...(1)$$

where x, y, z are some scalars.

(1) ⇒ $$\mathbf{r} \,.\, (\mathbf{b} \times \mathbf{c}) = x\mathbf{a} \,.\, (\mathbf{b} \times \mathbf{c}) + y\mathbf{b} \,.\, (\mathbf{b} \times \mathbf{c}) + z\mathbf{c} \,.\, (\mathbf{b} \times \mathbf{c})$$

$$= x\,[\mathbf{a\ b\ c}]$$

$\therefore$ $$x = \frac{\mathbf{r}\,.\,(\mathbf{b}\times\mathbf{c})}{[\mathbf{a\ b\ c}]} = (\mathbf{r}\,.\,\mathbf{a}')$$

Similarly $y = \mathbf{r}\,.\,\mathbf{b}',\quad z = \mathbf{r}\,.\,\mathbf{c}'$

$\therefore$ $$\mathbf{r} = (\mathbf{r}\,.\,\mathbf{a}')\,\mathbf{a} + (\mathbf{r}\,.\,\mathbf{b}')\,\mathbf{b} + (\mathbf{r}\,.\,\mathbf{c}')\,\mathbf{c}. \qquad ...(2)$$

Cor. 1. Similarly, we can get

$$\mathbf{r} = (\mathbf{r}\,.\,\mathbf{a})\,\mathbf{a}' + (\mathbf{r}\,.\,\mathbf{b})\,\mathbf{b}' + (\mathbf{r}\,.\,\mathbf{c})\,\mathbf{c}'$$

Cor. 2. Since the system **i, j, k** is self-reciprocal, (2) can be written as

$$\mathbf{r} = (\mathbf{r}\,.\,\mathbf{i})\,\mathbf{i} + (\mathbf{r}\,.\,\mathbf{j})\,\mathbf{j} + (\mathbf{r}\,.\,\mathbf{k})\,\mathbf{k}.$$

EXAMPLES

Example 1. *Prove that*

$$[\mathbf{a}\times\mathbf{p},\ \mathbf{b}\times\mathbf{q},\ \mathbf{c}\times\mathbf{r}] + [\mathbf{a}\times\mathbf{q},\ \mathbf{b}\times\mathbf{r},\ \mathbf{c}\times\mathbf{p}] + [\mathbf{a}\times\mathbf{r},\ \mathbf{b}\times\mathbf{p},\ \mathbf{c}\times\mathbf{q}] = 0.$$

(Kumaon 2001)

Solution. $$[\mathbf{a}\times\mathbf{p},\ \mathbf{b}\times\mathbf{q},\ \mathbf{c}\times\mathbf{r}] = (\mathbf{a}\times\mathbf{p})\,.\,[(\mathbf{b}\times\mathbf{q})\times(\mathbf{c}\times\mathbf{r})]$$
$$= (\mathbf{a}\times\mathbf{p})\,.\,[\{(\mathbf{b}\times\mathbf{q})\,.\,\mathbf{r}\}\,\mathbf{c} - \{(\mathbf{b}\times\mathbf{q})\,.\,\mathbf{c}\}\,\mathbf{r}]$$
$$= (\mathbf{a}\times\mathbf{p})\,.\,\{[\mathbf{b\ q\ r}]\,\mathbf{c} - [\mathbf{b\ q\ c}]\,\mathbf{r}\}$$
$$= [\mathbf{a\ p\ c}]\,[\mathbf{b\ q\ r}] - [\mathbf{a\ p\ r}]\,[\mathbf{b\ q\ c}] \qquad ...(1)$$

Also, $[\mathbf{a}\times\mathbf{q},\ \mathbf{b}\times\mathbf{r},\ \mathbf{c}\times\mathbf{p}]$
$$= (\mathbf{b}\times\mathbf{r})\,.\,[(\mathbf{c}\times\mathbf{p})\times(\mathbf{a}\times\mathbf{q})]$$
$$= (\mathbf{b}\times\mathbf{r})\,.\,[\{(\mathbf{c}\times\mathbf{p})\,.\,\mathbf{q}\}\,\mathbf{a} - \{(\mathbf{c}\times\mathbf{p})\,.\,\mathbf{a}\}\,\mathbf{q}]$$
$$= (\mathbf{b}\times\mathbf{r})\,.\,\{[\mathbf{c\ p\ q}]\,\mathbf{a} - [\mathbf{c\ p\ a}]\,\mathbf{q}\}$$
$$= [\mathbf{b\ r\ a}]\,[\mathbf{c\ p\ q}] - [\mathbf{b\ q\ r}]\,[\mathbf{a\ p\ c}] \qquad ...(2)$$

$[\mathbf{a}\times\mathbf{r},\ \mathbf{b}\times\mathbf{p},\ \mathbf{c}\times\mathbf{q}]$
$$= (\mathbf{c}\times\mathbf{q})\,.\,[(\mathbf{a}\times\mathbf{r})\times(\mathbf{b}\times\mathbf{p})]$$
$$= (\mathbf{c}\times\mathbf{q}\,.\,[\{(\mathbf{a}\times\mathbf{r})\,.\,\mathbf{p}\}\,\mathbf{b} - \{(\mathbf{a}\times\mathbf{r})\,.\,\mathbf{b}\}\,\mathbf{p}]$$
$$= [\mathbf{c\ q\ b}]\,[\mathbf{a\ r\ p}] - [\mathbf{c\ q\ b}]\,[\mathbf{a\ r\ b}] \qquad ...(3)$$

Adding (1), (2), (3) the required result follows.

Example 2. *Prove that*

$$[\mathbf{a}\times\mathbf{b},\ \mathbf{c}\times\mathbf{d},\ \mathbf{e}\times\mathbf{f}] = [\mathbf{a\ b\ d}]\,[\mathbf{c\ e\ f}] - [\mathbf{a\ b\ c}]\,[\mathbf{d\ e\ f}]$$
$$= [\mathbf{a\ b\ e}]\,[\mathbf{f\ c\ d}] - [\mathbf{a\ b\ f}]\,[\mathbf{e\ c\ d}]$$
$$= [\mathbf{c\ d\ a}]\,[\mathbf{b\ e\ f}] - [\mathbf{c\ d\ b}]\,[\mathbf{a\ e\ f}]$$

(Awadh 98, Rohilkhand 2006)

Solution. $[\mathbf{a}\times\mathbf{b},\ \mathbf{c}\times\mathbf{d},\ \mathbf{e}\times\mathbf{f}]$
$$= (\mathbf{a}\times\mathbf{b})\,.\,[(\mathbf{c}\times\mathbf{d})\times(\mathbf{e}\times\mathbf{f})]$$
$$= (\mathbf{a}\times\mathbf{b})\,.\,[\{(\mathbf{c}\times\mathbf{d})\,.\,\mathbf{f}\}\,\mathbf{e} - \{(\mathbf{c}\times\mathbf{d})\,.\,\mathbf{e}\}\,\mathbf{f}]$$
$$= (\mathbf{a}\times\mathbf{b})\,.\,\{[\mathbf{c\ d\ f}]\,\mathbf{e} - [\mathbf{c\ d\ e}]\,\mathbf{f}\}$$
$$= [\mathbf{a\ b\ e}]\,[\mathbf{c\ d\ f}] - [\mathbf{a\ b\ f}]\,[\mathbf{c\ d\ e}]$$
$$= [\mathbf{a\ b\ e}]\,[\mathbf{f\ c\ d}] - [\mathbf{a\ b\ f}]\,[\mathbf{e\ c\ d}]$$

Also,

$$
\begin{aligned}
[\mathbf{a} \times \mathbf{b},\ \mathbf{c} \times \mathbf{d},\ \mathbf{e} \times \mathbf{f}]
&= (\mathbf{c} \times \mathbf{d}) \,.\, [(\mathbf{e} \times \mathbf{f}) \times (\mathbf{a} \times \mathbf{b})] \\
&= (\mathbf{c} \times \mathbf{d}) \,.\, [\{(\mathbf{e} \times \mathbf{f}) \,.\, \mathbf{b}\}\, \mathbf{a} - \{(\mathbf{e} \times \mathbf{f}) \,.\, \mathbf{a}\}\, \mathbf{b}] \\
&= (\mathbf{c} \times \mathbf{d}) \,.\, \{[\mathbf{e}\ \mathbf{f}\ \mathbf{b}]\, \mathbf{a} - [\mathbf{e}\ \mathbf{f}\ \mathbf{a}]\, \mathbf{b}\} \\
&= [\mathbf{c}\ \mathbf{d}\ \mathbf{a}]\, [\mathbf{e}\ \mathbf{f}\ \mathbf{b}] - [\mathbf{c}\ \mathbf{d}\ \mathbf{b}]\, [\mathbf{e}\ \mathbf{f}\ \mathbf{a}] \\
&= [\mathbf{c}\ \mathbf{d}\ \mathbf{a}]\, [\mathbf{b}\ \mathbf{e}\ \mathbf{f}] - [\mathbf{c}\ \mathbf{d}\ \mathbf{b}]\, [\mathbf{a}\ \mathbf{e}\ \mathbf{f}]
\end{aligned}
$$

Similarly,

$$
\begin{aligned}
[\mathbf{a} \times \mathbf{b},\ \mathbf{c} \times \mathbf{d},\ \mathbf{e} \times \mathbf{f}]
&= (\mathbf{e} \times \mathbf{f}) \,.\, [(\mathbf{a} \times \mathbf{b}) \times (\mathbf{c} \times \mathbf{d})] \\
&= (\mathbf{e} \times \mathbf{f}) \,.\, [\{\mathbf{a} \times \mathbf{b}) \,.\, \mathbf{d}\}\, \mathbf{c} - \{(\mathbf{a} \times \mathbf{b}) \,.\, \mathbf{c}\}\, \mathbf{d}] \\
&= (\mathbf{e} \times \mathbf{f}) \,.\, \{[\mathbf{a}\ \mathbf{b}\ \mathbf{d}]\, \mathbf{c} - [\mathbf{a}\ \mathbf{b}\ \mathbf{c}]\, \mathbf{d}\} \\
&= [\mathbf{e}\ \mathbf{f}\ \mathbf{c}]\, [\mathbf{a}\ \mathbf{b}\ \mathbf{d}] - [\mathbf{e}\ \mathbf{f}\ \mathbf{d}]\, [\mathbf{a}\ \mathbf{b}\ \mathbf{c}] \\
&= [\mathbf{a}\ \mathbf{b}\ \mathbf{d}]\, [\mathbf{c}\ \mathbf{e}\ \mathbf{f}] - [\mathbf{a}\ \mathbf{b}\ \mathbf{c}]\, [\mathbf{d}\ \mathbf{e}\ \mathbf{f}]
\end{aligned}
$$

Example 3. *Prove that*

$$
[\mathbf{a}\ \mathbf{b}\ \mathbf{c}]^2 = \begin{vmatrix} \mathbf{a}\,.\,\mathbf{a} & \mathbf{a}\,.\,\mathbf{b} & \mathbf{a}\,.\,\mathbf{c} \\ \mathbf{b}\,.\,\mathbf{a} & \mathbf{b}\,.\,\mathbf{b} & \mathbf{b}\,.\,\mathbf{c} \\ \mathbf{c}\,.\,\mathbf{a} & \mathbf{c}\,.\,\mathbf{b} & \mathbf{c}\,.\,\mathbf{c} \end{vmatrix}
$$

Solution. In § 5.2.1, Page 119 we proved that

$$(\mathbf{a} \times \mathbf{b})^2 = \mathbf{a}^2\mathbf{b}^2 - (\mathbf{a} \,.\, \mathbf{b})^2.$$

In the following we shall make use of this result.

We have

$$
\begin{aligned}
[(\mathbf{a} \times \mathbf{b}) \times \mathbf{c}]^2 &= (\mathbf{a} \times \mathbf{b})^2\, \mathbf{c}^2 - (\mathbf{a} \times \mathbf{b} \,.\, \mathbf{c})^2 \qquad (\text{§ 5.2.1., P. 119}) \\
&= \{\mathbf{a}^2\mathbf{b}^2 - (\mathbf{a} \,.\, \mathbf{b})^2\}\, \mathbf{c}^2 - [\mathbf{a}\ \mathbf{b}\ \mathbf{c}]^2 \\
&= \mathbf{a}^2\mathbf{b}^2\mathbf{c}^2 - (\mathbf{a} \,.\, \mathbf{b})^2\, \mathbf{c}^2 - [\mathbf{a}\ \mathbf{b}\ \mathbf{c}]^2.
\end{aligned}
$$

Also

$$
\begin{aligned}
[(\mathbf{a} \times \mathbf{b}) \times \mathbf{c}]^2 &= [(\mathbf{a}.\mathbf{c})\, \mathbf{b} - (\mathbf{b} \,.\, \mathbf{c})\, \mathbf{a}]^2 \\
&= (\mathbf{a} \,.\, \mathbf{c})^2\, \mathbf{b}^2 - 2\, (\mathbf{a} \,.\, \mathbf{c})\, (\mathbf{b} \,.\, \mathbf{c})\, (\mathbf{a} \,.\, \mathbf{b}) + (\mathbf{b} \,.\, \mathbf{c})^2\, \mathbf{a}^2
\end{aligned}
$$

$$
\begin{aligned}
\therefore \quad & \mathbf{a}^2\mathbf{b}^2\mathbf{c}^2 - (\mathbf{a} \,.\, \mathbf{b})^2\, \mathbf{c}^2 - [\mathbf{a}\ \mathbf{b}\ \mathbf{c}]^2 \\
& \qquad = (\mathbf{a} \,.\, \mathbf{c})^2\, \mathbf{b}^2 - 2\, (\mathbf{a} \,.\, \mathbf{c})\, (\mathbf{b} \,.\, \mathbf{c})\, (\mathbf{a} \,.\, \mathbf{b}) + (\mathbf{b} \,.\, \mathbf{c})^2\, \mathbf{a}^2 \\
\Rightarrow \quad & [\mathbf{a}\ \mathbf{b}\ \mathbf{c}]^2 = \mathbf{a}^2\mathbf{b}^2\mathbf{c}^2 - (\mathbf{a} \,.\, \mathbf{b})^2\, \mathbf{c}^2 - (\mathbf{b} \,.\, \mathbf{c})^2\, \mathbf{a}^2 - (\mathbf{c} \,.\, \mathbf{a})^2\, \mathbf{b}^2 \\
& \qquad\qquad + 2\, (\mathbf{a} \,.\, \mathbf{b})\, (\mathbf{b} \,.\, \mathbf{c})\, (\mathbf{c} \,.\, \mathbf{a})
\end{aligned}
$$

$$
\begin{vmatrix} \mathbf{a}\,.\,\mathbf{a} & \mathbf{a}\,.\,\mathbf{b} & \mathbf{a}\,.\,\mathbf{c} \\ \mathbf{b}\,.\,\mathbf{a} & \mathbf{b}\,.\,\mathbf{b} & \mathbf{b}\,.\,\mathbf{c} \\ \mathbf{c}\,.\,\mathbf{a} & \mathbf{c}\,.\,\mathbf{b} & \mathbf{c}\,.\,\mathbf{c} \end{vmatrix}.
$$

Example 4. *Decompose a vector* **r** *as a linear combination of a vector* **a** *and another vector perpendicular to* **a** *and coplanar with* **r** *and* **a.**

Solution. The vector $(\mathbf{a} \times \mathbf{r}) \times \mathbf{a}$ is coplanar with **a** and **r** and perpendicular to *a*. Let

$$\mathbf{r} = x\mathbf{a} + y\,[(\mathbf{a} \times \mathbf{r}) \times \mathbf{a}]$$

Now we have

$$\mathbf{r}\,.\,\mathbf{a} = x\,\mathbf{a}\,.\,\mathbf{a} \quad \Rightarrow \quad x = (\mathbf{r}\,.\,\mathbf{a}) \,/\, (\mathbf{a}\,.\,\mathbf{a}),$$

$$\Rightarrow \quad \mathbf{r} \times \mathbf{a} = y\,[(\mathbf{a} \times \mathbf{r}) \times \mathbf{a}] \times \mathbf{a}$$

$$= y\,[(\mathbf{a}\,.\,\mathbf{a})\,\mathbf{r} - (\mathbf{a}\,.\,\mathbf{r})\,\mathbf{a}] \times \mathbf{a}$$

$$= y\,(\mathbf{a}\,.\,\mathbf{a})\,(\mathbf{r} \times \mathbf{a}).$$

$$\Rightarrow \quad y = 1/(\mathbf{a}\,.\,\mathbf{a}).$$

Thus,

$$\mathbf{r} = \frac{\mathbf{r}\,.\,\mathbf{a}}{\mathbf{a}\,.\,\mathbf{a}}\,\mathbf{a} + \frac{1}{\mathbf{a}\,.\,\mathbf{a}}(\mathbf{a} \times \mathbf{r}) \times \mathbf{a}.$$

Example 5. *Prove that*

$$2\,(\mathbf{a} \times \mathbf{b}) \times (\mathbf{c} \times \mathbf{d}) = \begin{vmatrix} -\mathbf{a} & -\mathbf{b} & \mathbf{c} & \mathbf{d} \\ a_1 & b_1 & c_1 & d_1 \\ a_2 & b_2 & c_2 & d_2 \\ a_3 & b_3 & c_3 & d_3 \end{vmatrix}$$

where $\mathbf{a} = a_1\mathbf{i} + a_2\mathbf{j} + a_3\mathbf{k}$ *etc.*

Solution. We have

$$(\mathbf{a} \times \mathbf{b}) \times (\mathbf{c} \times \mathbf{d}) = [\mathbf{a}\;\mathbf{b}\;\mathbf{d}]\,\mathbf{c} - [\mathbf{a}\;\mathbf{b}\;\mathbf{c}]\,\mathbf{d}$$

$$= \begin{vmatrix} a_1 & b_1 & d_1 \\ a_2 & b_2 & d_2 \\ a_3 & b_3 & d_3 \end{vmatrix} \mathbf{c} - \begin{vmatrix} a_1 & b_1 & c_1 \\ a_2 & b_2 & c_2 \\ a_3 & b_3 & c_3 \end{vmatrix} \mathbf{d}.$$

Also,

$$(\mathbf{a} \times \mathbf{b}) \times (\mathbf{c} \times \mathbf{d}) = [\mathbf{a}\;\mathbf{c}\;\mathbf{d}]\,\mathbf{b} - [\mathbf{b}\;\mathbf{c}\;\mathbf{d}]\,\mathbf{a}$$

$$= \begin{vmatrix} a_1 & c_1 & d_1 \\ a_2 & c_2 & d_2 \\ a_3 & c_3 & d_3 \end{vmatrix} \mathbf{b} - \begin{vmatrix} b_1 & c_1 & d_1 \\ b_2 & c_2 & d_2 \\ b_3 & c_3 & d_3 \end{vmatrix} \mathbf{a}.$$

Adding, we get the required result.

EXERCISES

1. Prove that

   (*i*) $(\mathbf{a} \times \mathbf{b}) \times (\mathbf{c} \times \mathbf{d}) + (\mathbf{a} \times \mathbf{c}) \times (\mathbf{d} \times \mathbf{b}) + (\mathbf{a} \times \mathbf{d}) \times (\mathbf{b} \times \mathbf{c}) = -2\,[\mathbf{b}\;\mathbf{c}\;\mathbf{d}]\,\mathbf{a}$ (*Garhwal 99*)

   (*ii*) $(\mathbf{b} \times \mathbf{c})\,.\,(\mathbf{a} \times \mathbf{d}) + (\mathbf{c} \times \mathbf{a})\,.\,(\mathbf{b} \times \mathbf{d}) + (\mathbf{a} \times \mathbf{b})\,.\,(\mathbf{c} \times \mathbf{d}) = 0.$ (*Patna 2003*)

2. If **a** and **b** lie in a plane normal to the plane containing **c** and **d**, then show that $(\mathbf{a} \times \mathbf{b})\,.\,(\mathbf{c} \times \mathbf{d}) = 0$.

3. If the four vectors **a, b, c, d** are coplanar, then show that

$$(\mathbf{a} \times \mathbf{b}) \times (\mathbf{c} \times \mathbf{d}) = 0.$$

4. If **a, b, c** and **a′, b′, c′** are reciprocal system of vectors, prove that

(*i*) $\mathbf{a} \cdot \mathbf{a}' + \mathbf{b} \cdot \mathbf{b}' + \mathbf{c} \cdot \mathbf{c}' = 3.$

(*ii*) $\mathbf{a}' \times \mathbf{b}' + \mathbf{b}' \times \mathbf{c}' + \mathbf{c}' \times \mathbf{a}' = \dfrac{\mathbf{a}+\mathbf{b}+\mathbf{c}}{[\mathbf{a}\ \mathbf{b}\ \mathbf{c}]}$, $[\mathbf{a}\ \mathbf{b}\ \mathbf{c}] \neq 0.$

(Kumaon 98, 2000, Avadh 99)

5. Prove that the set of vectors reciprocal to the set of vectors $2\mathbf{i} + 3\mathbf{j} - \mathbf{k}$, $\mathbf{i} - \mathbf{j} - \mathbf{k}$, $-\mathbf{i} + 2\mathbf{j} + 2\mathbf{k}$ is *(Rohilkhand 96, 2003)*

$$\frac{1}{3}(2\mathbf{i}+\mathbf{k}),\ \frac{1}{3}(-8\mathbf{i}+7\mathbf{j}-7\mathbf{k}),\ \frac{1}{3}(-7\mathbf{i}+3\mathbf{j}-5\mathbf{k}).$$

7.7. SOLUTION OF VECTOR EQUATIONS

7.7.1. *To solve for* **r**

$$\mathbf{r} \times \mathbf{b} = \mathbf{a} \times \mathbf{b}, \quad ...(i)$$

where **a, b, c** *are two given vectors.*

Let **r** be any solution of the equation.

Rewriting the given equation as

$$(\mathbf{r} - \mathbf{a}) \times \mathbf{b} = 0.$$

we see that $(\mathbf{r} - \mathbf{a})$ is parallel to **b**, so that we must have

$$\mathbf{r} - \mathbf{a} = t\mathbf{b} \quad \Rightarrow \quad \mathbf{r} = \mathbf{a} + t\mathbf{b}, \quad ...(ii)$$

where t is a scalar. Also it may be seen that $\mathbf{a} + t\mathbf{b}$ satisfies (*i*) for every value of the scalar t. Thus, (*ii*) gives the general solution of the given equation.

Geometrically, we know that the points whose position vectors are given by (*ii*) for different values of the scalar t lie on a straight line which passes through the point, with position vector **a**, and is parallel to the vector **b**. (Refer § 2.4, Ch. 2, P. 32)

7.7.2. *To solve for* **r**

$$\mathbf{r} \times \mathbf{b} = \mathbf{a},$$

where **a, b** *are two given vectors such that* **a** *is perpendicular to* **b**.

$$\mathbf{a}, \quad \mathbf{b}, \quad \mathbf{a} \times \mathbf{b},$$

so that every vector is expressible as a linear combination of the same.

Suppose that a solution of a given equation is

$$\mathbf{r} = x\mathbf{a} + y\mathbf{b} + z\mathbf{a} \times \mathbf{b}.$$

Substituting in a given equation, we obtain

$$(x\mathbf{a} + y\mathbf{b} + z\mathbf{a} \times \mathbf{b}) \times \mathbf{b} = \mathbf{a}$$

$$\Rightarrow \quad x\mathbf{a} \times \mathbf{b} + z\,\{(\mathbf{a} \cdot \mathbf{b})\,\mathbf{b} - (\mathbf{b} \cdot \mathbf{b})\,\mathbf{a}\} = \mathbf{a}$$

$$\Rightarrow \quad -\{1 + z\,(\mathbf{b} \cdot \mathbf{b})\}\,\mathbf{a} + x\mathbf{a} \times \mathbf{b} = 0, \text{ for } \mathbf{a} \cdot \mathbf{b} = 0.$$

$$\Rightarrow \quad 1 + z\,(\mathbf{b} \cdot \mathbf{b}) = 0, \quad x = 0.$$

a, $\mathbf{a} \times \mathbf{b}$, being non-collinear vectors. Thus, we have

$$\mathbf{r} = y\mathbf{b} - \frac{1}{\mathbf{b} \cdot \mathbf{b}}\,\mathbf{a} \times \mathbf{b}. \quad ...(i)$$

Substituting (*i*) in the given equation, we may verify that this is a solution for every value of the scalar y. It follows that

$$\mathbf{r} = -\frac{1}{\mathbf{b}\,.\,\mathbf{b}}\,\mathbf{a} \times \mathbf{b} + y\mathbf{b},$$

is the *general* solution of the equation; y being the parameter.

Geometrically speaking, the points whose position vectors satisfy the given equation lie on the straight line which passes through the point with position vector $-\frac{1}{\mathbf{b}\,.\,\mathbf{b}}(\mathbf{a} \times \mathbf{b})$ and is parallel to the vector **b**.

7.7.3. *To solve simultaneously for* **r**

$$\mathbf{r} \times \mathbf{b} = \mathbf{c} \times \mathbf{b} \qquad ...(i)$$

$$\mathbf{r}\,.\,\mathbf{a} = 0 \qquad ...(ii)$$

provided that **a** *is not perpendicular to* **b**.

Suppose that **r** is a solution of the given equations. Rewriting (*i*) as

$$(\mathbf{r} - \mathbf{c}) \times \mathbf{b} = 0,$$

we see that **r** – **c** and **b** are collinear so that

$$\mathbf{r} - \mathbf{c} = t\mathbf{b} \quad \Leftrightarrow \quad \mathbf{r} = \mathbf{c} + t\mathbf{b},$$

where t is a scalar. Substituting in (*ii*), we obtain

$$(\mathbf{c} + t\mathbf{b})\,.\,\mathbf{a} = 0$$

$$\Leftrightarrow \quad \mathbf{c}.\mathbf{a} + t\,\mathbf{b}\,.\,\mathbf{a} = 0 \;\Rightarrow\; t = -\frac{\mathbf{c}\,.\,\mathbf{a}}{\mathbf{b}\,.\,\mathbf{a}}, \text{ for } \mathbf{b}\,.\,\mathbf{a} \neq 0.$$

Thus, $$\mathbf{r} = \mathbf{c} - \left(\frac{\mathbf{c}\,.\,\mathbf{a}}{\mathbf{b}\,.\,\mathbf{a}}\right)\mathbf{b}. \qquad ...(iii)$$

We may also easily verify that, **r**, given by (*iii*) does satisfy the given equations.

Hence,

$$\mathbf{r} = \mathbf{c} - \frac{\mathbf{c}.\mathbf{a}}{\mathbf{b}.\mathbf{a}}\,\mathbf{b},$$

is a solution and the only solution of the given equations.

It may be seen that **r** . **a** = 0 represents the plane which passes through the origin and is normal to the vector **a**.

Thus, the solution, in question, represents the point of intersection of a line and a plane.

7.7.4. *To solve for* **r**

$$k\mathbf{r} + \mathbf{r} \times \mathbf{a} = \mathbf{b}, \qquad ...(i)$$

where, k is a given non-zero scalar and **a**, **b** *are two given vectors.*

Suppose that **r** is a solution. Expressing **r** as a linear combination of the non-coplanar vectors **a**, **b**, **a** × **b**, we write

$$\mathbf{r} = x\mathbf{a} + y\mathbf{b} + z\mathbf{a} \times \mathbf{b}.$$

Substituting in the given equation, we obtain

$$k(x\mathbf{a}+y\mathbf{b}+z\,\mathbf{a}\times\mathbf{b})+y\,\mathbf{b}\times\mathbf{a}+z\{(\mathbf{a}\cdot\mathbf{a})\,\mathbf{b}-(\mathbf{a}\cdot\mathbf{b})\,\mathbf{a}\}=\mathbf{b}$$

$$\Rightarrow\ \{kx-z(\mathbf{a}\cdot\mathbf{b})\}\,\mathbf{a}+(ky+z\mathbf{a}^2-1)\,\mathbf{b}+(kz-y)\,\mathbf{a}\times\mathbf{b}=0$$

$$\Rightarrow\ kx-z(\mathbf{a}\cdot\mathbf{b})=0,\quad ky+z\mathbf{a}^2-1=0,\quad kz-y=0;$$

a, b, a × b being non-coplanar vectors. It follows that

$$x=\frac{\mathbf{a}\cdot\mathbf{b}}{k(k^2+\mathbf{a}^2)},\ y=\frac{k}{k^2+\mathbf{a}^2},\ z=\frac{1}{k^2+\mathbf{a}^2}.$$

It follows that

$$\mathbf{r}=\frac{1}{k^2+\mathbf{a}^2}\left[\frac{\mathbf{a}\cdot\mathbf{b}}{k}\,\mathbf{a}+k\mathbf{b}+\mathbf{a}\times\mathbf{b}\right]. \qquad ...(ii)$$

Also we may easily verify that **r** given by (*ii*) satisfies the given equation. Hence, this is a solution and the only solution.

EXERCISES

1. Solve simultaneously $\mathbf{r}\cdot\mathbf{n}_1=1,\ \mathbf{r}\cdot\mathbf{n}_2=1$.
2. Find the condition for the equations

$$\mathbf{r}\times\mathbf{a}=\mathbf{b},\quad \mathbf{r}\times\mathbf{c}=\mathbf{d}$$

to be consistent. Assuming the condition for consistency to be satisfied, solve the equations.

OBJECTIVE QUESTIONS

For each of the following questions, four alternatives are given for the answer. Only one of them is correct. Choose the correct alternative.

1. If $\mathbf{a}=2\mathbf{i}+3\mathbf{j}-\mathbf{k}$, $\mathbf{b}=-\mathbf{i}+2\mathbf{j}-4\mathbf{k}$ and $\mathbf{c}=\mathbf{i}+\mathbf{j}+\mathbf{k}$, then $(\mathbf{a}\times\mathbf{b})\cdot(\mathbf{c}\times\mathbf{d})$ is equal to

   (*a*) 60 (*b*) 64 (*c*) 74 (*d*) – 74
2. The product $(a\mathbf{i}+b\mathbf{j})\times(\mathbf{i}-\mathbf{j}+\mathbf{k})\cdot(\mathbf{i}+\mathbf{j}-\mathbf{k})\times(-\mathbf{i}+\mathbf{j}+\mathbf{k})$ is

   (*a*) 0 (*b*) $a+b$ (*c*) $b-a$ (*d*) $-2a$
3. If $\mathbf{p}=\dfrac{\mathbf{a}\times\mathbf{b}}{[\mathbf{a}\,\mathbf{b}\,\mathbf{c}]}$, $\mathbf{q}=\dfrac{\mathbf{c}\times\mathbf{a}}{[\mathbf{a}\,\mathbf{b}\,\mathbf{c}]}$, $\mathbf{r}=\dfrac{\mathbf{a}\times\mathbf{b}}{[\mathbf{a}\,\mathbf{b}\,\mathbf{c}]}$, where **a b c** are three non-coplanar vectors then the value of the expression $(\mathbf{a}+\mathbf{b}+\mathbf{c})\cdot(\mathbf{p}+\mathbf{q}+\mathbf{r})$ is

   (*a*) 3 (*b*) 2 (*c*) 1 (*d*) 0
4. In Q. 3, value of $(\mathbf{a}+\mathbf{b})\cdot\mathbf{p}+(\mathbf{b}+\mathbf{c})\cdot\mathbf{q}+(\mathbf{c}+\mathbf{a})\cdot\mathbf{r}$ is

   (*a*) 0 (*b*) 1 (*c*) 2 (*d*) 3
5. If **a, b, c** be three non-coplanar vectors for which $[\mathbf{a}\,\mathbf{b}\,\mathbf{c}]\neq$ and **a′**, **b′**, **c′** constitute the reciprocal system of vectors, then any vector

r can be expressed as

(*a*) $\mathbf{r} = (\mathbf{r} \times \mathbf{a}) \cdot \mathbf{a}' + (\mathbf{r} \times \mathbf{b}) \cdot \mathbf{b}' + (\mathbf{r} \times \mathbf{c}) \cdot \mathbf{c}'$

(*b*) $\mathbf{r} = (\mathbf{r} \times \mathbf{a}') \cdot \mathbf{a} + (\mathbf{r} \times \mathbf{b}') \cdot \mathbf{b} + (\mathbf{r} \times \mathbf{c}') \cdot \mathbf{c}$

(*c*) $\mathbf{r} = (\mathbf{r} \cdot \mathbf{a})\, \mathbf{a}' + (\mathbf{r} \cdot \mathbf{b})\, \mathbf{b}' + (\mathbf{r} \cdot \mathbf{c})\, \mathbf{c}'$

(*d*) $\mathbf{r} = (\mathbf{r} \cdot \mathbf{a}')\, \mathbf{a} + (\mathbf{r} \cdot \mathbf{b}')\, \mathbf{b} + (\mathbf{r} \cdot \mathbf{c}')\, \mathbf{c}$

6. Value of $\begin{vmatrix} \mathbf{a}\cdot\mathbf{a} & \mathbf{a}\cdot\mathbf{b} & \mathbf{a}\cdot\mathbf{c} \\ \mathbf{b}\cdot\mathbf{a} & \mathbf{b}\cdot\mathbf{b} & \mathbf{b}\cdot\mathbf{c} \\ \mathbf{c}\cdot\mathbf{a} & \mathbf{c}\cdot\mathbf{b} & \mathbf{c}\cdot\mathbf{c} \end{vmatrix}$ is

(*a*) $[\mathbf{a}\ \mathbf{b}\ \mathbf{c}]$ (*b*) $[\mathbf{a}\ \mathbf{b}\ \mathbf{c}]^2$

(*c*) $[\mathbf{a}\ \mathbf{b}\ \mathbf{c}]^3$ (*d*) $3\,[\mathbf{a}\ \mathbf{b}\ \mathbf{c}]$

7. If four vectors **a, b, c, d** are coplanar, then

(*a*) $(\mathbf{a} \times \mathbf{b}) \cdot (\mathbf{c} \times \mathbf{d}) = 0$ (*b*) $(\mathbf{a} \times \mathbf{c}) \cdot (\mathbf{b} \times \mathbf{d}) = 0$

(*c*) $(\mathbf{a} \times \mathbf{b}) \times (\mathbf{c} \times \mathbf{d}) = 0$ (*d*) $(\mathbf{a} \times \mathbf{c}) \times (\mathbf{b} \times \mathbf{d}) = 0$

8. The value of $(\mathbf{a} \times \mathbf{b}) \times (\mathbf{b} \times \mathbf{c})$ is

(*a*) $\mathbf{a}\,[\mathbf{a}\ \mathbf{b}\ \mathbf{c}]$ (*b*) $\mathbf{b}\,[\mathbf{a}\ \mathbf{b}\ \mathbf{c}]$

(*c*) $\mathbf{c}\,[\mathbf{a}\ \mathbf{b}\ \mathbf{c}]$ (*d*) **0**

9. The value of

$(\mathbf{a} \times \mathbf{b}) \cdot (\mathbf{c} \times \mathbf{d}) + (\mathbf{b} \times \mathbf{c}) \cdot (\mathbf{a} \times \mathbf{d}) + (\mathbf{c} \times \mathbf{a}) \cdot (\mathbf{b} \times \mathbf{d})$ is

(*a*) 0 (*b*) 3 (*c*) 2 (*d*) – 3

10. Value of $(\mathbf{a} \times \mathbf{b}) \cdot (\mathbf{b} \times \mathbf{c}) + (\mathbf{c} \times \mathbf{a})$ is

(*a*) $[\mathbf{a}\ \mathbf{b}\ \mathbf{c}]$ (*b*) $[\mathbf{a}\ \mathbf{b}\ \mathbf{c}]^2$

(*c*) $[\mathbf{a}\ \mathbf{b}\ \mathbf{c}]^3$ (*d*) None of these

ANSWERS

1. (*d*) **2.** (*d*) **3.** (*a*) **4.** (*d*) **5.** (*d*)

6. (*b*) **7.** (*c*) **8.** (*b*) **9.** (*a*) **10.** (*b*)

APPENDIX II

Some Properties of Tetrahedra

1. Centre and radius of the circumsphere of a tetrahedron. *If OABC is a tetrahedron and*

$$\vec{OA} = \mathbf{a},\ \vec{OB} = \mathbf{b},\ \vec{OC} = \mathbf{c},$$

then the radius ρ *of the circumsphere of the tetrahedron is given by*

$$\begin{vmatrix} \frac{1}{2}\mathbf{a}^2 & \mathbf{a}\,.\,\mathbf{a} & \mathbf{a}\,.\,\mathbf{b} & \mathbf{a}\,.\,\mathbf{c} \\ \frac{1}{2}\mathbf{b}^2 & \mathbf{b}\,.\,\mathbf{a} & \mathbf{b}\,.\,\mathbf{b} & \mathbf{b}\,.\,\mathbf{c} \\ \frac{1}{2}\mathbf{c}^2 & \mathbf{c}\,.\,\mathbf{a} & \mathbf{c}\,.\,\mathbf{b} & \mathbf{c}\,.\,\mathbf{c} \\ \rho^2 & \frac{1}{2}\mathbf{a}^2 & \frac{1}{2}\mathbf{b}^2 & \frac{1}{2}\mathbf{c}^2 \end{vmatrix} = 0 \qquad \text{...(I)}$$

and the position vector **r** *of the circumcentre is given by*

$$\begin{vmatrix} \mathbf{r} & \mathbf{a} & \mathbf{b} & \mathbf{c} \\ \frac{1}{2}\mathbf{a}^2 & \mathbf{a}\,.\,\mathbf{a} & \mathbf{a}\,.\,\mathbf{b} & \mathbf{a}\,.\,\mathbf{c} \\ \frac{1}{2}\mathbf{b}^2 & \mathbf{b}\,.\,\mathbf{a} & \mathbf{b}\,.\,\mathbf{b} & \mathbf{b}\,.\,\mathbf{c} \\ \frac{1}{2}\mathbf{c}^2 & \mathbf{c}\,.\,\mathbf{a} & \mathbf{c}\,.\,\mathbf{b} & \mathbf{c}\,.\,\mathbf{c} \end{vmatrix} = 0. \qquad \text{...(II)}$$

We have

$$\rho^2 = \mathbf{r}^2 = (\mathbf{r} - \mathbf{a})^2 = (\mathbf{r} - \mathbf{b})^2 = (\mathbf{r} - \mathbf{c})^2 \qquad \text{...}(i)$$

These give

$$\mathbf{a}^2 = 2\mathbf{r}\,.\,\mathbf{a},\ \mathbf{b}^2 = 2\mathbf{r}\,.\,\mathbf{b},\ \mathbf{c}^2 = 2\mathbf{r}\,.\,\mathbf{c}. \qquad \text{..}(ii)$$

Let the expression for **r** in terms of **a, b, c** be

$$\mathbf{r} = x\mathbf{a} + y\mathbf{b} + z\mathbf{c}. \qquad \text{...}(iii)$$

Multiplying scalarly with **a**, **b** and **c** successively and using the relations (*ii*), we get

$$\frac{1}{2}\mathbf{a}^2 = x\mathbf{a}\cdot\mathbf{a} + y\mathbf{a}\cdot\mathbf{b} + z\mathbf{a}\cdot\mathbf{c}, \qquad ...(iv)$$

$$\frac{1}{2}\mathbf{b}^2 = x\mathbf{b}\cdot\mathbf{a} + y\mathbf{b}\cdot\mathbf{b} + z\mathbf{b}\cdot\mathbf{c}, \qquad ...(v)$$

$$\frac{1}{2}\mathbf{c}^2 = x\mathbf{c}\cdot\mathbf{a} + y\mathbf{c}\cdot\mathbf{b} + z\mathbf{c}\cdot\mathbf{c}. \qquad ...(vi)$$

Again multiplying (*iii*) scalarly with **r** and using the relations (*ii*) and the relation $\mathbf{r}^2 = \rho^2$, we get

$$\rho^2 = \frac{1}{2}x\mathbf{a}^2 + \frac{1}{2}y\mathbf{b}^2 + \frac{1}{2}z\mathbf{c}^2. \qquad ...(vii)$$

Eliminating x, y, z from (*iv*), (*v*), (*vi*) and (*vii*), we get I and elimi-nating x, y, z from (*iii*), (*iv*), (*v*) and (*vi*), we get II.

2. *If a tetrahedron OABC is such that the mid-points of its six edges lie on a sphere of radius, r, then*

(I) *the opposite edges of the tetrahedron are perpendicular.*

(II) *centre of the sphere is the centroid of the tetrahedron.*

(III) $GA^2 + GB^2 + GC^2 + GD^2 = 12r^2$*; G, being the centroid.*

(IV) $AB^2 + CD^2 = BC^2 + AD^2 = CA^2 + BD^2 = 16r^2$.

Take the centre O of the sphere through the six mid-points of the edges as the origin of reference. Let

$$\overrightarrow{OA} = \mathbf{a},\ \overrightarrow{OB} = \mathbf{b},\ \overrightarrow{OC} = \mathbf{c},\ \overrightarrow{OD} = \mathbf{d},$$

The sphere passes through the points with position vectors

$$\frac{1}{2}(\mathbf{a}+\mathbf{b}),\ \frac{1}{2}(\mathbf{c}+\mathbf{d}),\ \frac{1}{2}(\mathbf{b}+\mathbf{c}),\ \frac{1}{2}(\mathbf{a}+\mathbf{d}),\ \frac{1}{2}(\mathbf{c}+\mathbf{a}),\ \frac{1}{2}(\mathbf{b}+\mathbf{d}).$$

Thus, we have

$$4r^2 = (\mathbf{a}+\mathbf{b})^2 = (\mathbf{c}+\mathbf{d})^2 = (\mathbf{b}+\mathbf{c})^2$$
$$= (\mathbf{a}+\mathbf{d})^2 = (\mathbf{c}+\mathbf{a})^2 = (\mathbf{b}+\mathbf{d})^2 \qquad ...(i)$$

From (*i*), we have

$$(\mathbf{a}+\mathbf{b})^2 + (\mathbf{c}+\mathbf{d})^2 = (\mathbf{b}+\mathbf{c})^2 + (\mathbf{a}+\mathbf{d})^2$$

$$\Rightarrow \qquad \mathbf{a}\cdot\mathbf{b} + \mathbf{c}\cdot\mathbf{d} = \mathbf{b}\cdot\mathbf{c} + \mathbf{a}\cdot\mathbf{d}$$

$$\Rightarrow \qquad (\mathbf{a}-\mathbf{c})\cdot(\mathbf{b}-\mathbf{d}) = 0 \ \Rightarrow\ AC \perp BD.$$

Similarly, $BC \perp AD$, $CA \perp AD$. Hence, the result (I).

Again, from (*i*),

$$(\mathbf{a}+\mathbf{b})^2 = (\mathbf{c}+\mathbf{d})^2$$

$$\Rightarrow \qquad (\mathbf{a}+\mathbf{b}+\mathbf{c}+\mathbf{d})\cdot[(\mathbf{a}+\mathbf{b}) - (\mathbf{c}+\mathbf{d})] = 0$$

Similarly,

$$(\mathbf{c}+\mathbf{a})^2 = (\mathbf{b}+\mathbf{d})^2 \ \Rightarrow\ (\mathbf{a}+\mathbf{b}+\mathbf{c}+\mathbf{d})\cdot[(\mathbf{c}+\mathbf{a}) - (\mathbf{b}+\mathbf{d})] = 0.$$

$$(\mathbf{b}+\mathbf{c})^2 = (\mathbf{a}+\mathbf{d})^2 \ \Rightarrow\ (\mathbf{a}+\mathbf{b}+\mathbf{c}+\mathbf{d})\cdot[(\mathbf{b}+\mathbf{c}) - (\mathbf{a}+\mathbf{d})] = 0.$$

The vectors

$$(\mathbf{a}+\mathbf{d}) - (\mathbf{c}+\mathbf{d}),\ (\mathbf{b}+\mathbf{c}) - (\mathbf{a}+\mathbf{d}),\ (\mathbf{c}+\mathbf{a}) - (\mathbf{b}+\mathbf{d})$$

cannot be coplanar, lying as they do along the lines joining the mid-points of the pairs of opposite edges of the tetrahedron. Therefore

$$\mathbf{a} + \mathbf{b} + \mathbf{c} + \mathbf{d} = 0$$

so that the centre of the circumsphere coincides with the centroid of the tetrahedron. Hence, the result (II).

From (*i*), by addition

$$24r^2 = 3\,\Sigma\mathbf{a}^2 + 2\,\Sigma\mathbf{a}\,.\,\mathbf{b}.$$

Also

$$\Sigma\mathbf{a} = 0 \;\Rightarrow\; (\Sigma\mathbf{a})^2 = 0 \;\Rightarrow\; \Sigma\mathbf{a}^2 = -\,2\,\Sigma\mathbf{a}\,.\,\mathbf{b}.$$

Thus, we have

$$24r^2 = 3\,\Sigma\mathbf{a}^2 - \Sigma\mathbf{a}^2 = 2\,\Sigma\mathbf{a}^2 \;\Rightarrow\; \Sigma\mathbf{a}^2 = 12r^2.$$

$$\Rightarrow \qquad \Sigma GA^2 = 12r^2.$$

Hence, the result (III).

Again

$$(\mathbf{a} + \mathbf{b})^2 + (\mathbf{c} + \mathbf{d})^2 = 8r^2$$

$$\Rightarrow \qquad 2\mathbf{a}\,.\,\mathbf{b} + 2\mathbf{c}\,.\,\mathbf{d} = 8r^2 - \Sigma\mathbf{a}^2 = -\,4r^2.$$

Thus, we have

$$(\mathbf{a} - \mathbf{b})^2 + (\mathbf{c} - \mathbf{d})^2 = \Sigma\mathbf{a}^2 - (2\mathbf{a}\,.\,\mathbf{b} + 2\mathbf{c}\,.\,\mathbf{d})$$

$$= 12r^2 + 4r^2 = 16r^2$$

$$\Rightarrow \qquad AB^2 + CD^2 = 16r^2.$$

Hence, the result (IV).

3. *The six mid-points of the six edges of a tetrahedron lie on a sphere, if the pairs of opposite edges of the tetrahedron are perpendicular to each other.*

Take the origin of reference at the centroid of the tetrahedron. Denoting the position vectors of the vertices *A, B, C, D* by **a, b, c, d,** we are given that

$$\mathbf{a} + \mathbf{b} + \mathbf{c} + \mathbf{d} = 0 \qquad ...(i)$$

$$\mathbf{a}\,.\,\mathbf{b} + \mathbf{c}\,.\,\mathbf{d} = \mathbf{a}\,.\,\mathbf{c} + \mathbf{b}\,.\,\mathbf{d} = \mathbf{b}\,.\,\mathbf{c} - \mathbf{a}\,.\,\mathbf{d} \qquad ...(ii)$$

From (*i*)

$$\mathbf{a} + \mathbf{b} = -\,(\mathbf{c} + \mathbf{d}) \;\Rightarrow\; (\mathbf{a} + \mathbf{b})^2 = (\mathbf{c} + \mathbf{d})^2.$$

Similarly

$$(\mathbf{a} + \mathbf{c})^2 = (\mathbf{b} + \mathbf{d})^2,\;\; (\mathbf{b} + \mathbf{c})^2 = (\mathbf{a} + \mathbf{d})^2. \qquad ...(iii)$$

Again we have

$$(\mathbf{a} + \mathbf{b})^2 + (\mathbf{c} + \mathbf{d})^2 = \Sigma\mathbf{a}^2 + 2\,(\mathbf{a}\,.\,\mathbf{b} + \mathbf{c}\,.\,\mathbf{d})$$

$$= \Sigma\mathbf{a}^2 + 2\,(\mathbf{a}\,.\,\mathbf{c} + \mathbf{b}\,.\,\mathbf{d}); \qquad \text{By } (ii)$$

$$= (\mathbf{a} + \mathbf{c})^2 + (\mathbf{b} + \mathbf{d})^2.$$

Similarly

$$(\mathbf{a} + \mathbf{b})^2 = (\mathbf{c} + \mathbf{d})^2 = (\mathbf{b} + \mathbf{c})^2 + (\mathbf{a} + \mathbf{d})^2. \qquad ...(iv)$$

From (*iii*) and (*iv*), it follows that

$$(\mathbf{a}+\mathbf{b})^2 = (\mathbf{c}+\mathbf{d})^2 = (\mathbf{b}+\mathbf{c})^2 = (\mathbf{a}+\mathbf{d})^2 = (\mathbf{c}+\mathbf{a})^2 + (\mathbf{b}+\mathbf{d})^2.$$

Thus, the middle points lie on a sphere with its centre at the centroid which has been taken as the origin.

This result is the converse of the result (I) of 2.

4. Isosceles Tetrahedra. Def. *A tetrahedron, the pairs of whose opposite edges are equal in length is called an isosceles tetrahedron.*

Theorem. *The shortest distances between pairs of opposite edges of an isosceles tetrahedron lie along the joins of their middle points and the three shortest distances bisect each other at right angles.*

Take the vertex O of the isosceles tetrahedron $OABC$ as the origin of reference. Let

$$\vec{OA} = \mathbf{a},\ \vec{OB} = \mathbf{b},\ \vec{OC} = \mathbf{c}.$$

Now

$$\left.\begin{aligned} OA = BC &\Rightarrow \mathbf{a}^2 = \mathbf{b}^2 + \mathbf{c}^2 - 2\mathbf{b}\,.\,\mathbf{c} \Rightarrow \mathbf{b}\,.\,\mathbf{c} = \frac{1}{2}(\mathbf{b}^2 + \mathbf{c}^2 - \mathbf{a}^2) \\ OB = CA &\Rightarrow \mathbf{b}^2 = \mathbf{c}^2 + \mathbf{a}^2 - 2\mathbf{c}\,.\,\mathbf{a} \Rightarrow \mathbf{c}\,.\,\mathbf{a} = \frac{1}{2}(\mathbf{c}^2 + \mathbf{a}^2 - \mathbf{b}^2) \\ OC = AB &\Rightarrow \mathbf{c}^2 = \mathbf{a}^2 + \mathbf{b}^2 - 2\mathbf{a}\,.\,\mathbf{b} \Rightarrow \mathbf{a}\,.\,\mathbf{b} = \frac{1}{2}(\mathbf{a}^2 + \mathbf{b}^2 - \mathbf{c}^2) \end{aligned}\right\} \quad ...(i)$$

The vectors

$$\frac{1}{2}(\mathbf{b}+\mathbf{c}) - \frac{1}{2}\mathbf{a},\ \frac{1}{2}(\mathbf{c}+\mathbf{a}) - \frac{1}{2}\mathbf{b},\ \frac{1}{2}(\mathbf{a}+\mathbf{b}) - \frac{1}{2}\mathbf{c}$$

lie along the joins of the mid-points of the pairs of opposite edges

$$OA, BC; \quad OB, CA; \quad OC, AB,$$

respectively.

The join of the mid-points of OA and BC will be perpendicular to OA and BC, if

$$\left[\frac{1}{2}(\mathbf{b}+\mathbf{c}) - \frac{1}{2}\mathbf{a}\right]\cdot\mathbf{a} = 0 \text{ and } \left[\frac{1}{2}(\mathbf{b}+\mathbf{c}) - \frac{1}{2}\mathbf{a}\right]\cdot(\mathbf{b}-\mathbf{c}) = 0$$

$$\Rightarrow \quad \mathbf{a}^2 = \mathbf{a}\,.\,\mathbf{b} + \mathbf{a}\,.\,\mathbf{c};\ \mathbf{b}^2 - \mathbf{c}^2 = \mathbf{a}\,.\,\mathbf{b} - \mathbf{a}\,.\,\mathbf{c}. \quad ...(ii)$$

It can be at once seen that the relations (*ii*) are simple consequences of the relations (*i*). Thus, the shortest distance between OA and OB lies along the join of their middle points.

Similarly we may show that the shortest distances between the other pairs of opposite edges lie along the joins of their mid-points.

The three shortest distance lines lie along the vectors

$$\mathbf{b}+\mathbf{c}-\mathbf{a},\ \ \mathbf{c}+\mathbf{a}-\mathbf{b},\ \ \mathbf{a}+\mathbf{b}-\mathbf{c}$$

which join the mid-points of the opposite edges.

Now we have

$$(\mathbf{b} + \mathbf{c} - \mathbf{a}) \cdot (\mathbf{c} + \mathbf{a} - \mathbf{b}) = \mathbf{c}^2 - (\mathbf{b} - \mathbf{a})^2$$

$$= \mathbf{c}^2 - \mathbf{b}^2 - \mathbf{a}^2 + 2\mathbf{a} \cdot \mathbf{b} = 0, \quad \text{By } (i)$$

Thus, we see that the lines of shortest distance are at right angles to each other.

Surely they also all concur at the centroid.

5. In-sphere of a tetrahedron. In-centre and In-radius. It will be shown that there exists a point I inside any given tetrahedron equidistant from the four faces of the same. The sphere with centre I and radius equal to its distance from any face will touch the four faces of the tetrahedron internally. This sphere is known as the *In-sphere* and its centre I is known as the *In-centre* of the tetrahedron. The radius of the in-sphere is called the *In-radius.*

Let *OABC* be any tetrahedron and let **a, b, c** be the position vectors of *A, B, C* with reference to *O* as origin.

The equations of the three planes

OBC, OCA, OAB

through the vertex *O* are

$$\mathbf{r} \cdot \mathbf{b} \times \mathbf{c} = 0, \quad \mathbf{r} \cdot \mathbf{c} \times \mathbf{a} = 0, \quad \mathbf{r} \cdot \mathbf{a} \times \mathbf{b} = 0.$$

The bisecting planes of the angles between the planes *OBC, OCA* are

$$\frac{\mathbf{r} \cdot \mathbf{b} \times \mathbf{c}}{|\mathbf{b} \times \mathbf{c}|} \pm \frac{\mathbf{r} \cdot \mathbf{c} \times \mathbf{a}}{|\mathbf{c} \times \mathbf{a}|} = 0.$$

Of these two, the equation

$$\frac{\mathbf{r} \cdot \mathbf{b} \times \mathbf{c}}{|\mathbf{b} \times \mathbf{c}|} - \frac{\mathbf{r} \cdot \mathbf{c} \times \mathbf{a}}{|\mathbf{c} \times \mathbf{a}|} = 0 \quad \text{...}(i)$$

represents the internal bisecting plane, for the points *A, B* lie on opposite sides of it. By substituting **a, b** for **r** respectively in (*i*), we obtain expressions having opposite signs.

The other two internal bisecting planes are

$$\frac{\mathbf{r} \cdot \mathbf{c} \times \mathbf{a}}{|\mathbf{c} \times \mathbf{a}|} - \frac{\mathbf{r} \cdot \mathbf{a} \times \mathbf{b}}{|\mathbf{a} \times \mathbf{b}|} = 0 \quad \text{...}(ii)$$

$$\frac{\mathbf{r} \cdot \mathbf{a} \times \mathbf{b}}{|\mathbf{a} \times \mathbf{b}|} - \frac{\mathbf{r} \cdot \mathbf{b} \times \mathbf{c}}{|\mathbf{b} \times \mathbf{c}|} = 0. \quad \text{...}(iii)$$

For the sake of brevity, we write α, β, γ for

$$\frac{1}{2}|\mathbf{b} \times \mathbf{c}|, \ \frac{1}{2}|\mathbf{c} \times \mathbf{a}|, \ \frac{1}{2}|\mathbf{a} \times \mathbf{b}|,$$

respectively so that they denote the areas of the triangular faces, *OBC, OCA, OAB* respectively.

The three planes (*i*), (*ii*) and (*iii*) obviously meet in a line through *O*. The line of intersection of these planes is parallel to the vector

$$\left(\frac{\mathbf{b}\times\mathbf{c}}{\alpha}-\frac{\mathbf{c}\times\mathbf{a}}{\beta}\right)\times\left(\frac{\mathbf{c}\times\mathbf{a}}{\beta}-\frac{\mathbf{a}\times\mathbf{b}}{\gamma}\right)=\frac{[\mathbf{a}\ \mathbf{b}\ \mathbf{c}]}{\alpha\,\beta\,\gamma}\{\alpha\mathbf{a}+\beta\mathbf{b}+\gamma\mathbf{c}\}.$$

Thus, the line of intersection of these planes is

$$\mathbf{r} = t\,(\alpha\mathbf{a} + \beta\mathbf{b} + \gamma\mathbf{c}).$$

From this, we easily deduce that the point I where this line meets the plane *ABC* whose equation is

$$\mathbf{r} \cdot (\mathbf{b} \times \mathbf{c} + \mathbf{c} \times \mathbf{a} + \mathbf{a} \times \mathbf{b}) = [\mathbf{a}\ \mathbf{b}\ \mathbf{c}]$$

is

$$\frac{\alpha\mathbf{a}+\beta\mathbf{b}+\gamma\mathbf{c}}{\alpha+\beta+\gamma}$$

$$\Rightarrow \quad \overrightarrow{OI_1}=\frac{\alpha\,\overrightarrow{OA}+\beta\,\overrightarrow{OB}+\gamma\,\overrightarrow{OC}}{\alpha+\beta+\gamma}.$$

We now suppose that the position vectors of the vertices *A, B, C, D* with reference to any again *O* are

$$\mathbf{a},\ \mathbf{b},\ \mathbf{c},\ \mathbf{d}$$

respectively. The position vector of I_1 being

$$(\alpha\mathbf{a} + \beta\mathbf{b} + \gamma\mathbf{c}) \,/\, (\alpha + \beta + \gamma),$$

the point dividing $O.\ I_1$ in the ratio $\alpha + \beta + \gamma : \delta$ is

$$\frac{\alpha\mathbf{a}+\beta\mathbf{b}+\gamma\mathbf{c}+\delta\mathbf{d}}{\alpha+\beta+\gamma+\delta} \qquad ...(iv)$$

δ denoting the area of the triangle *ABC*.

By symmetry, we see that the point (*iv*) lies on the line of intersection of the internal bisecting planes of the planes through every vertex.

Thus, the planes bisecting internally the angles between pairs of planes of the tetrahedron meet in the point given by (*iv*) which is the *in-centre*.

In-radius. The distance of the point (*iv*) from the face *ABC* is the in-radius. Consider the face *ABC* whose equation is

$$\mathbf{r} \cdot (\mathbf{b} \times \mathbf{c} + \mathbf{c} \times \mathbf{a} + \mathbf{a} \times \mathbf{b}) = [\mathbf{a}\ \mathbf{b}\ \mathbf{c}].$$

Thus, the in-radius is

$$\frac{\dfrac{\alpha\mathbf{a}+\beta\mathbf{b}+\gamma\mathbf{c}+\delta\mathbf{d}}{\alpha+\beta+\gamma+\delta}\cdot(\mathbf{b}\times\mathbf{c}+\mathbf{c}\times\mathbf{a}+\mathbf{a}\times\mathbf{b})-[\mathbf{a}\ \mathbf{b}\ \mathbf{c}]}{|\,\mathbf{b}\times\mathbf{c}+\mathbf{c}\times\mathbf{a}+\mathbf{a}\times\mathbf{b}\,|}$$

$$=\frac{\alpha\,|\,[\mathbf{a}\ \mathbf{b}\ \mathbf{c}]+[\mathbf{a}\ \mathbf{c}\ \mathbf{d}]+[\mathbf{a}\ \mathbf{b}\ \mathbf{d}]-[\mathbf{b}\ \mathbf{c}\ \mathbf{d}]\,|}{(\alpha+\beta+\gamma+\delta)\,2\alpha}$$

$$=\frac{6V}{2S}=\frac{3V}{S},$$

where V denotes the volume and S, the sum of the areas of the faces of the tetrahedron.

The reader may compare this formula with that for the In-radius of a triangle.

6. Regular tetrahedron. Def. *A tetrahedron whose edges are all equal in length is called a Regular tetrahedron.*

The faces of a regular tetrahedron are equilateral triangles so that the angle between any two concurrent edges of a regular tetrahedron is 60°.

The following results will be established :

I. *The angle between any two plane faces of a regular tetrahedron is* $\cos^{-1}\frac{1}{3}$.

II. *The angle between any edge and a face not containing the edge is* $\cos^{-1}\left(\sqrt{\frac{1}{3}}\right)$.

III. *Any two opposite edges are perpendicular to each other.*

IV. *The distance of any vertex from the opposite face is* $\sqrt{\frac{2}{3}}k$; *k being the length of any edge.*

I. Let $OABC$ be a regular tetrahedron. Take O as the origin of reference. Let

$$\overrightarrow{OA} = \mathbf{a},\ \overrightarrow{OB} = \mathbf{b},\ \overrightarrow{OC} = \mathbf{c}.$$

and let k be the length of each of the edges,

We have

$$|\mathbf{a}| = |\mathbf{b}| = |\mathbf{c}| = k,$$

$$|\mathbf{a}\cdot\mathbf{b}| = |\mathbf{b}\cdot\mathbf{c}| = |\mathbf{c}\cdot\mathbf{a}|$$

$$= k^2\cos 60° = \frac{1}{2}k^2.$$

$$\mathbf{a}\cdot\mathbf{a} = \mathbf{b}\cdot\mathbf{b} = \mathbf{c}\cdot\mathbf{c} = k^2.$$

The equation of the planes OAB and OBC are

$$\mathbf{r}\cdot\mathbf{a}\times\mathbf{b} = 0,\quad \mathbf{r}\cdot\mathbf{b}\times\mathbf{c} = 0.$$

Now the angle between two planes is equal to the angle between the normals to the same. Thus, if θ denotes the angle between these two planes, we have

$$\cos\theta = \frac{(\mathbf{a}\times\mathbf{b})}{|\mathbf{a}\times\mathbf{b}|}\cdot\frac{(\mathbf{b}\times\mathbf{c})}{|\mathbf{b}\times\mathbf{c}|}.$$

Now $\quad (\mathbf{a}\times\mathbf{b})\cdot(\mathbf{b}\times\mathbf{c}) = (\mathbf{a}\cdot\mathbf{b})(\mathbf{b}\cdot\mathbf{c}) - (\mathbf{b}\cdot\mathbf{b})(\mathbf{a}\cdot\mathbf{c}) = \frac{-1}{4}k^4$

$$|\mathbf{a}\times\mathbf{b}| = |\mathbf{a}|\,|\mathbf{b}|\sin 60° = \frac{\sqrt{3}}{2}k^2 = |\mathbf{b}\times\mathbf{c}|.$$

$$\therefore \qquad \cos\theta = -\frac{1}{3}.$$

Thus, the acute angle between any two plane faces is $\cos^{-1}\frac{1}{3}$.

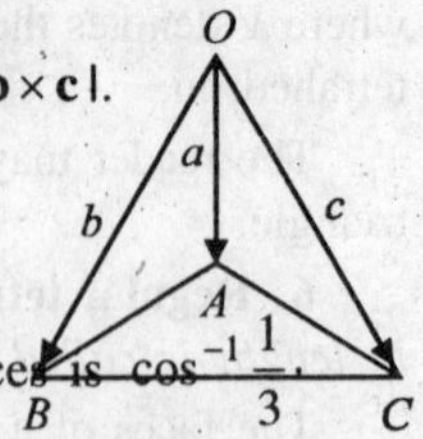

Fig. 7.1.

II. Consider the face *ABC* and the edge *OA* whose equations are

$$\mathbf{r}\,.\,[\mathbf{b}\times\mathbf{c}+\mathbf{c}\times\mathbf{a}+\mathbf{a}\times\mathbf{b}] = [\mathbf{a}\;\mathbf{b}\;\mathbf{c}], \quad \mathbf{r} = t\mathbf{a}$$

respectively.

The angle between a line and a plane is equal to the complement of the angle between the line and the normal to the plane. Thus, if θ denotes the angle between the face and the edge, we have

$$\sin\theta = \frac{(\mathbf{b}\times\mathbf{c}+\mathbf{c}\times\mathbf{a}+\mathbf{a}\times\mathbf{b})\,.\,\mathbf{a}}{|\mathbf{b}\times\mathbf{c}+\mathbf{c}\times\mathbf{a}+\mathbf{a}\times\mathbf{b}|\;|\mathbf{a}|}$$

$$= \frac{[\mathbf{a}\;\mathbf{b}\;\mathbf{c}]}{|\mathbf{b}\times\mathbf{c}+\mathbf{c}\times\mathbf{a}+\mathbf{a}\times\mathbf{b}|\;|\mathbf{a}|} \qquad \ldots(i)$$

Now $\quad [\mathbf{a}\;\mathbf{b}\;\mathbf{c}]^2 = \begin{vmatrix} \mathbf{a}.\mathbf{a} & \mathbf{a}.\mathbf{b} & \mathbf{a}.\mathbf{c} \\ \mathbf{b}.\mathbf{a} & \mathbf{b}.\mathbf{b} & \mathbf{b}.\mathbf{c} \\ \mathbf{c}.\mathbf{a} & \mathbf{c}.\mathbf{b} & \mathbf{c}.\mathbf{c} \end{vmatrix} = \frac{1}{2}k^4 \qquad$ (Cor. P. 179)

Also $\quad [\mathbf{b}\times\mathbf{c}+\mathbf{c}\times\mathbf{a}+\mathbf{a}\times\mathbf{b}]$

is twice the area of the triangle *ABC* which is equilateral with each side *k* so that this is

$$= \frac{\sqrt{3}}{2}k^2.$$

Substituting in (*i*), we obtain

$$\sin\theta = \sqrt{\frac{2}{3}} \Rightarrow \cos\theta = \sqrt{\frac{1}{3}} \Rightarrow \theta = \cos^{-1}\left(\sqrt{\frac{1}{3}}\right).$$

III. We have $\overrightarrow{OA}\cdot\overrightarrow{BC} = \mathbf{a}\,.\,(\mathbf{c}-\mathbf{b}) = \mathbf{a}\,.\,\mathbf{c}-\mathbf{a}\,.\,\mathbf{b} = 0$ so that $OA \perp BC$.

Similarly,

$$OB \perp CA \quad \text{and} \quad OC \perp AB.$$

IV. The equation of the plane *ABC* is

$$\mathbf{r}\,.\,[\mathbf{b}\times\mathbf{c}+\mathbf{c}\times\mathbf{a}+\mathbf{a}\times\mathbf{b}] = [\mathbf{a}\;\mathbf{b}\;\mathbf{c}],$$

so that the distance of the vertex *O* from this plane

$$= \frac{[\mathbf{a}\;\mathbf{b}\;\mathbf{c}]}{|\mathbf{b}\times\mathbf{c}+\mathbf{c}\times\mathbf{a}+\mathbf{a}\times\mathbf{b}|} = \frac{2}{3}k.$$

7. Orthocentric tetrahedra. A *tetrahedron is called orthocentric if it* *is such that the altitude from its vertices to the opposite faces are concurrent.*

The following results will be established :

I. *If in a tetrahedron OABC, the altitudes from A and B to the opposite faces intersect then the edge AB is perpendicular to the opposite edge OC.*

II. *If the altitudes from A and B intersect and those from B and C also intersect then all the four altitudes are concurrent and the pairs of opposite edges are at right angles to each other.*

Take O as the origin of reference. Let

$$\overrightarrow{OA} = \mathbf{a},\ \overrightarrow{OB} = \mathbf{b},\ \overrightarrow{OC} = \mathbf{c}.$$

I. The equations of the altitudes from A and B to the opposite faces are

$$\mathbf{r} = \mathbf{a} + t\,(\mathbf{b} \times \mathbf{c}), \quad \mathbf{r} = \mathbf{b} + p\,(\mathbf{c} \times \mathbf{a}). \qquad ...(i)$$

Now

$$(\mathbf{a} - \mathbf{b}) \cdot (\mathbf{b} \times \mathbf{c}) \times (\mathbf{c} \times \mathbf{a}) = (\mathbf{a} + \mathbf{b}) \cdot [\mathbf{a}\ \mathbf{b}\ \mathbf{c}]$$
$$= [\mathbf{a}\ \mathbf{b}\ \mathbf{c}]\ (\mathbf{a} - \mathbf{b}) \cdot \mathbf{c}$$

so that the altitudes (*i*) will be coplanar if, and only if

$$(\mathbf{a} - \mathbf{b}) \cdot \mathbf{c} = 0, \quad \Rightarrow \quad AB \perp OC.$$

II. Assuming the condition as satisfied, we find the position vector of the point of intersection of the altitudes from A and B.

For this purpose, we express the vectors **a, b** as linear combination of the non-coplanar vectors **b × c, c × a, a × b.**

Let

$$\mathbf{a} = l_1 \mathbf{b} \times \mathbf{c} + m_1 \mathbf{c} \times \mathbf{a} + n_1 \mathbf{a} \times \mathbf{b}, \quad \mathbf{b} = l_2 \mathbf{b} \times \mathbf{c} + m_2 \mathbf{c} \times \mathbf{a} + n_2 \mathbf{a} \times \mathbf{b},$$

so that, as may be easily shown,

$$l_1 = \frac{\mathbf{a} \cdot \mathbf{a}}{[\mathbf{a}\ \mathbf{b}\ \mathbf{c}]},\ m_1 = \frac{\mathbf{a} \cdot \mathbf{b}}{[\mathbf{a}\ \mathbf{b}\ \mathbf{c}]},\ n_1 = \frac{\mathbf{a} \cdot \mathbf{c}}{[\mathbf{a}\ \mathbf{b}\ \mathbf{c}]},$$

$$l_2 = \frac{\mathbf{b} \cdot \mathbf{a}}{[\mathbf{a}\ \mathbf{b}\ \mathbf{c}]},\ m_2 = \frac{\mathbf{b} \cdot \mathbf{b}}{[\mathbf{a}\ \mathbf{b}\ \mathbf{c}]},\ n_2 = \frac{\mathbf{b} \cdot \mathbf{c}}{[\mathbf{a}\ \mathbf{b}\ \mathbf{c}]}.$$

At the point of intersection of the lines (*i*), we have

$$l_1 (\mathbf{b} \times \mathbf{c}) + m_1 (\mathbf{c} \times \mathbf{a}) + n_1 (\mathbf{a} \times \mathbf{b}) + t\,(\mathbf{b} \times \mathbf{c})$$
$$= l_2 (\mathbf{b} \times \mathbf{c}) + m_2 (\mathbf{c} \times \mathbf{a}) + n_2 (\mathbf{a} \times \mathbf{b}) + p\,(\mathbf{c} \times \mathbf{a})$$

$$\Rightarrow \qquad l_1 + t = l_2, \quad m_1 = m_2 + p, \quad n_1 = n_2.$$

Here $n_1 = n_2$ is equivalent to (*ii*), *viz.*, $(\mathbf{a} - \mathbf{b}) \cdot \mathbf{c} = 0$.

$$\therefore \qquad t = l_2 - l_1 = \frac{\mathbf{b} \cdot \mathbf{a} - \mathbf{a} \cdot \mathbf{a}}{[\mathbf{a}\ \mathbf{b}\ \mathbf{c}]},\ p = m_1 - m_2 = \frac{\mathbf{a} \cdot \mathbf{b} - \mathbf{b} \cdot \mathbf{b}}{[\mathbf{a}\ \mathbf{b}\ \mathbf{c}]}.$$

Thus, the point of intersection of the altitudes from A and B is

$$\frac{1}{[\mathbf{a}\ \mathbf{b}\ \mathbf{c}]} \{(\mathbf{b} \cdot \mathbf{a})(\mathbf{b} \times \mathbf{c}) + (\mathbf{a} \cdot \mathbf{b})(\mathbf{c} \times \mathbf{a}) + (\mathbf{a} \cdot \mathbf{c})(\mathbf{a} \times \mathbf{b})\} \qquad ...(iii)$$

Again suppose that the altitudes from B and C also intersect so that $(\mathbf{b}-\mathbf{c}) \cdot \mathbf{a} = 0$. The point of intersection of these altitudes is

$$\frac{1}{[\mathbf{a}\,\mathbf{b}\,\mathbf{c}]}\{(\mathbf{a}\cdot\mathbf{b})(\mathbf{b}\times\mathbf{c})+(\mathbf{b}\cdot\mathbf{c})(\mathbf{c}\times\mathbf{a})+(\mathbf{c}\cdot\mathbf{a})(\mathbf{a}\times\mathbf{b})\} \quad ...(iv)$$

As $\mathbf{a}\cdot\mathbf{b} = \mathbf{b}\cdot\mathbf{c} = \mathbf{c}\cdot\mathbf{a}$, the points (*iii*) and (*iv*) are the same and the common point is

$$\frac{k}{[\mathbf{a}\,\mathbf{b}\,\mathbf{c}]}\{\mathbf{b}\times\mathbf{c}+\mathbf{c}\times\mathbf{a}+\mathbf{a}\times\mathbf{b}\} \quad ...(v)$$

where $k = \mathbf{a}.\mathbf{b} = \mathbf{b}\cdot\mathbf{c} = \mathbf{c}\cdot\mathbf{a}$.

From the symmetry of (v), it is clear that all four altitudes concur at this point. Also the conditions

$$\mathbf{a}.\mathbf{b} = \mathbf{b}.\mathbf{c} = \mathbf{c}.\mathbf{a}$$

imply that the pairs of opposite edges of the tetrahedron are perpendicular to each other.

Also the point of concurrence is called the **Orthocentre** of the tetrahedron.

8. Equifacial Tetrahedra. Def. *A tetrahedron whose faces have equal areas is called equifacial.*

The following results will be established :

I. *If the areas of two faces OAB and OAC of a tetrahedron OABC be equal, then the line of shortest distance between OA and BC is the line through the middle point of BC perpendicular to OA.*

II. *If the areas of two faces OAB and OAC are equal and those of the faces BCA, BCO are equal then the line of shortest distance between OA and BC is the join of their middle points : also the faces equal in area are congruent.*

III. *If the areas of three triangular faces through any vertex O are equal, then the centroid and in-centre are collinear with O.*

IV. *The faces of an equifacial tetrahedron are congruent triangles. Every equifacial tetrahedron is isosceles.*

I. Take O as origin of reference. Let

$$\vec{OA} = \mathbf{a},\ \vec{OB} = \mathbf{b},\ \vec{OC} = \mathbf{c}.$$

Vector areas of the faces OAB and OAC are

$$\frac{1}{2}\vec{OA}\times\vec{OB} \text{ and } \frac{1}{2}\vec{OA}\times\vec{OC}$$

so that we are given that

$$|\mathbf{a}\times\mathbf{b}| = |\mathbf{a}\times\mathbf{c}|$$

$$\Rightarrow \quad (\mathbf{a}\times\mathbf{b})\cdot(\mathbf{a}\times\mathbf{b}) = (\mathbf{a}\times\mathbf{c})\cdot(\mathbf{a}\times\mathbf{c})$$

$$\Rightarrow \quad (\mathbf{a}\times\mathbf{b}-\mathbf{a}\times\mathbf{c})\cdot(\mathbf{a}\times\mathbf{b}+\mathbf{a}\times\mathbf{c}) = 0$$

$$\Rightarrow \quad [\mathbf{a}\times(\mathbf{b}-\mathbf{c})]\cdot[\mathbf{a}\times(\mathbf{b}+\mathbf{c})] = 0 \quad ...(i)$$

Now $\mathbf{a} \times (\mathbf{b} - \mathbf{c})$ is parallel to the line of shortest distance between *OA* and *BC*.

Also $\frac{1}{2}(\mathbf{b}+\mathbf{c})$ is the position vector of the mid-point *D* of *BC*.

As $\vec{OA} = \mathbf{a}$ and $\vec{OD} = \frac{1}{2}(\mathbf{b}+\mathbf{c})$,

the vector $\mathbf{a} \times (\mathbf{b} + \mathbf{c})$ is normal to the plane *OAD*.

By (*i*), the line of shortest distance between *OA* and *BC* lies in the plane *OAD*. Also it meets *BC*. Hence, we have the result I.

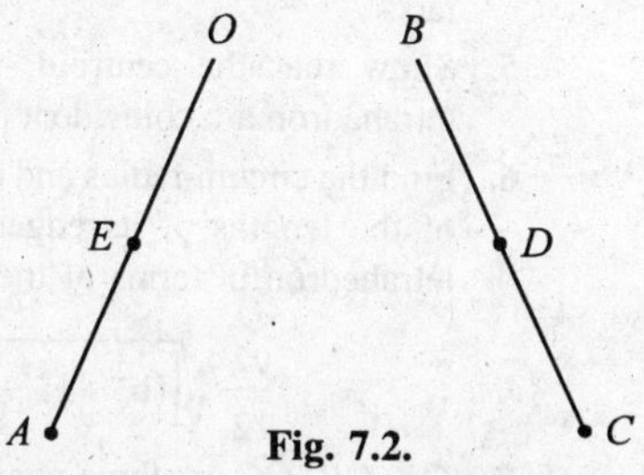

Fig. 7.2.

II. By I, the line of shortest distance passes through the mid-points of *OA* and *BC* and hence it is their join. Let *E* be the mid-point of *OA*.

The figure is symmetrical above *DE* so that

$$OB = AC \text{ and } OC = AB.$$

Hence, the triangles *OAB*, *OAC* are congruent and the triangles *BCO* and *BCA* are congruent.

III. If α, β, γ, δ denote the areas of the faces opposite to the vertices *A, B, C, O* the position vector of the in-centre is

$$\frac{\alpha\mathbf{a}+\beta\mathbf{b}+\gamma\mathbf{c}+\delta\mathbf{d}}{\alpha+\beta+\gamma+\delta} = \frac{3\alpha\,(\mathbf{a}+\mathbf{b}+\mathbf{c})}{3\alpha+\delta}, \qquad \text{(5 Page 193)}$$

for $\alpha = \beta = \gamma$.

Also the centroid is

$$\frac{\mathbf{a}+\mathbf{b}+\mathbf{c}+\mathbf{0}}{4} = \frac{\mathbf{a}+\mathbf{b}+\mathbf{c}}{4}$$

Thus,

$$\vec{OG} = \frac{\mathbf{a}+\mathbf{b}+\mathbf{c}}{4},$$

$$\vec{OI} = \frac{3\alpha\,(\mathbf{a}+\mathbf{b}+\mathbf{c})}{3\alpha+\delta} = \frac{12\alpha}{3\alpha+\delta}\vec{OG}.$$

It follows that *O, I, G* are collinear.

IV is an immediate consequence of II.

EXERCISES

1. *OABC* is an orthocentric tetrahedron and α, β, γ are the angles between the edges *OA, OB, OC* taken in pairs, show that

$$\frac{OA}{\cos\alpha} = \frac{OB}{\cos\beta} = \frac{OC}{\cos\gamma}$$

2. Find the circum-radius and in-radius of a regular tetrahedron in terms of

the length of each edge.

3. Show that a tetrahedron which is isosceles and orthocentric is regular.
4. Show that the feet of the altitudes from the vertices to the opposite faces of an orthocentric tetrahedron are the orthocentres of the corresponding faces.
5. Show that the centroid, circum-centre and in-centre of an isosceles tetrahedron are coincident points.
6. Find the circum-radius and the in-radius of an isosceles tetrahedron in terms of the lengths of its edges. Also show that the volume of an isosceles tetrahedron in terms of the lengths of the edges is

$$\frac{\sqrt{2}}{12}\sqrt{\left[(\mathbf{b}^2+\mathbf{c}^2-\mathbf{a}^2)(\mathbf{c}^2+\mathbf{a}^2-\mathbf{b}^2)(\mathbf{a}^2+\mathbf{b}^2-\mathbf{c}^2)\right]}.$$

7. *OA, OB, OC* are three concurrent lines; find the axis of the right circular cone touching the planes *OBC, OCA, OAB* internally.
8. Show that in any tetrahedron, the line joining the middle points of any one pair of opposite edges is perpendicular to the shortest distance lines between either of the two other pairs of opposite edges.
9. The altitudes from the vertices *A* and *B* to the opposite faces of a tetrahedron intersect; show that *AB* is perpendicular to *CD* and the altitudes from the vertices *C, D to* the opposite faces also intersect. Prove the converse also.
10. *OABC* and *OA′B′C′* are two tetrahedra such that the plane through *OA* perpendicular to the face *B′OC′* and the two other similar planes intersect in a line; show that the plane through *OA′* perpendicular to the face *BOC* and two other similar planes also intersect in a line.
11. *ABCD* is a tetrahedron : h_1, h_2, h_3, h_4 are the perpendicular distances of the vertices from the opposite faces and s_1, s_2, s_3 are the shortest distances between the three pairs of opposite edges; prove that

$$\frac{1}{S_1^2}+\frac{1}{S_2^2}+\frac{1}{S_3^2}=\frac{1}{h_1^2}+\frac{1}{h_2^2}+\frac{1}{h_3^2}+\frac{1}{h_4^2}.$$

12. Show that the centre of the sphere through four points *A, B, C, D* with position vectors **a, b, c, d** is

$$-\frac{\mathbf{a}^2\,\Delta\,\overrightarrow{BCD}+\mathbf{b}^2\,\Delta\,\overrightarrow{CAD}+\mathbf{c}^2\,\Delta\,\overrightarrow{ABD}-\mathbf{d}^2\,\Delta\,\overrightarrow{ABC}}{\overrightarrow{AB}\times\overrightarrow{AC}\cdot\overrightarrow{AD}}.$$

13. Show that the circum-centre of the tetrahedron *OABC* is

$$\frac{\mathbf{a}^2(\mathbf{b}\times\mathbf{c})+\mathbf{b}^2(\mathbf{c}\times\mathbf{a})+\mathbf{c}^2(\mathbf{a}\times\mathbf{b})}{2\,[\mathbf{a}\ \mathbf{b}\ \mathbf{c}]},$$

where $\overrightarrow{OA}=\mathbf{a},\ \overrightarrow{OB}=\mathbf{b},\ \overrightarrow{OC}=\mathbf{c}.$

8

Statics with Vectors

Introduction. A study of Theoretical Statics with the help of Vector Algebra will be undertaken in this chapter.

In this connection it is important to remember that a force cannot be completely characterised by a vector and that we require two vectors for the purpose. As against this, a Couple is completely characterisable by one vector.

The study will be based on the **Principles of Transmissibility of Forces and the Parallelogram Law of Forces.**

8.1. FORCES REPRESENTED BY LINE VECTORS

A force acting on a rigid body is completely specified by its (*i*) *Magnitude,* (*ii*) *Line of action,* (*iii*) *Point of application, and* (*iv*) *Sense along the line of action.*

As a consequence of the principle of **Transmissibility of forces,** it can be shown that the effect of a force acting on a *rigid* body is unaltered if its point of application is changed to any other point on the line of action of the force without any change in its magnitude and sense. Thus, while specifying a force acting on a rigid body, we need only state its magnitude, line of action and sense. Unlike vectors however, the particular line of action of a force is material and the effects of two forces acting along *different* parallel lines are different even though they may have the same magnitude and sense.

Thus, a force cannot be thought of as being completely representable by a vector. For the representation of forces, we need the notion of *Line Vectors* which requires a more restricted definition of equality than that for vectors (§ 1.2.2, page 3).

Line Vectors. Def. *A directed line segment is called a line vector such that two directed line segments having the same magnitude,* **support** *and sense are equal line vectors.*

Thus, two directed line segments may be equal vectors but unequal line vectors. For the equality of line vectors, it is necessary that their supports are same and not *just* parallel as for the equality of vectors.

While two vectors are equal if the corresponding directed line segments have the same length, same or parallel supports and the same sense, the two line vectors are equal if the corresponding directed line segments have the same length, same supports and the same sense.

A line vector is also sometimes called a *Localised vector* and for the sake of distinction, the vectors, so called, are often termed **Free Vectors.**

Forces represented by line vectors. From the foregoing, it is clear that a force acting on rigid body can be represented by a line vector such that two equivalent forces are represented by equal line vectors and *vice-versa.*

Two forces may be said to be equivalent if they have the same effect.

8.2. STUDY OF STATICS WITH VECTORS

In view of the preceding, we see that an Analytical Vectorial study of Statics requires the setting up of an *Algebra of Line Vectors.* It can be shown, however, that a line vector can itself be specified by two vectors so that, on our part, we shall avoid direct reference to the notion of line vectors and show that whereas a force cannot be represented by a single vector it can be specified by two vectors, to be called

(*i*) *Vector of the Force,*

(*ii*) *Moment Vector of the Force about a given point.*

The Algebra of Vectors developed in the preceding chapters will thus form the basis of the Vectorial study of Statics.

8.2.1. Vectors of a Force

A vector whose

(*i*) length represents the magnitude of the force according to some scale,

(*ii*) support is the same or parallel to that of the line of action of the force,

(*iii*) sense is the same as that of the force,

is called the vector of the given force or *Force Vector.* The *Vector of a force* will be denoted by the same symbol as denotes the force.

It will be seen that correspondence between Forces and Vectors is *not* one-one; to each force there corresponds a single vector but to a given vector there will correspond several, not necessarily equivalent forces. Forces corresponding to the same vector will have the same magnitude, the same or *parallel* lines of action and the same sense; they may have, however, different points of application.

A force is completely specified by its Vector and its point of application.

Just as the magnitude of a force gives us only an incomplete idea of the force, so the vector of a force also gives an incomplete idea, even though this tells us more than the magnitude of the force.

8.2.2. Moment Vector about a given Point

Let **F** be any given force and let P be any point on its line of action.

Take any arbitrary point O and let

$$\overrightarrow{OP} = \mathbf{r}.$$

Then the *vector*

$$\mathbf{M} = \mathbf{r} \times \mathbf{F}$$

is called the Moment vector of the force about **O**. The symbol **F**, here denotes the vector of the force.

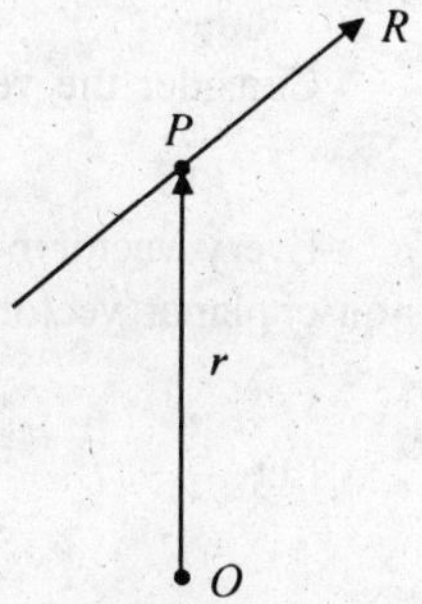

Fig. 8.1.

The moment vector about O is independent of the choice of the point P on the line of action of the force. Thus, if Q be any other point on the line of action of the force, we have

$$\overrightarrow{OQ} \times \mathbf{F} = (\overrightarrow{OP} + \overrightarrow{PQ}) \times \mathbf{F}$$

$$= \overrightarrow{OP} \times \mathbf{F} + \overrightarrow{PQ} \times \mathbf{F}$$

$$= \overrightarrow{OP} \times \mathbf{F};$$

$\overrightarrow{OP} \times \mathbf{F}$ being the zero vector, for $\overrightarrow{PQ}$ and **F** are parallel vectors.

Of course the moment vector of a force does depend upon the point, O and to lay emphasis on this point, whenever necessary, we may write, O – moment vector instead of just moment vector.

The two vectors **F** and **M** are clearly perpendicular to each other so that we have

$$\mathbf{F} \cdot \mathbf{M} = 0.$$

8.2.3. Vanishing of the Moment Vector

The moment vector of a non-zero force about a point O is zero if, and only if, the point O lies on the line of action of the force.

8.3. FORCE DETERMINED BY ITS FORCE VECTOR AND MOMENT VECTOR

We shall now show that the line of action of a force is known as soon as we know **F** and **M**.

Let **F**, **M** be two vectors such that

$$\mathbf{F} \cdot \mathbf{M} = 0.$$

Consider the vector equation

$$\mathbf{r} \times \mathbf{F} = \mathbf{M}. \qquad \text{...}(i)$$

Every vector **r** is uniquely expressible as a linear combination of the non-coplanar vectors

$$\mathbf{M},\ \mathbf{F},\ \mathbf{F} \times \mathbf{M}.$$

Let

$$\mathbf{r} = \lambda_1\mathbf{M} + \lambda_2\mathbf{F} + \lambda_3\mathbf{F} \times \mathbf{M}. \qquad \text{...}(ii)$$

Substituting in (*i*) we get

$$\lambda_1\mathbf{M} \times \mathbf{F} + \lambda_3(\mathbf{F} \times \mathbf{M}) \times \mathbf{F} = \mathbf{M}$$

$$\Rightarrow \quad \lambda_1[\mathbf{M} \times \mathbf{F} + \lambda_3]\ (\mathbf{F} \cdot \mathbf{F})\ \mathbf{M} - \mathbf{M} \cdot \mathbf{F})\ \mathbf{F}] = \mathbf{M}$$

$$\Rightarrow \quad \lambda_1\mathbf{M} \times \mathbf{F} + [\lambda_2(\mathbf{F} \cdot \mathbf{F}) - 1]\ \mathbf{M} = 0, \quad \text{for } \mathbf{F} \cdot \mathbf{M} = 0,$$

$$\Leftrightarrow \quad \lambda_1 = 0, \quad \lambda_3(\mathbf{F} \cdot \mathbf{F}) - 1 = 0.$$

Thus, $\lambda_1 = 0$ and $\lambda_3 = 1/\mathbf{F} \cdot \mathbf{F}$.

Substituting in (*ii*), we get

$$\mathbf{r} = \frac{1}{\mathbf{F} \cdot \mathbf{F}}(\mathbf{F} \times \mathbf{M}) + \lambda_2\mathbf{F} \qquad \text{...}(iii)$$

Also we may easily see that the vector **r** given by (*iii*) satisfy (*i*) for every value of the scalar λ_2.

Now the equation (*iii*) represents a straight line parallel to the force vector **F** such that for any point on this line, the equation (*i*) is true.

Hence (*iii*) is the equation of the line of action of the force which has **F** and **M** as Force Vector and Moment Vector respectively. Thus, the force is completely determined.

8.4. VECTOR SUM AND MOMENT VECTOR SUM

As in the case of a single force, we can associate two vectors to any system of forces such that, so far as the effect on a rigid body is concerned, they determined the system completely.

Consider any system of forces

$$\mathbf{F}_1,\ \mathbf{F}_2,\ \mathbf{F}_3, \ldots$$

acting on a rigid body and let

$$\mathbf{r}_1, \ \mathbf{r}_2, \ \mathbf{r}_3, \ldots$$

be the position vectors of any points on the lines of action of the forces with reference to any point O.

The force vectors of the forces are also denoted by

$$\mathbf{F}_1, \ \mathbf{F}_2, \ \mathbf{F}_3, \ldots$$

We write

$$\mathbf{R} = \mathbf{F}_1 + \mathbf{F}_2 + \mathbf{F}_3 + \ldots$$

$$\mathbf{G} = \mathbf{r}_1 \times \mathbf{F}_1 + \mathbf{r}_2 \times \mathbf{F}_2 + \mathbf{r}_3 \times \mathbf{F}_3 + \ldots$$

so that the vectors **R** and **G** are the sums of the vectors of the forces and of the moment vectors of the forces about O. For the sake of brevity, these will be referred to as

Vector sum and **Moment sum**

respectively.

8.4.1. Change in G with change in origin

Let O, O', be any two points and

$$\mathbf{r}_1, \mathbf{r}_2 \ldots, \mathbf{r}_4 \ldots; \quad \mathbf{r}_1', \mathbf{r}_2' \ldots, \mathbf{r}_6' \ldots$$

be the position vectors of the points

$$P_1, P_2 \ldots, P_i \ldots,$$

on the lines of action of the given forces with respect to O and O' respectively.

We have

$$\overrightarrow{OP_i} = \mathbf{r}_i, \ \overrightarrow{O'P_i} = \mathbf{r}'_i$$

$$\Rightarrow \qquad \mathbf{r}'_i = \overrightarrow{O'P_i}$$

$$= \overrightarrow{O'O} + \overrightarrow{OP_i} = \overrightarrow{OP_i} - \overrightarrow{OQ'} = \mathbf{r}_i - \mathbf{s},$$

where $\mathbf{s} = \overrightarrow{OO}$

If **G**, **G**′ denote the moment sums about O and O', we have

$$\mathbf{G}' = \Sigma\,(\mathbf{r}_i' \times \mathbf{F}_i)$$

$$= \Sigma\,(\mathbf{r}_i - \mathbf{s}) \times \mathbf{F}_i = \Sigma\,(\mathbf{r}_i \times \mathbf{F}_i) - \Sigma \mathbf{s} \times \mathbf{F}_i$$

$$= \Sigma\,(\mathbf{r}_i \times \mathbf{F}_i) - \Sigma\,(\mathbf{s} \times \mathbf{F}_i)$$

$$= \mathbf{G} - \mathbf{s} \times \Sigma \mathbf{F}_i = \mathbf{G} - \mathbf{s} \times \mathbf{R}.$$

Thus, the *moment sum* **G**′ *about any point* O' *is*

$$\mathbf{G} - \mathbf{s} \times \mathbf{R},$$

where **G** *is the moment sum about O and* **s** *is the position vector of O′ relative to O.*

Cor. 1. The moment sums **G, G′** are the same for two points O, O' when OO' is parallel to the vector sum **R**, for in that case

$$\mathbf{s} \times \mathbf{R} = 0.$$

Cor. 2. If the vector sum is zero and the moment sum about one point is zero then the moment sum about every point is zero.

8.5. REDUCTION OF A SYSTEM OF FORCES

An important problem of Statics consists in the determination of systems of forces which are equivalent to any given system so far as the effect on a rigid body is concerned and which are at the same time simpler to deal with than the given system. It is also of great importance to determine analytical conditions which characterise equivalent systems of forces. In this connection the two vectors **R** and **G** will be seen to play a very important part.

The two principles of reduction. The following two principles are fundamental for determining systems equivalent to a given system :

I. *Principle of transmissibility of forces* as a result of which a force can be thought of as acting at any point of its line of action.

II. *Parallelogram law of forces* as a result of which two forces $\mathbf{F}_1$ and $\mathbf{F}_2$ acting at any point O are equivalent to a single force called their esultant, whose vector is $\mathbf{F}_1 + \mathbf{F}_2$ and which also acts as O.

A system S_2 of forces which arises from another system S_1 be any series of applications of the above two principles is equivalent to S_1, in as much as the effects of the two are the same.

By successive applications of the Parallelogram law of forces, it can be shown that

(*i*) *if* $\mathbf{F}_1$, $\mathbf{F}_2$, $\mathbf{F}_3$ *be three forces acting at any point O and*

$$\overrightarrow{OA_1} = \mathbf{F}_1,\ \overrightarrow{OA_2} = \mathbf{F}_2,\ \overrightarrow{OA_3} = \mathbf{F}_3,$$

then the forces are equivalent to a single force whose line of action passes through O and whose vector is where OP is the diagonal of the parallelopiped constructed with

$$OA_1,\ OA_2,\ OA_3,$$

as adjacent sides.

(*ii*) *A system of concurrent three forces is equivalent to a single force whose point of application is the point of concurrence and whose vector is the sum of the vectors of the given forces.*

8.5.1. Zero Force

Def. *A force whose vector is the zero vector is called Zero force.*

Thus, a force whose magnitude is zero is the zero force.

Equilibrium. *A system of forces equivalent to the Zero force is said to be in equilibrium.*

Negative of a force. The force whose magnitude and line of action are the same as those of a force **F** but whose sense is opposite to that of **F** is said to be the negative of **F** and is denoted by – **F**.

The two forces **F**, – **F** form a system in equilibrium.

The moment vector of **F**, about any point is the negative of that of, – **F**, about the same point.

Theorem. *The vector sum and the moment sum about any point are the same for two equivalent systems.*

All that we need to show is that both the vector sum and the moment sum remain unaltered as a result of application of each of the two principles of reduction.

The vector sum is obviously invariant for each of the two principles.

For the moment sum, we prove the following :

The sum of the moment vectors of two forces acting at a point is equal to the moment vector of their resultant about the same point. (*Patna 2003*)

(Varignon's Theorem)

Let $\mathbf{F}_1$, $\mathbf{F}_2$, be two forces acting at any point *A*. The sum of the moment vectors about any point *O*

$$= \vec{OA} \times \mathbf{F}_1 + \vec{OA} \times \mathbf{F}_2$$

$$= \vec{OA} \times (\mathbf{F}_1 + \mathbf{F}_2)$$

= moment vector of the resultant about *O*.

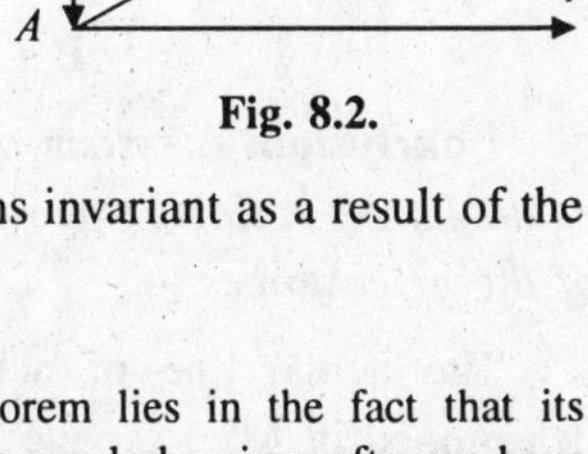

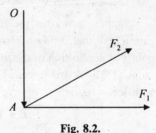

Fig. 8.2.

Also, obviously, the moment vector remains invariant as a result of the application of the first principle.

Hence the theorem.

Note. The importance of the Varignon's theorem lies in the fact that its converse is also true, but the proof of the converse can only be given after we have considered some further developments of the subject. (Refer § 8.9, Page 214)

8.6. REDUCTION OF A SYSTEM CONSISTING OF TWO PARALLEL FORCES

Couples. Let *P, Q* be two given parallel forces. Take two points *A, B* on their lines of action.

Introduce two forces **F**, – **F** at A and B acting along the join of A and B.

The forces **P** and **F** have a resultant which passes through A and whose vector is **P** + **F**.

The forces **Q** and – **F** have a resultant which passes through B and whose vector is **Q** – **F**.

Now two cases are possible : *the lines of action of these two latter forces* **P** + **F** *and* **Q** – **F** *meet or are parallel.*

Firstly, we investigate the condition for the lines of action to be parallel.

Let

$$\mathbf{P} = a\mathbf{S}, \quad \mathbf{Q} = b\mathbf{S},$$

where a, b are scalars and **S** is a unit vector.

$$\therefore \qquad \mathbf{P} + \mathbf{F} = a\mathbf{S} + \mathbf{F}, \quad \mathbf{Q} - \mathbf{F} = b\mathbf{S} - \mathbf{F}.$$

For parallelism

$$(a\mathbf{S} + \mathbf{F}) \times (b\mathbf{S} - \mathbf{F}) = 0$$

$$\Rightarrow \qquad (a + b)\, \mathbf{F} \times \mathbf{S} = 0.$$

As $\mathbf{F} \times \mathbf{S} \neq 0$, we have $a + b = 0$, *i.e.*, the two forces are equal in magnitude and opposite in sense.

A system of two such forces is called a *Couple.*

The discussion of couples will be taken up in § 8.7.

Suppose now that the lines of action of **P** + **F** and **Q** – **F** are not parallel, so that, being coplanar, they will meet at some point, say O. The points of application of the forces at A and B may now be transferred to the point O. These two forces at O are equivalent to a single force through O with force vector

$$\mathbf{P} + \mathbf{F} + \mathbf{Q} - \mathbf{F} = \mathbf{P} + \mathbf{Q},$$

Conclusion. *A system of two parallel forces which is not-a couple is equivalent to a single force whose vector is equal to the sum of the vectors of the given forces.*

The actual line of action of the single equivalent force will be determined in § 8.15, page 219.

8.7. REDUCTION TO THREE FORCES

Any given system of forces is equivalent to a system of three forces whose points of application are three given points.

Let L, M, N be the three given points.

Choose points

$$A_1, A_2, A_3, \ldots$$

on the lines of action of the given forces in such a manner that none of these points lies on the plane determined by the points L, M, N so that these are the points other than those where this plane meets the lines of action of the given forces.

Consider any force $\mathbf{F}_1$ acting at A_1.

Let

$$\overrightarrow{A_1B_1} = \mathbf{F}_1.$$

There exists a parallelopiped whose one diagonal is along the line A_1B_1 of application of the force $\mathbf{F}_1$ and whose three edges through A_1 lie along A_1L, A_1M and A_1N. Thus there exist three concurrent forces acting along A_1L, A_1M, A_1N equivalent to $\mathbf{F}_1$. We suppose that the points of application of these forces are transferred to L, M, N.

Let each force of the system be treated in a like manner.

Thus, we arrive at three systems of concurrent forces separately concurrent at L, M and N. Replacing these concurrent forces by their resultants, we arrive at the theorem stated.

8.8. CONDITION FOR EQUILIBRIUM

Theorem. *If for a system of forces, the vector sum and the moment sum about a point are both zero, then the system is in equilibrium.*

By Cor. 2, § 8.4.1., page 213 moment sum about every point is zero.

We replace the given system by an equivalent system of three forces having any three points L, M, N as their points of application. Let $\mathbf{F}_1$, $\mathbf{F}_2$, $\mathbf{F}_3$ be these three forces.

The vector sum and the moment sum are both zero for the new system also. (§ 8.5, page 213)

Take any three points A, B, C on the lines of action of the forces.

Equating to zero the moment sum about A, we obtain

$$\overrightarrow{AB} \times \mathbf{F}_2 + \overrightarrow{AC} \times \mathbf{F}_3 = \mathbf{0},$$

so that the vectors $\overrightarrow{AB} \times \mathbf{F}_2$ and $\overrightarrow{AC} \times \mathbf{F}_3$ are collinear. Thus, there exists a linear normal to the lines AB, AC and the lines of action of the forces $\mathbf{F}_2$ and $\mathbf{F}_3$. Accordingly the lines of action of the forces lie in a plane determined by the intersecting lines AB, AC. In other words, the lines of action of the forces $\mathbf{F}_2$ and $\mathbf{F}_3$ are coplanar and the plane containing the same passes through the point A of the line of action of $\mathbf{F}_1$. As the point A is arbitrary, we deduce that the three lines of action of the three forces are coplanar.

Now two cases arise according as some two of the three lines of action are non-parallel or all three are parallel.

In the first case, suppose that the lines of action of $\mathbf{F}_1$ and $\mathbf{F}_2$ are non-parallel so that they meet at a point say, O. Equating to zero the moment sum about O, we see that O lies on the line of action $\mathbf{F}_3$ also. Thus, the three lines of action are concurrent in this case. These three concurrent forces are equivalent to a single force whose line of action passes through O and whose vector is equal to the sum of the vectors of the forces. The latter being zero, we deduce that the system is in equilibrium.

Suppose now that the *lines of action of three forces are all parallel.*

As the vector sum of the forces is zero, the three parallel forces cannot all have the same sense.

Suppose that **P** and **Q** are having the same sense opposite to that of **R.**

The forces **P** and **Q** are equivalent to a single force parallel to the given force with vector $\mathbf{P} + \mathbf{Q}$.

Equating to zero the moment about any point A on the line of action of $\mathbf{P} + \mathbf{Q}$, we see that the point A lies on the line of action of **R.** Thus, R and $\mathbf{P} + \mathbf{Q}$ have the same line of action. Also $\mathbf{P} + \mathbf{Q} + \mathbf{R} = 0$. Hence, the system is in equilibrium.

8.9. CRITERIA FOR EQUIVALENT SYSTEMS

Theorem. *Two systems of forces having the same vector sum and the same moment sum about a point are equivalent.*

Let there be two systems of forces

$$\mathbf{F}_1, \mathbf{F}_2, \ldots\ldots\ldots\ldots$$

$$\mathbf{F}_1', \mathbf{F}_2', \ldots\ldots\ldots\ldots$$

having the same vector sum **R** and the same moment sum **G** about a point **O**.

Consider now a new system equivalent to the first obtained by introducing the pairs of forces

$$\mathbf{F}_1', -\mathbf{F}_1'; \ \mathbf{F}_2', -\mathbf{F}_2' \ldots\ldots$$

to the same. For the part of this new system consisting of the forces

$$\mathbf{F}_1, \quad \mathbf{F}_2, \ldots\ldots\ldots\ldots$$

$$-\mathbf{F}_1', -\mathbf{F}_2', \ldots\ldots\ldots\ldots$$

the vector sum and the moment sum about O are both zero and as such this part is in equilibrium. Thus, we are left with the forces

$$\mathbf{F}_1', \mathbf{F}_2', \ldots\ldots\ldots$$

forming a system equivalent to the first.

Hence the result.

8.10. COUPLES

Def. *A system consisting of a pair of equal unlike parallel forces is called a Couple.*

The vector sum of the two forces of a couple is clearly zero.

About moment sum, we have the following important result :

The moment sum of the two forces of a couple is the same about every point.

Let the lines of action of the forces of the couple be two parallel lines. Let *A, C* be any two points on these parallel lines. The two forces have equal magnitudes but opposite sense.

Thus, if **F** denote the vector of one force, then – **F** denotes that of the other.

The moment sum of the forces of the couple about any point *O*

$$= \overrightarrow{OA} \times \mathbf{F} + \overrightarrow{OC} \times (-\mathbf{F})$$

$$= (\overrightarrow{OA} - \overrightarrow{OC}) \times \mathbf{F} = \overrightarrow{CA} \times \mathbf{F},$$

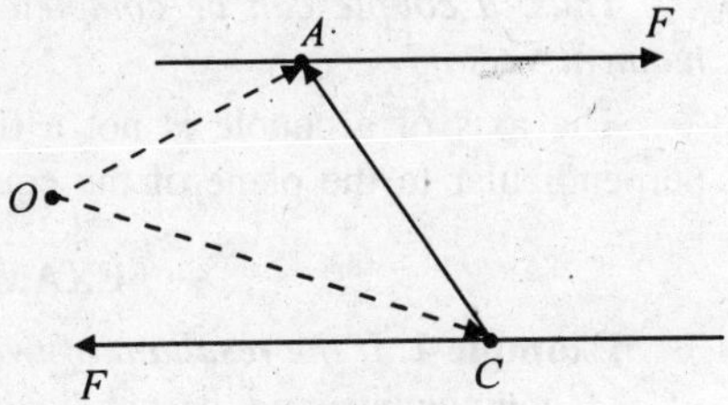

Fig. 8.3.

so that the moment sum is independent of the point *O*. Here *A, C* may be thought of as any two points on the lines of action of the two forces.

The result obtained above is contained in the general theorem that if the vector sum of the forces of a system is zero, then the moment sum about every point is the same.

Moment of a Couple. *The constant moment sum of the forces of a couple is called the moment of the couple.*

Axis of a Couple. *Any line perpendicular to the plane of a couple is called the axis of the couple* so that the axis of a couple is parallel to its moment vector.

Thus, the moment of a couple is a vector perpendicular to the plane of the couple and its magnitude is the product of the magnitude of either force with the perpendicular distance between the lines of action of the forces.

8.10.1. Equivalent Couples

Theorem. *Two couples with equal moments are equivalent*, for the vector and the moment sum are the same for each couple (§ 8.6, Page 209).

Two couples with equal moments necessarily lie in the same or parallel planes.

8.10.2. Composition of Couples

Theorem. *Two couples with moments* $\mathbf{M}_1$ *and* $\mathbf{M}_2$ *are equivalent to a single couple with moment* $\mathbf{M}_1 + \mathbf{M}_2$.

This result is also an immediate consequence of the Theorem of § 8.9.

8.10.3. Couples Represented by Vectors

We have seen that two couples with equal moments are equivalent and as such a couple is completely known by its moment which is a vector.

Also a system consisting of two couples is equivalent to a third couple whose moment is the vector sum of the moments of the two component couples.

Finally any vector equal to the moment vector of a couple is also the moment vector of the couple.

Thus, a couple can be completely characterised by a vector, viz., its moment vector.

The axis of a couple is not a fixed line; it can, in fact, be any line perpendicular to the plane of the couple.

EXAMPLES

Example 1. *If the resultant of two forces be equal in magnitude to one of the components and perpendicular to it in direction, find the other component.*

Solution. Let P and Q be the forces inclined at an angle θ to each other. Let $\mathbf{i}$ and $\mathbf{j}$ be the direction of P and perpendicular to it respectively. Along $\mathbf{j}$ the resultant of forces P and Q is acting.

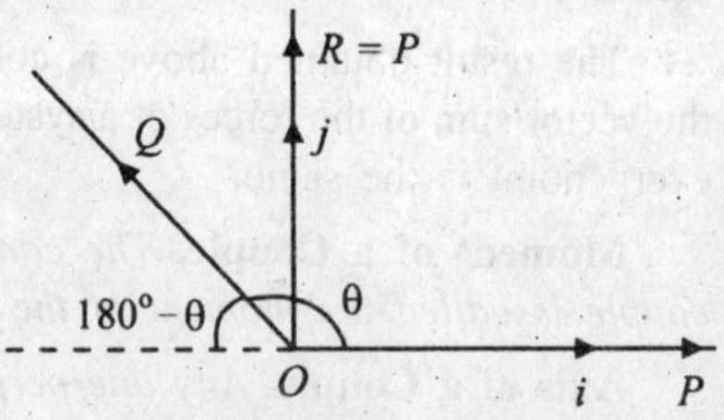

Fig. 8.4.

$$\therefore \quad P\mathbf{i} + \{-Q\cos(180° - \theta)\,\mathbf{i} + Q\sin(180° - \theta)\,\mathbf{j}\} = P\mathbf{j}$$

$$\Rightarrow \quad (P + Q\cos\theta)\,\mathbf{i} + (Q\sin\theta - P)\,\mathrm{j} = 0$$

Since $\mathbf{i}$ and $\mathbf{j}$ are non-collinear vectors, hence

$$P + Q\cos\theta = 0 \quad \text{and} \quad Q\sin\theta - P = 0$$

$$\therefore \quad Q\sin\theta = P \quad \text{and} \quad Q\cos\theta = -P$$

$\Rightarrow \qquad \tan\theta = -1 \quad \Rightarrow \quad \theta = 135°$

and $\qquad Q^2 \sin^2\theta + Q^2\cos^2\theta = P^2 + P^2 \quad \Rightarrow \quad Q = \sqrt{2}P$

Example 2. *Forces P, Q act at O and have a resultant R. If any transversal cuts their lines of action at A, B, C respectively, show that*

$$P/OA + Q/OB = R/OC.$$

Solution. We have

$$\mathbf{P} + \mathbf{Q} = \mathbf{R} \qquad ...(1)$$

Let a transversal cuts the lines of action of P, Q and R at A, B, C. Let ON is $\perp$ to AB. Let $\hat{\mathbf{n}}$ be a unit vector along ON and $\angle BON = \alpha$, $\angle AON = \beta$ and $\angle CON = \gamma$.

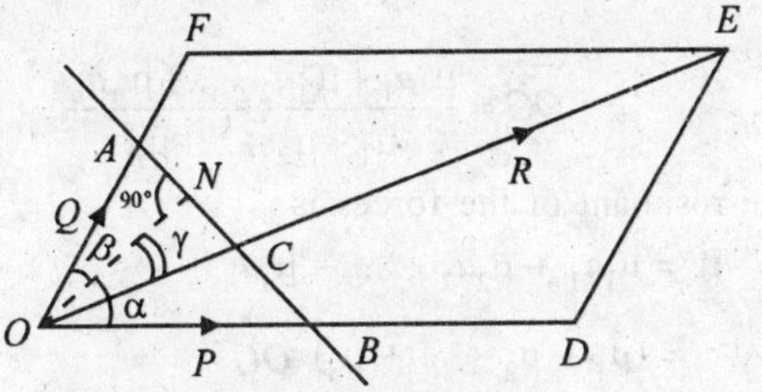

Fig. 8.5.

(1) $\Rightarrow \qquad \mathbf{P}\cdot\hat{\mathbf{n}} + \mathbf{Q}\cdot\hat{\mathbf{n}} = \mathbf{R}\cdot\hat{\mathbf{n}}$

$\Rightarrow \qquad P\cos\alpha + Q\cos\beta = R\cos\gamma$

From $\Delta\, AON$, $\Delta\, BON$ and $\Delta\, CON$,

$$\cos\alpha = \frac{ON}{OA}, \quad \cos\beta = \frac{ON}{OB}, \quad \cos\gamma = \frac{ON}{OC}$$

$$\therefore \qquad P\left(\frac{ON}{OA}\right) + Q\left(\frac{ON}{OB}\right) = R\left(\frac{ON}{OC}\right)$$

$$\Rightarrow \qquad \frac{P}{OA} + \frac{Q}{OB} = \frac{R}{OC}.$$

Example 3. *The forces P and Q act at a point O. Their resultant R is such that R = P. If P is doubled, show that the new resultant is at right angles to Q.*

Solution. Given $\mathbf{P} + \mathbf{Q} = \mathbf{R}$

Let $\mathbf{S}$ be the resultant of $2\mathbf{P}$ and Q.

$\therefore \qquad 2\mathbf{P} + \mathbf{Q} = \mathbf{S}$

Now $\qquad \mathbf{S}\cdot\mathbf{Q} = (2\mathbf{P} + \mathbf{Q})\cdot\mathbf{Q}$

$\qquad = (\mathbf{R} + \mathbf{P})\cdot(\mathbf{R} - \mathbf{P})$

$\qquad = R^2 - P^2 = 0.$

$\Rightarrow$ $\mathbf{S}$ is perpendicular to $\mathbf{Q}$.

Example 4. *A particle is acted on by a number of centres of forces some of which attract and some repel, the forces in each case varying as the distance. The intensities for different centres are different. Prove that the resultant passes through a fixed point for all positions of the particle.*

Solution. Let O be the origin. Let the given particle be situated at O. Let the position vectors of various centres of forces $A_1, A_2, ..., A_n$ be $\boldsymbol{a}_1, \boldsymbol{a}_2, ..., \boldsymbol{a}_n$. Hence, the forces acting on the particle can be taken as $\mu_1\boldsymbol{a}_1, \mu_2\boldsymbol{a}_2, ..., \mu_n\boldsymbol{a}_n$, where $\mu_1, \mu_2, ..., \mu_n$ are intensities of the centres and may be positive or negative accordingly as the centres attract or repel.

Let G be the centroid of $\boldsymbol{a}_1, \boldsymbol{a}_2, ..., \boldsymbol{a}_n$ with associated numbers $\mu_1, \mu_2, ..., \mu_n$. Then

$$\overrightarrow{OG} = \frac{\mu_1\boldsymbol{a}_1 + \mu_2\boldsymbol{a}_2 + ... + \mu_n\boldsymbol{a}_n}{\mu_1 + \mu_2 + ... + \mu_n}$$

Then **R**, the resultant of the forces is

$$\mathbf{R} = \mu_1\boldsymbol{a}_1 + \mu_2\boldsymbol{a}_2 + ... + \mu_n\boldsymbol{a}_n$$

$$= (\mu_1 + \mu_2 + ... + \mu_n)\ \overrightarrow{OG}$$

Hence, the resultant is a force passing through G, which is independent of the origin. Hence, the resultant passes through the fixed point G for all positions of the particle.

Example 5. *ABCDEF is a regular hexagon and O any point in the plane. Show that the forces represented both in magnitude and direction by OA, OC, OE, BO, DO, FO are in equilibrium.*

Solution. We have

$$\overrightarrow{OA}+\overrightarrow{OC}+\overrightarrow{OE}+\overrightarrow{BO}+\overrightarrow{DO}+\overrightarrow{FO} = (\overrightarrow{DO}+\overrightarrow{OC})+(\overrightarrow{FO}+\overrightarrow{OA})+(\overrightarrow{BO}+\overrightarrow{OE})$$

$$= \overrightarrow{DC}+\overrightarrow{FA}+\overrightarrow{BE}$$

$$= \frac{1}{2}\overrightarrow{EB}+\frac{1}{2}\overrightarrow{EB}-\overrightarrow{EB}$$

$$= \overrightarrow{EB}-\overrightarrow{EB} = 0$$

$\Rightarrow$ the given forces are in equilibrium.

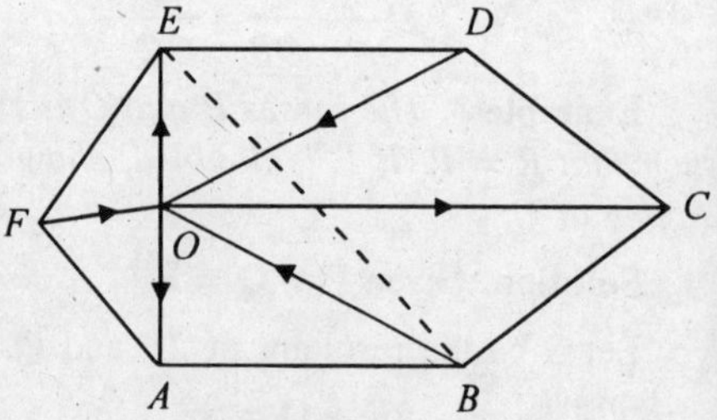

Fig. 8.6.

Example 6. *Find the vector moment of the three forces* **i** + 2**j** – 3**k**, 2**i** + 3**j** + 4**k** *and* – **i** – **j** + **k** *acting on a particle at point P (0, 1, 2) about A (1, – 2, 0).*

Solution. The resultant of three forces

$$\mathbf{F} = (\mathbf{i} + 2\mathbf{j} - 3\mathbf{k}) + (2\mathbf{i} + 3\mathbf{j} + 4\mathbf{k}) + (-\mathbf{i} - \mathbf{j} + \mathbf{k})$$
$$= 2\mathbf{i} + 4\mathbf{j} + 2\mathbf{k},$$

acting at P.

$$\overrightarrow{AP} = (0, 1, 2) - (1, -2, 0) = -\mathbf{i} + 3\mathbf{j} + 2\mathbf{k}$$

$\therefore$ Moment of $\mathbf{F}$ acting at P about A

$$= \overrightarrow{AP} \times \mathbf{F} = (-\mathbf{i} + 3\mathbf{j} + 2\mathbf{k}) \times (2\mathbf{i} + 4\mathbf{j} + 2\mathbf{k})$$

$$= \begin{vmatrix} \mathbf{i} & \mathbf{j} & \mathbf{k} \\ -1 & 3 & 2 \\ 2 & 4 & 2 \end{vmatrix} = -2\mathbf{i} + 6\mathbf{j} - 10\mathbf{k}.$$

EXERCISES

1. ABC is a triangle and D, E, F are the mid-points of the sides. Forces represented by AD, $\frac{2}{3}BE$ and $\frac{1}{3}CF$ act on a particle at a point where AD and BE meet. Show that the resultant is represented both in magnitude and direction by $\frac{1}{2}AC$ and the line of action of the resultant divides BC in the ratio 2 : 1.
2. $ABCD$ is a quadrilateral and E and F are the mid-points of the sides AB and CD. Show that the resultant of forces PA, PB, PC, PD is 4 PG where G is the middle point of EF, wherever the point P may be. When will these forces be in equilibrium ?
3. Three forces P, Q, R act along three straight lines OA, OB, OC. If their resultant is parallel to the plane ABC, prove that
$$(P/OA) + (Q/OB) + (R/OC) = 0.$$
4. Forces act at the vertices of a tetrahedron along the lines joining them to the centroids of opposite faces and proportional to the lengths of the same. Show that the system is in equilibrium.
5. Forces act at the circumcentre O of a triangle ABC equal to $\overrightarrow{OA}$, $\overrightarrow{OB}$ and $\overrightarrow{OC}$. Show that their resultant is $\overrightarrow{OH}$ where H is the orthocentre of $\Delta\, ABC$.
6. Three forces $\overrightarrow{OA}, \overrightarrow{OB}, \overrightarrow{OC}$ diverge from the point O and three forces $\overrightarrow{AO'}, \overrightarrow{BO'}, \overrightarrow{CO'}$ converge to the point O'. Show that the resultant of six forces is represented in direction and magnitude by $3\overrightarrow{OO'}$.

7. H is the orthocentre of $\Delta\, ABC$. Show that the resultant of forces acting on a particle and represented by AH, HB, HC is represented by the diameter through A of the circumcentre of $\Delta\, ABC$.

8. Show that the moment about the point $\mathbf{i} + 2\mathbf{j} - \mathbf{k}$ of a force represented by $\mathbf{i} + 2\mathbf{j} + \mathbf{k}$ acting through the point $2\mathbf{i} + 3\mathbf{j} + \mathbf{k}$ is $-3\mathbf{i} + \mathbf{j} + \mathbf{k}$.

9. A force with components $(-7, 4, 5)$ acts at the point $(2, 4, -3)$. Find its moment about the origin. Find also its moment about the line $L : x = y = z$; the positive direction on the line being that in which x increases.

10. Prove that the sum of moment of two intersecting forces about a point in their plane is equal to the moment of their resultant about that point.

11. Define the moment of a vector about a point. Find the moment about a corner of a cube of three unit vectors converging on the opposite corner along three edges. Show that the sum of these moments is zero.

12. $ABCD$ is a quadrilateral. Find by the vector method the position of a point O inside the quadrilateral such that the forces represented by $\overrightarrow{OA}$, $\overrightarrow{OB}$, $\overrightarrow{OC}$ and $\overrightarrow{OD}$ may be in equilibrium.

8.11. REDUCTION TO SIMPLER SYSTEMS

8.11.1. Theorem

Every system of forces acting on a rigid body is equivalent to a single force acting at an arbitrary point together with a couple whose moment is the sum of the moments of the forces about the point.

Let O be any arbitrary point and let $\mathbf{R}$ be the vector sum and $\mathbf{G}$, the moment sum of the system about O.

Then the system consisting of

(*i*) the force whose vector is $\mathbf{R}$ and whose point of application is O, and

(*ii*) a couple whose moment is $\mathbf{G}$.

is clearly equivalent to the given system, for both the systems have the same vector sum and the same moment sum about a point O.

8.11.2. Poinsot's Central Axis. Wrench

Theorem. *Every system of forces is equivalent to a force and a couple such that the line of action of the force is the axis of the couple.*

Suppose that such a reduction is possible so that there exists a line such that the system is equivalent to a force along the line and a couple whose axis is the line. Then *the moment sum of the system about any point of this line is parallel to the vector sum of the system.*

Now the moment sum about any point O' is

$$\mathbf{G} - \mathbf{r} \times \mathbf{R} \text{ where } \overrightarrow{OO'} = \mathbf{r}.$$

This will be parallel to **R**, if

$$(\mathbf{G} - \mathbf{r} \times \mathbf{R}) \times \mathbf{R} = 0 \qquad ...(i)$$

$$\Rightarrow \quad \mathbf{G} \times \mathbf{R} - (\mathbf{r} \,.\, \mathbf{R})\, \mathbf{R} + (\mathbf{R} \,.\, \mathbf{R})\, \mathbf{r} = 0$$

$$\Rightarrow \quad \mathbf{r} = -\frac{\mathbf{G} \times \mathbf{R}}{\mathbf{R} \,.\, \mathbf{R}} + t\mathbf{R}, \qquad ...(ii)$$

where t is a scalar. Conversely, it may be easily seen that **r** given by (*ii*) satisfies (*i*) for every value of the scalar t.

The points given by (*ii*) for different values of the scalar t, lie on a line.

Thus, we have obtained a line whose vector equation is (*ii*) such that for every point O' on the line, the moment sum about O' is parallel to the vector sum.

This establishes the existence and uniqueness of a line such that the system is equivalent to a force along the line and a couple whose axis is the line.

This line is known as the *Poinsot's Central axis.*

Its equation is

$$\mathbf{r} = -\frac{\mathbf{G} \times \mathbf{R}}{\mathbf{R} \,.\, \mathbf{R}} + t\mathbf{R} = \frac{\mathbf{G} \times \mathbf{R}}{\mathbf{R} \,.\, \mathbf{R}} + t\mathbf{R};$$

R being the vector sum and **G** the moment sum about the origin of reference.

Wrench. Def. *A system consisting of a force and a couple such that the line of the force is the axis of the couple is called a Wrench.*

Thus, what we have proved above is equivalent to saying that *any given system of forces is equivalent to a wrench.*

8.11.3. Moment of the Couple of the Wrench equivalent to a given System

The moment sum about any point

$$-\frac{\mathbf{G} \times \mathbf{R}}{\mathbf{R} \,.\, \mathbf{R}} + t\mathbf{R}$$

of the central axis

$$= \mathbf{G} - \left(-\frac{\mathbf{G} \times \mathbf{R}}{\mathbf{R} \cdot \mathbf{R}} + t\mathbf{R} \right) \times \mathbf{R}$$

$$= \mathbf{G} + \frac{1}{\mathbf{R} \,.\, \mathbf{R}} [(\mathbf{G} \,.\, \mathbf{R})\, \mathbf{R} - (\mathbf{R} \,.\, \mathbf{R})\, \mathbf{G}]$$

$$= \frac{\mathbf{G} \,.\, \mathbf{R}}{\mathbf{R} \,.\, \mathbf{R}} \mathbf{R} = p\mathbf{R}, \text{ say,}$$

where p is a scalar.

8.11.4. Intensity and Pitch of a Wrench

The magnitude $|\mathbf{R}|$ of the force $\mathbf{R}$ of a wrench is called its *Intensity*.

The moment of the couple of a wrench is necessarily the product of the force vector $\mathbf{R}$ of the wrench by a scalar. This scalar is called the *Pitch* of the wrench. It is usual to denote a wrench whose force is $\mathbf{R}$ and pitch p by the symbol

$$(\mathbf{R}, p\mathbf{R})$$

so that $p\mathbf{R}$ is the couple of the wrench.

It will be useful to remember that the wrench equivalent to a system of forces for which the vector sum is $\mathbf{R}$ and moment sum about the origin of reference O is $\mathbf{G}$ is

$$(\mathbf{R}, p\mathbf{R})$$

where

$$p = \frac{\mathbf{G} \cdot \mathbf{R}}{\mathbf{R} \cdot \mathbf{R}}.$$

Also the equation of the axis of the wrench is

$$\mathbf{r} = -\frac{\mathbf{G} \times \mathbf{R}}{\mathbf{R} \cdot \mathbf{R}} + t\mathbf{R}.$$

8.11.5. Degenerate and Non-degenerate Wrenches

Def. *A wrench is said to be degenerate if either the force or the couple of the wrench is zero.*

If the force and the couple be both zero, then the system is in equilibrium.

8.12. INVARIANTS OF A SYSTEM OF FORCES

Let $\mathbf{R}$ be the vector sum and $\mathbf{G}$ the moment sum about a point O of a given system of forces.

Clearly $\mathbf{R}$ is invariant in as much as it does not depend upon the choice of the point O. The scalar $\mathbf{R} \cdot \mathbf{R}$ is also thus an invariant.

In addition to the above, there is also another scalar invariant, *viz.*,

$$\mathbf{G} \cdot \mathbf{R},$$

i.e., the scalar product of the moment sum and the vector sum.

The moment sum $\mathbf{G}'$ about a point O' such that

$$\overrightarrow{OO'} = \mathbf{r},$$

is given by

$$\mathbf{G}' = \mathbf{G} - \mathbf{r} \times \mathbf{R}.$$

$$\Rightarrow \quad \mathbf{G}' \cdot \mathbf{R} = (\mathbf{G} - \mathbf{r} \times \mathbf{R}) \cdot \mathbf{R} = \mathbf{G} \cdot \mathbf{R}.$$

The invariance of **G . R** is also an immediate consequence of the invariance of the pitch **G . R / R . R** of the unique equivalent wrench and of the invariance of **R . R**, as seen above.

Parameter. The invariant **G . R** is called Parameter of the system and it plays an important part in investigation concerning systems of forces.

8.12.1. Geometrical Interpretation of the Parameter G . R

Firstly we consider a system consisting of two forces $\mathbf{F}_1$, $\mathbf{F}_2$. Let $\mathbf{r}_1$, $\mathbf{r}_2$ be the position vectors of any two points *A, B* on the lines of action of the forces. For these two forces, we have

$$\mathbf{R} = \mathbf{F}_1 + \mathbf{F}_2, \quad \mathbf{G} = \mathbf{r}_1 \times \mathbf{F}_1 + \mathbf{r}_2 \times \mathbf{F}_2.$$

$$\Rightarrow \qquad \mathbf{R} \cdot \mathbf{G} = (\mathbf{F}_1 + \mathbf{F}_2) \cdot (\mathbf{r}_1 \times \mathbf{F}_1 + \mathbf{r}_2 \times \mathbf{F}_2)$$

$$= (\mathbf{F}_2 \times \mathbf{F}_1) \cdot (\mathbf{r}_1 - \mathbf{r}_2).$$

where

$$\mathbf{r}_2 - \mathbf{r}_1 = \overrightarrow{AB}.$$

If the two forces be denoted by the line vectors $\overrightarrow{AC}$, $\overrightarrow{BD}$ then

$$\mathbf{R} \cdot \mathbf{G} = \overrightarrow{BD} \times \overrightarrow{AC} \cdot \overrightarrow{AB},$$

so that the parameter of two forces is six times the algebraic value of the volume of the tetrahedron formed by the line vectors representing the given forces as a pair of opposite edges.

Any system of forces. Consider now a system consisting of any number of forces

$$\mathbf{F}_1, \mathbf{F}_2, \ldots\ldots$$

Let $\overrightarrow{A_1B_1}$, $\overrightarrow{A_2B_2}, \ldots$ be the line vectors denoting the forces. For this system,

$$\mathbf{R} \cdot \mathbf{G} = (\mathbf{F}_1 + \mathbf{F}_2 + \ldots) \cdot (\overrightarrow{OA_1} \times \mathbf{F}_1 + \overrightarrow{OA_2} \times \mathbf{F}_2 + \ldots)$$

$$= \Sigma\,(\mathbf{F}_j \times \mathbf{F}_i),\ (\overrightarrow{OA_i} - \overrightarrow{OA_j}),$$

where the summation extends to every pair of forces.

$$\therefore \quad \mathbf{R} \cdot \mathbf{G} = \Sigma\,(\mathbf{F}_i \times \mathbf{F}_j) \cdot (\overrightarrow{A_i\ A_j}) = \Sigma\,(\overrightarrow{A_iB_i} \times \overrightarrow{A_jB_j} \cdot \overrightarrow{A_jA_i}),$$

so that the parameter of any system of forces is six times the algebraic sum of the volumes of the tetrahedra formed by pairs of line vectors denoting forces of the system as pairs of opposite edges.

The above geometric manner of the representation of **R . G** gives another proof of invariance of the same.

8.13. PARAMETERS FOR SOME SPECIAL SYSTEMS OF FORCES

It should be noticed that the parameter **R . G** is zero, if **R** = 0 or **G** = 0 or R is perpendicular to **G.**

I. *The parameter for a system of coplanar forces is zero.*

The moment vector of every force of the system about a point *O* of the plane is perpendicular to the plane and as such the moment sum **G** of the forces, about *O* is perpendicular to the plane. Also the vector **R** is parallel to the plane. Thus, **G . R** = 0.

II. *The parameter for a system of parallel forces is zero.*

The moment vector about any point *O* of a force of the system lies in the plane through *O* perpendicular to the force. As the forces are parallel, the plane π through *O* perpendicular to any one force is also perpendicular to the others. Thus, the moment sum **G** lies in the plane π.

Also the vector sum **R** is perpendicular to the plane π. Thus, **G . R** = 0.

III. *The parameter of a system of concurrent forces is zero.*

The proof follows from the fact that the moment vector of each force about the point of the concurrence *O* is zero and as a consequence their sum **G** is zero.

IV. *The parameter of a system consisting of any number of couples is zero.*

The proof follows from the fact that for such a system the vector sum **R** is zero.

V. *The parameter of a system consisting of two non-zero forces lying along non-coplanar lines cannot be zero.*

Let $\mathbf{F}_1$, $\mathbf{F}_2$ be two forces acting along two skew lines *AB, CD*. We have,

$$\mathbf{R} = \mathbf{F}_1 + \mathbf{F}_2$$

$$\mathbf{G} = \overrightarrow{AC} \times \mathbf{F}_2,$$

where **G** denotes moment sum about *A*.

Thus,

$$\mathbf{GR} = (\overrightarrow{AC} \times \mathbf{F}_2) \cdot (\mathbf{F}_1 + \mathbf{F}_2)$$

$$= \overrightarrow{AC} \times \mathbf{F}_2 \cdot \mathbf{F}_1 \neq 0,$$

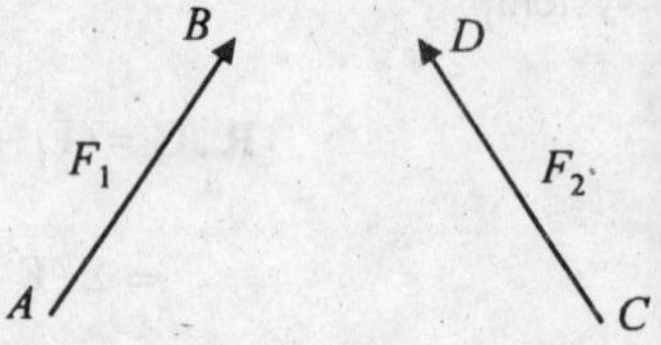

Fig. 8.7.

for the lines of action of $\mathbf{F}_2$, $\mathbf{F}_1$ are not coplanar.

8.14. CRITERIA FOR THE NATURE OF A SYSTEM OF FORCES

Theorem.

I. If $\mathbf{R} \cdot \mathbf{G} \neq 0$, the system is equivalent to a non-degenerate wrench.

II. If $\mathbf{R} \cdot \mathbf{G} = 0$ then the system is

(*i*) equivalent to a couple of moment **G** if $\mathbf{R} = 0$ but $\mathbf{G} \neq 0$,

(*ii*) equivalent to a single force if $\mathbf{R} \neq 0$,

(*iii*) in equilibrium if $\mathbf{R} = 0$ as well as $\mathbf{G} = 0$.

The above criteria are immediate consequences of the result that any given systems of forces is equivalent to a wrench

$$(\mathbf{R}, p\mathbf{R}) \quad \text{where} \quad p = \mathbf{G} \cdot \mathbf{R} / \mathbf{R} \cdot \mathbf{R} ;$$

R being the vector sum and **G** the moment sum of the forces of the system about a point.

Cor. A system of coplanar forces must be equivalent to a single force or a couple, unless it be in equilibrium, for the parameter of such a system is zero.

Similarly, a system of parallel forces, unless it be in equilibrium, is equivalent to a single force or a couple.

8.15. CENTRE OF A SYSTEM OF PARALLEL FORCES

Let

$$\mathbf{r}_1, \mathbf{r}_2, \ldots$$

be the position vectors of any points on the lines of action of the parallel forces

$$\mathbf{F}_1, \mathbf{F}_2, \ldots$$

Let **a** denote a unit vector parallel to the forces. We write

$$\mathbf{F}_1 = p_1\mathbf{a}, \quad \mathbf{F}_2 = p_2\mathbf{a}, \quad \text{etc.,}$$

so that $|\, p_1 \,|$, $|\, p_2 \,|$, etc., denote the magnitudes of the forces.

We have

$$\mathbf{R} = \mathbf{F}_1 + \mathbf{F}_2, \ldots$$
$$= (p_1 + p_2 + \ldots)\, \mathbf{a} = (\Sigma p_i)\, \mathbf{a}.$$
$$\mathbf{G} = \mathbf{r}_1 \times p_1\mathbf{a} + \mathbf{r}_2 \times p_2\mathbf{a} + \ldots$$
$$= p_1\mathbf{r}_1 \times \mathbf{a} + p_2\mathbf{r}_2 \times \mathbf{a} + \ldots$$
$$= (\Sigma p_i\, \mathbf{r}_i) \times \mathbf{a}.$$

If $\Sigma p_i = 0$, *i.e.*, $\mathbf{R} = 0$, the system is equivalent to a couple unless **G** is also zero, in which case the system will be in equilibrium.

Suppose now that $\Sigma p_i \neq 0$. We shall obtain points about which the moment sum is zero. Let **r** be the position vector of such a point.

The moment sum of the system about a point with position vector **r** is

$$= \mathbf{G} - \mathbf{r} \times \mathbf{R} = (\Sigma p_i \mathbf{r}_i) \times \mathbf{a} - (\Sigma p_i)\, \mathbf{r} \times \mathbf{a}$$
$$= [\Sigma p_i \mathbf{r}_i - (\Sigma p_i)\, \mathbf{r}] \times \mathbf{a}.$$

Equating it to zero, we see that **r** is given by

$$\Sigma p_i \mathbf{r}_i - (\Sigma p_i)\, \mathbf{r} = t\, \mathbf{a},$$

where t is a scalar parameter,

$$\Rightarrow \qquad \mathbf{r} = \frac{\Sigma p_i \mathbf{r}_i}{\Sigma p_i} - \frac{t}{\Sigma p_i}\, \mathbf{a}$$

which is the equation of the line of action of the resultant. Clearly this line passes through a point

$$\frac{\Sigma p_i \mathbf{r}_i}{\Sigma p_i}$$

which is independent of the actual direction of the forces and depends only on the magnitudes and the points of application of the forces.

The point is known as the *Centre of parallel forces.* Clearly this is the centroid of the weighted points

$$\mathbf{r}_1,\ \mathbf{r}_2,\ \ldots$$

the respective weights being

$$p_1,\ p_2,\ \ldots$$

8.16. MOMENT OF FORCE ABOUT A LINE

We have so far been concerned with moments of forces about points but we shall now introduce the concept of the moment of a force about a line.

Let **F** be any given force and, **1**, any directed line.

Def. *The moment of a force* **F** *about a directed line* **1** *is the projection on* **1** *of the moment vector of the force about any point on* **1**.

Thus, if O, P be two points on the line **1** such that the vector $\overrightarrow{OP}$ is of unit length, then the moment of **F** about the line **1** is the scalar triple product

$$\overrightarrow{OA} \times \mathbf{F} \cdot \overrightarrow{\mathbf{OP}}, \qquad \ldots(i)$$

where A is any point on the line of action of the force **F**.

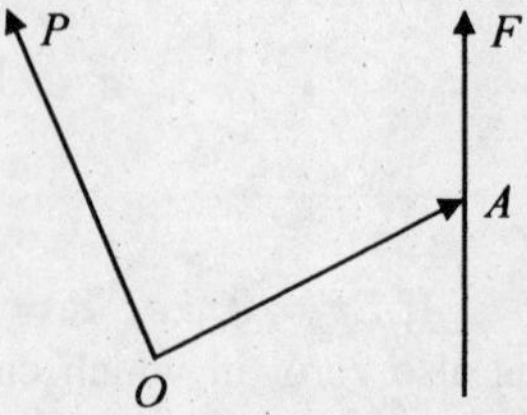

Fig. 8.8.

It may be easily seen that the moment of **F** about **1** is independent of the choice of the point O on the line **1**.

It should be carefully noted that the moments of forces about points are vectors whereas those about lines are scalars.

8.16.1. Geometrical Representation of the Moment of a Force about a Line

It may be easily seen from (*i*) above that *the moment of a force represented by a line vector* $\overrightarrow{AB}$ *about a directed line* **1** *is represented by six times the volume of the tetrahedron constructed with AB and any segment of unit length along* **I** *as a pair of opposite edges.*

It also follows from above that if α be the angle between the line of action of **F** and the line **1** and, c, be the length of shortest distance between the same, then the moment of **F** about **1** is

$$|\mathbf{F}|\, c \sin \alpha.$$

8.16.2. Vanishing of the Moment of a Force about a Line.

Theorem. *The moment of a non-zero force about a line is zero if, and only if, the line of action of the force is coplanar with the line for the moment*

$$\overrightarrow{OA} \times \mathbf{F} \cdot \overrightarrow{OP}$$

of a non-zero, **F** about the line $\overrightarrow{OP}$ will be zero if, and only if, the three vectors $\overrightarrow{OA}$, **F** and $\overrightarrow{OP}$ are all parallel to the same plane; *i.e.*, the line of action of **F** and the line OP are coplanar.

8.16.3. Moment of a System of Forces about a Line is, by definition, the sum of the moments of the forces of the system about the line. Thus, if $\overrightarrow{OP}$ be a unit vector along the given line and $\mathbf{F}_1$, $\mathbf{F}_2$, ... be the forces of the system and $\mathbf{r}_1$, $\mathbf{r}_2$, ... the position vectors of any points on their lines of action with respect to O then the moment of the system about the line

$$= \mathbf{r}_1 \times \mathbf{F}_1 \cdot \overrightarrow{OP} + \mathbf{r}_2 \times \mathbf{F}_2 \cdot \overrightarrow{OP} + \dots$$

$$= (\mathbf{r}_1 \times \mathbf{F}_1 + \mathbf{r}_2 \times \mathbf{F}_2 + \dots) \cdot \overrightarrow{OP} = \mathbf{G} \cdot \overrightarrow{OP},$$

so that the moment of a system about any line is equal to the projection on the line of the moment sum of the system about any point of the line.

From this it follows that the moment about a line of a system remains unchanged when we pass from a given system to another equivalent system.

8.17. NULL LINES. NULL PLANES AND NULL POINTS

8.17.1. Null Lines

Def. *Any line about which the moment of a system is zero is called a Null line.*

Clearly any given line will be a null line if the moment sum of the system about a point of the line is perpendicular to the line.

Null lines through a given point. Any line perpendicular to the moment sum about the point is a null line and as such we see that the *null lines through a point lie in a plane which passes through the point and is perpendicular to the moment sum about the point.*

This plane is called the *Null plane of the point.*

If **a** be the position vector of the given point, then the moment sum of the system about the point is $\mathbf{G} - \mathbf{a} \times \mathbf{R}$ and accordingly the equation of the null plane of the point is

$$(\mathbf{r} - \mathbf{a}) \cdot (\mathbf{G} - \mathbf{a} \times \mathbf{R}) = 0$$

$$\Rightarrow \quad \mathbf{r} \cdot (\mathbf{G} - \mathbf{a} \times \mathbf{R}) = \mathbf{a} \cdot \mathbf{G}.$$

Also a point is called the *null point of its null plane.*

It is clear that a null line lies in the null plane of every point of the line.

8.17.2. Theorem

If the null plane of a point A passes through a point B then the null plane of B passes through A.

As the null plane of *A* passes through *B*, the line *AB, i.e., BA* is a null line and as such it lies in the null plane of the point *B*. Hence, the null plane of *B* passes through *A*.

Analytically, we may see that the condition for the null plane

$$\mathbf{r} \cdot (\mathbf{G} - \mathbf{a} \times \mathbf{R}) = \mathbf{a} \cdot \mathbf{G}$$

of a point with position vector **a** to pass through the point with position vector **b** is

$$\mathbf{b} \cdot (\mathbf{G} - \mathbf{a} \times \mathbf{R}) = \mathbf{a} \cdot \mathbf{G}$$

$$\Rightarrow \quad \mathbf{b} \cdot \mathbf{G} - \mathbf{a} \cdot \mathbf{G} = [\mathbf{b}\ \mathbf{a}\ \mathbf{R}] \qquad ...(i)$$

which is also the condition for the null plane of **b**, to pass through **a**. In fact, interchanging **a** and **b** in (*i*), we arrive against at (*i*).

Cor. *Null planes of collinear points have a common line of intersection.*

8.17.3. Theorem

If a system of forces is equivalent to two forces then every line intersecting the lines of action of the two forces is a null line of the system.

This follows from the fact that the moment of each of the two forces about every line intersecting their lines of action is zero.

8.18. CONJUGATE FORCES

Theorem. *Any system of forces is, in general, equivalent to two forces one of which acts along a given line.*

Let OA be the given line.

The system is equivalent to a force $\mathbf{R}$ whose line of action OB passes through O and a couple of moment $\mathbf{G}$.

We suppose that

(*i*) OA is not a null line, (*ii*) $\mathbf{R} \neq \mathbf{0}$, and (*iii*) OB does not coincide with OA.

The null line of O is the plane of the couple $\mathbf{G}$.

Let the plane through OA and OB meet the null plane of O in a line OC.

The force $\mathbf{R}$ along OB is equivalent to two forces along OA, OC. Let these forces be $\mathbf{F}_1$, $\mathbf{F}_2$ respectively.

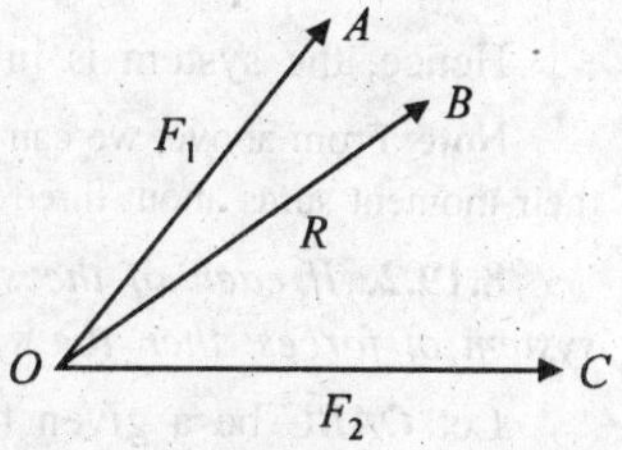

Fig. 8.9.

The couple $\mathbf{G}$ can be replaced by two parallel forces one of which is $-\mathbf{F}_2$ acting along CO and the other acts along another line in the plane of $\mathbf{G}$, *i.e.*, in the null plane of O. The two forces along OC and CO are in equilibrium and as such may be removed.

Thus, we are left with a force along the given line OA and another force whose line of action LM lies in the null plane of O.

As the system is equivalent to two forces along the lines OA and LM, every line intersecting these lines, is a null line of the system. Thus, the null plane of every point on one of these lines passes through the other.

8.18.1. Conjugate Forces and Conjugate Lines

Def. Two forces equivalent to a system of forces are called conjugate forces and their lines of action are called conjugate lines.

The line conjugate to any given line can be obtained as the line of intersection of the null planes of any two points of the given line.

8.19. OTHER FORMS OF CRITERIA FOR EQUILIBRIUM

8.19.1. *If the moment sum of a system about each of three non-collinear points is zero, then the system is in equilibrium.*

Let O, A, B be three non-collinear points. Let

$$\overrightarrow{OA} = \mathbf{a}, \ \overrightarrow{OB} = \mathbf{b}.$$

Let **R** be the vector sum and **G**, the moment sum about O. Then the moment sums about A and B are

$$\mathbf{G} - \mathbf{a} \times \mathbf{R}, \quad \mathbf{G} - \mathbf{b} \times \mathbf{R}.$$

Equating to zero the moment sums about O, A, B, we have

$$\mathbf{0} = \mathbf{G} = \mathbf{G} - \mathbf{a} \times \mathbf{R} = \mathbf{G} - \mathbf{b} \times \mathbf{R},$$

$$\Rightarrow \quad \mathbf{G} = \mathbf{0}, \ \mathbf{a} \times \mathbf{R} = \mathbf{0}, \ \mathbf{b} \times \mathbf{R} = \mathbf{0}.$$

Now **R** cannot be parallel to both the vectors **a** and **b**. Thus,

$$\mathbf{R} = \mathbf{0}.$$

Also

$$\mathbf{G} = \mathbf{0}.$$

Hence, the system is in equilibrium.

Note. From above, we can deduce that two systems of forces are equivalent if their moment sums about three non-collinear points are equal.

8.19.2. *If each of the six edges of a tetrahedron is a null line of a system of forces, then the system is in equilibrium.*

Let $OABC$ be a given tetrahedron and let

$$\overrightarrow{OA} = \mathbf{a}, \ \overrightarrow{OB} = \mathbf{b}, \ \overrightarrow{OC} = \mathbf{c}.$$

Let **R** be the vector sum and **G**, the moment sum about O of the system of forces.

Equating to zero the moments about the lines OA, OB, OC, we obtain

$$\mathbf{G} \cdot \frac{\mathbf{a}}{|\mathbf{a}|} = 0, \ \mathbf{G} \cdot \frac{\mathbf{b}}{|\mathbf{b}|} = 0, \ \mathbf{G} \cdot \frac{\mathbf{c}}{|\mathbf{c}|} = 0,$$

$$\Rightarrow \quad \mathbf{G} \cdot \mathbf{a} = 0, \quad \mathbf{G} \cdot \mathbf{b} = 0, \quad \mathbf{G} \cdot \mathbf{c} = 0. \qquad ...(i)$$

Again, as the moment sums about A, B, C are

$$\mathbf{G} - \mathbf{a} \times \mathbf{R}, \quad \mathbf{G} - \mathbf{b} \times \mathbf{R}, \quad \mathbf{G} - \mathbf{c} \times \mathbf{R},$$

we have, equating to zero the moments about AB, BC, CA,

$$(\mathbf{G} - \mathbf{a} \times \mathbf{R}) \cdot \frac{\mathbf{b} - \mathbf{a}}{|\mathbf{b} - \mathbf{a}|} = 0, \ (\mathbf{G} - \mathbf{b} \times \mathbf{R}) \cdot \frac{\mathbf{c} - \mathbf{b}}{|\mathbf{c} - \mathbf{b}|} = 0,$$

$$(\mathbf{G} - \mathbf{c} \times \mathbf{R}) \cdot \frac{\mathbf{a} - \mathbf{c}}{|\mathbf{a} - \mathbf{c}|} = 0 \qquad ...(ii)$$

The vector **a**, **b**, **c** being non-coplanar, we have, from (*i*)

$$\mathbf{G} = \mathbf{0}. \qquad \text{(§ 2.7, Page 45)}$$

Now the equations (*ii*) give

$$\mathbf{a} \times \mathbf{R} \cdot (\mathbf{b} - \mathbf{a}) = 0, \quad \mathbf{b} \times \mathbf{R} \cdot (\mathbf{c} - \mathbf{b}) = 0, \quad \mathbf{c} \times \mathbf{R} \cdot (\mathbf{a} - \mathbf{c}) = 0.$$

which are equivalent to

$$\mathbf{R} \cdot \mathbf{a} \times \mathbf{b} = 0, \quad \mathbf{R} \cdot \mathbf{b} \times \mathbf{c} = 0, \quad \mathbf{R} \cdot \mathbf{c} \times \mathbf{a} = 0.$$

The vectors $\mathbf{a} \times \mathbf{b}$, $\mathbf{b} \times \mathbf{c}$, $\mathbf{c} \times \mathbf{a}$ being non-coplanar, we obtain

$$\mathbf{R} = \mathbf{0}. \qquad \text{(§ 2.7, Page 45)}$$

Hence, the system is in equilibrium.

Note. The result obtained above enables us to state the criterion for the equivalence of two systems of forces in terms of moments about lines as follows :

Two systems of forces are equivalent if the moments of the two systems about each of the six edges of a tetrahedron are the same.

8.20. FORMULATION IN TERMS OF RECTANGULAR CARTESIAN AXES

Let

$$\mathbf{F}_1, \mathbf{F}_2, \ldots, \mathbf{F}_p, \ldots$$

be forces acting on a rigid body and let

$$P_1, P_2, \ldots, P_p, \ldots$$

be any points on their lines of action.

Let X_p, Y_p, Z_p, be the resolved parts of $\mathbf{F}_p$ along the three co-ordinate axes and (x_p, y_p, z_p) the co-ordinates of the point P_p. We have

$$\mathbf{F}_p = X_p\mathbf{i} + Y_p\mathbf{j} + Z_p\mathbf{k},$$

$$\mathbf{r}_p = x_p\mathbf{i} + y_p\mathbf{j} + z_p\mathbf{k},$$

where $\mathbf{r}_p$ is the position vector of the point $\mathbf{P}_p$ with reference to the origin $\mathbf{0}$.

Moment of $\mathbf{F}_p$ *about the origin*

$$= \mathbf{r}_p \times \mathbf{F}_p$$

$$= (x_p\mathbf{i} + y_p\mathbf{j} + z_p\mathbf{k}) \times (Z_p\mathbf{i} + Y_p\mathbf{j} + Z_p\mathbf{k})$$

$$= (y_pZ_p - z_pY_p)\,\mathbf{i} + (z_pX_p - x_pZ_p)\,\mathbf{j} + (x_pY_p - x_pX_p)\,\mathbf{k}$$

Moment of $\mathbf{F}_p$ *about the coordinate axes.* We have :

$$\mathbf{r}_p \times \mathbf{F}_p \cdot \mathbf{i} = y_pZ_p - z_pY_p, \quad \mathbf{r}_p \times \mathbf{F}_p \cdot \mathbf{j} = z_pX_p - x_pZ_p,$$

$$\mathbf{r}_p \times \mathbf{F}_p \cdot \mathbf{k} = x_pY_p - y_pX_p$$

$$\Rightarrow \qquad y_pZ_p - z_pY_p, \quad z_pX_p - x_pZ_p, \quad x_pY_p - y_pX_p$$

are the moments of $\mathbf{F}_p$ about the co-ordinate axes.

The vector sum and the moment sum about O,

$$\mathbf{R} = \mathbf{i}\,\Sigma X_p + \mathbf{j}\,\Sigma Y_p + \mathbf{k}\,\Sigma Z_p$$
$$= \mathbf{i}X + \mathbf{j}Y + \mathbf{k}Z, \text{ say.}$$
$$\mathbf{G} = \mathbf{i}\,\Sigma\,(y_p Z_p - z_p Y_p) + \mathbf{j}\,\Sigma\,(z_p X_p - x_p Z_p) + \mathbf{k}\,\Sigma\,(x_p Y_p - y_p X_p)$$
$$= \mathbf{i}L + \mathbf{j}M + \mathbf{k}N.$$

Here L, M, N are the moments of the system about the co-ordinate axes.

Moment sum about a point, A (a_1, a_2, a_3).

The position vector of the point A being

$$\mathbf{a} = a_1\mathbf{i} + a_2\mathbf{j} + a_3\mathbf{k},$$

the moment sum about A

$$= \mathbf{G} - \mathbf{a} \times \mathbf{R}$$
$$= (L\mathbf{i} + M\mathbf{j} + N\mathbf{k}) - (a_1\mathbf{i} + a_2\mathbf{j} + a_3\mathbf{k}) \times (X\mathbf{i} + Y\mathbf{j} + Z\mathbf{k})$$
$$= [L - (a_2 Z - a_3 Y)]\,\mathbf{i} + [M - (a_3 X - a_1 Z)]\,\mathbf{j} + [N - (a_1 Y - a_2 X)]\,\mathbf{k}$$

Thus, the moments about the lines through A parallel to the co-ordinate axes are

$$L - (a_2 Z - a_3 Y), \quad M - (a_3 X - a_1 Z), \quad N - (a_1 Y - a_2 X).$$

Parameter of the system :

$$\mathbf{R}\,.\,\mathbf{G} = LX + MY + NZ.$$

Equations of the central axis. The vector equation of the axis is

$$\mathbf{r} = -\frac{\mathbf{G} \times \mathbf{R}}{\mathbf{R}\,.\,\mathbf{R}} + t\mathbf{R}$$

$$\Rightarrow \quad x\mathbf{i} + y\mathbf{j} + z\mathbf{k} = -\frac{(L\mathbf{i} + M\mathbf{j} + N\mathbf{k}) \times (X\mathbf{i} + Y\mathbf{j} + Z\mathbf{k})}{X^2 + Y^2 + Z^2} + t\,(X\mathbf{i} + Y\mathbf{j} + Z\mathbf{k})$$

$$= -\frac{1}{X^2 + Y^2 + Z^2}\{(MZ - NY)\,\mathbf{i} + (NX - LZ)\,\mathbf{j} + (LY - MX)\,\mathbf{k}\} + t\,(X\mathbf{i} + Y\mathbf{j} + Z\mathbf{k})$$

Thus,
$$\frac{x + \dfrac{MZ - NY}{\Sigma X^2}}{X} = \frac{y + \dfrac{NX - LZ}{\Sigma X^2}}{Y} = \frac{\dfrac{LY - MX}{\Sigma X^2}}{Z}$$

are the cartesian equations of the central axis.

The pitch $= \dfrac{\mathbf{R}\,.\,\mathbf{G}}{\mathbf{R}\,.\,\mathbf{R}} = \dfrac{LX + MY + NZ}{X^2 + Y^2 + Z^2}.$

The equations of the null plane of a point A (a_1, a_2, a_3).

If **a** denote the position vector of the point **A**, the vector equation of the null plane of A is

$$\mathbf{r} \cdot (\mathbf{G} - \mathbf{a} \times \mathbf{R}) = \mathbf{a} \cdot \mathbf{G}.$$

Making substitutions, we obtain

$$[L - (a_2Z - a_3Y)]\, x + [M - (a_3X - a_1Z)]\, y + [N - (a_1Y - a_2X)]\, z = a_1L + a_2M + a_3N$$

as the required cartesian equation of the null plane.

Conditions for equilibrium :

$$X = 0,\ Y = 0,\ Z = 0;\ L = 0,\ M = 0,\ N = 0.$$

Conditions for the system to be equivalent to a single force

$$LX + MY + NZ = 0,\ X^2 + Y^2 + Z^2 \neq 0.$$

Conditions for the system to be equivalent to a single couple

$$LX + MY + NZ = 0,\ X^2 + Y^2 + Z^2 = 0,\ L^2 + M^2 + N^2 \neq 0.$$

EXAMPLES

Example 1. *One system of force reduces to a force* **R** *acting at a point O and a couple* **G** *and a second system to a force* **R′** *at O and a couple* **G′**. *Prove that if each system is reduced to a wrench and if*

$$\mathbf{R} \cdot \mathbf{G}' + \mathbf{R}' \cdot \mathbf{G} = (p + p')\, \mathbf{R} \cdot \mathbf{R}'.$$

where p and p′ are the pitches of the two wrenches then the axes of the two wrenches are coplanar.

Solution. With O as the origin of reference, the equations of the axes of the two wrenches are

$$\mathbf{r} = \frac{\mathbf{R} \times \mathbf{G}}{\mathbf{R} \cdot \mathbf{R}} + t\mathbf{R},\quad \mathbf{r} = \frac{\mathbf{R}' \times \mathbf{G}'}{\mathbf{R}' \cdot \mathbf{R}'} + k\mathbf{R}'.$$

They will be coplanar if the scalar triple product

$$\mathbf{R} \times \mathbf{R}' \cdot \left(\frac{\mathbf{R} \times \mathbf{G}}{\mathbf{R} \cdot \mathbf{R}} - \frac{\mathbf{R}' \times \mathbf{G}'}{\mathbf{R}' \cdot \mathbf{R}'}\right) = 0$$

$$\Rightarrow \quad \frac{(\mathbf{R} \times \mathbf{R}') \cdot (\mathbf{R} \times \mathbf{G})}{\mathbf{R} \cdot \mathbf{R}} - \frac{(\mathbf{R} \times \mathbf{R}') \cdot (\mathbf{R}' \times \mathbf{G}')}{\mathbf{R}' \cdot \mathbf{R}'} = 0$$

$$\Rightarrow \quad \frac{(\mathbf{R}' \cdot \mathbf{R}')(\mathbf{R}' \cdot \mathbf{G}) - (\mathbf{R} \cdot \mathbf{G})(\mathbf{R}' \cdot \mathbf{R})}{\mathbf{R} \cdot \mathbf{R}}$$

$$- \frac{(\mathbf{R}' \cdot \mathbf{R})(\mathbf{R}' \cdot \mathbf{G}') - (\mathbf{R} \cdot \mathbf{G}')(\mathbf{R}' \cdot \mathbf{R}')}{\mathbf{R}' \cdot \mathbf{R}'} = 0.$$

Putting

$$p = \frac{\mathbf{R} \cdot \mathbf{G}}{\mathbf{R} \cdot \mathbf{R}'} \cdot p' = \frac{\mathbf{R}' \cdot \mathbf{G}'}{\mathbf{R} \cdot \mathbf{R}'}$$

we obtain

$$\mathbf{R}' \cdot \mathbf{G} - p\,(\mathbf{R}' \cdot \mathbf{R}) - (\mathbf{R} \cdot \mathbf{R}')\,p' + \mathbf{R} \cdot \mathbf{G}' = 0,$$

$$\Rightarrow \quad \mathbf{R} \cdot \mathbf{G}' + \mathbf{R}' \cdot \mathbf{G} = (p + p')\,(\mathbf{R} \cdot \mathbf{R}'),$$

which is the required condition.

Example 2. *Forces act at the centroids of the faces of a tetrahedron along the outward drawn normals to the faces and proportional to their areas; show that these forces are in equilibrium.*

Solution. Let $OABC$ be the tetrahedron. Take O as the origin of reference.

Let

$$\vec{OA} = \mathbf{a}, \ \vec{OB} = \mathbf{b}, \ \vec{OC} = \mathbf{c}.$$

The centroids of the faces are

$$\frac{1}{3}(\mathbf{a} + \mathbf{b} + \mathbf{c}),$$

$$\frac{1}{3}(\mathbf{b} + \mathbf{c}), \ \frac{1}{3}(\mathbf{a} + \mathbf{c}), \ \frac{1}{3}(\mathbf{a} + \mathbf{b}),$$

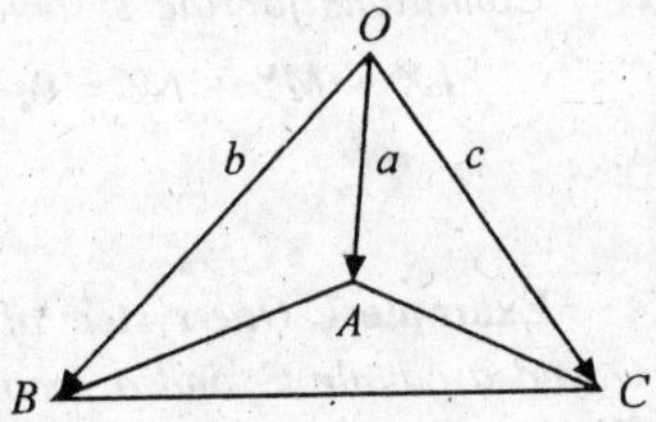

Fig. 8.10.

Outward drawn normal to the plane ABC is parallel to the vector

$$\vec{AB} \times \vec{AC} = (\mathbf{b} - \mathbf{a}) \times (\mathbf{c} - \mathbf{a})$$
$$= \mathbf{b} \times \mathbf{c} + \mathbf{c} \times \mathbf{a} + \mathbf{a} \times \mathbf{b}$$

Also the outward drawn normal vectors to the faces

$$OBA, \ OAC, \ OCB$$

are parallel to the vectors

$$\mathbf{b} \times \mathbf{a}, \ \mathbf{a} \times \mathbf{c}, \ \mathbf{c} \times \mathbf{b}$$

respectively. The magnitudes of each of these outward drawn normal vectors is equal to twice the area of the corresponding face.

Thus, omitting the constant of proportionality, we have the forces

$$\mathbf{b} \times \mathbf{a}, \ \mathbf{a} \times \mathbf{c}, \ \mathbf{c} \times \mathbf{b}, \ \ \mathbf{b} \times \mathbf{c} + \mathbf{c} \times \mathbf{a} + \mathbf{a} \times \mathbf{b}$$

acting at the points

$$\frac{1}{3}(\mathbf{b} + \mathbf{a}), \ \frac{1}{3}(\mathbf{a} + \mathbf{c}), \ \frac{1}{3}(\mathbf{c} + \mathbf{b}), \ \frac{1}{3}(\mathbf{a} + \mathbf{b} + \mathbf{c})$$

respectively.

Now we have

$$\mathbf{R} = 0$$

and
$$\mathbf{G} = \frac{1}{3}(\mathbf{b}+\mathbf{a})\times(\mathbf{b}\times\mathbf{a}) + \frac{1}{3}(\mathbf{a}+\mathbf{c})\times(\mathbf{a}\times\mathbf{c}) + \frac{1}{3}(\mathbf{c}+\mathbf{b})\times(\mathbf{c}\times\mathbf{b})$$
$$+\frac{1}{3}(\mathbf{a}+\mathbf{b}+\mathbf{c})\times(\mathbf{b}\times\mathbf{c}+\mathbf{c}\times\mathbf{a}+\mathbf{a}\times\mathbf{b})$$
$$= \frac{1}{3}[\mathbf{a}\times(\mathbf{b}\times\mathbf{c}) + \mathbf{b}\times(\mathbf{c}\times\mathbf{a}) + \mathbf{c}\times(\mathbf{a}\times\mathbf{b})] = 0.$$

Hence, the system is in equilibrium.

Example 3. *Equal forces act along the edges BC, CA, AB, DA, DB, DC of a regular tetrahedron; show that the central axis of the system is perpendicular from D to the plane ABC and the pitch of the equivalent wrench is* $\frac{1}{2\sqrt{2}}k$, *where k is an edge of the tetrahedron.*

Solution. Take D as the origin of reference.

Let

$$\overrightarrow{DA} = \mathbf{a},\ \overrightarrow{DB} = \mathbf{b},\ \overrightarrow{DC} = \mathbf{c},$$

$$\Rightarrow \quad \overrightarrow{BC} = \mathbf{c}-\mathbf{b},\ \overrightarrow{CA} = \mathbf{a}-\mathbf{c},\ \overrightarrow{AB} = \mathbf{b}-\mathbf{a}.$$

If we neglect the factor of proportionality, we may represent the forces themselves by

$$\mathbf{a},\ \mathbf{b},\ \mathbf{c},\ \mathbf{c}-\mathbf{b},\ \mathbf{a}-\mathbf{c},\ \mathbf{b}-\mathbf{a}.$$

The points of application of these forces may be taken to be

$$D,\ D,\ D,\ B,\ C,\ A$$

whose position vectors are **0, 0, 0, b, c, a.**

We have

$$\mathbf{R} = \mathbf{a}+\mathbf{b}+\mathbf{c}+\mathbf{c}-\mathbf{b}+\mathbf{a}-\mathbf{c}+\mathbf{b}-\mathbf{a} = \mathbf{a}+\mathbf{b}+\mathbf{c},$$

$$\mathbf{G} = \mathbf{0}\times\mathbf{a}+\mathbf{0}\times\mathbf{b}+\mathbf{0}\times\mathbf{c}+\mathbf{b}\times(\mathbf{c}-\mathbf{b})+\mathbf{c}\times(\mathbf{a}-\mathbf{c})+\mathbf{a}\times(\mathbf{b}-\mathbf{a})$$
$$= \mathbf{a}\times\mathbf{b}+\mathbf{b}\times\mathbf{c}+\mathbf{c}\times\mathbf{a}.$$

We represent $\mathbf{a}+\mathbf{b}+\mathbf{c}$ in terms of the vectors

$$\mathbf{b}\times\mathbf{c},\quad \mathbf{c}\times\mathbf{a},\quad \mathbf{a}\times\mathbf{b}.$$

Let

$$\mathbf{a}+\mathbf{b}+\mathbf{c} = l\,\mathbf{b}\times\mathbf{c} + m\,\mathbf{c}\times\mathbf{a} + n\,\mathbf{a}\times\mathbf{b}.$$

Multiplying scalarly by **a, b, c** respectively, we obtain

$$l = \frac{\mathbf{a}.\mathbf{a}+\mathbf{a}.\mathbf{b}+\mathbf{a}.\mathbf{c}}{[\mathbf{a}\,\mathbf{b}\,\mathbf{c}]},\quad m = \frac{\mathbf{b}.\mathbf{a}+\mathbf{b}.\mathbf{b}+\mathbf{b}.\mathbf{c}}{[\mathbf{a}\,\mathbf{b}\,\mathbf{c}]},\quad n = \frac{\mathbf{c}.\mathbf{a}+\mathbf{c}.\mathbf{b}+\mathbf{c}.\mathbf{c}}{[\mathbf{a}\,\mathbf{b}\,\mathbf{c}]}.$$

The tetrahedron being regular of edge k, we have

$$\mathbf{a}\,.\,\mathbf{a}=\mathbf{b}\,.\,\mathbf{b}=\mathbf{c}\,.\,\mathbf{c}=k^2,\ \mathbf{a}\,.\,\mathbf{b}=\mathbf{b}\,.\,\mathbf{c}=\mathbf{c}\,.\,\mathbf{a}=\frac{1}{2}k^2$$

$$[\mathbf{a}\,\mathbf{b}\,\mathbf{c}]^2=\begin{vmatrix}\mathbf{a}\,.\,\mathbf{a} & \mathbf{a}\,.\,\mathbf{b} & \mathbf{a}\,.\,\mathbf{c}\\ \mathbf{b}\,.\,\mathbf{a} & \mathbf{b}\,.\,\mathbf{b} & \mathbf{b}\,.\,\mathbf{c}\\ \mathbf{c}\,.\,\mathbf{a} & \mathbf{c}\,.\,\mathbf{b} & \mathbf{c}\,.\,\mathbf{c}\end{vmatrix}=\frac{1}{2}k^6$$

so that $l = m = n = \dfrac{2\sqrt{2}}{k}$.

$$\therefore \qquad \mathbf{R}=\frac{2\sqrt{2}}{k}=(\mathbf{b}\times\mathbf{c}+\mathbf{c}\times\mathbf{a}+\mathbf{a}\times\mathbf{b})=\frac{2\sqrt{2}}{k}\mathbf{G}$$

$$\Rightarrow \qquad \mathbf{G}=\frac{k}{2\sqrt{2}}\mathbf{R}.$$

Now it is known that the vector $\mathbf{b} \times \mathbf{c} + \mathbf{c} \times \mathbf{a} + \mathbf{a} \times \mathbf{b}$ is normal to the plane ABC.

Thus, $\mathbf{R}$ and $\mathbf{G}$ about the point D have the same direction and are in fact both perpendicular to the plane ABC.

Thus, the line through D normal to the plane ABC is the central axis of the system.

Also $\mathbf{G}=\dfrac{k}{2\sqrt{2}}\mathbf{R}$ so that the pitch $=\dfrac{k}{2\sqrt{2}}$.

Example 4. *Forces act along the edges BC, CA, AB, AO, BO, CO of a regular tetrahedron OABC such that the forces acting along any pair of opposite edges are of the same magnitude; show that the pitch of the resultant wrench is* $\dfrac{k}{2\sqrt{2}}$ *where k is the length of each edge.*

Solution. Take O as the origin of reference and let

$$\vec{OA}=\mathbf{a},\ \vec{OB}=\mathbf{b},\ \vec{OC}=\mathbf{c}.$$

The forces are

$$-\lambda_1\mathbf{a},\ \lambda_1(\mathbf{c}-\mathbf{b});\ -\lambda_2\mathbf{b},\ \lambda_2(\mathbf{a}-\mathbf{c});\ -\lambda_3\mathbf{c},\ \lambda_3(\mathbf{b}-\mathbf{a})$$

acting at the points

$$O,\ B;\ \ O,\ C;\ \ O,\ A$$

with position vectors

$$\mathbf{0},\ \mathbf{b};\ \ \mathbf{0},\ \mathbf{c};\ \ \mathbf{0},\ \mathbf{a};$$

where λ_1/k, λ_2/k, λ_3/k, are the magnitudes of the forces along the opposite pairs of edges. We have

$$\begin{cases} \mathbf{R} = (\lambda_2 - \lambda_3 - \lambda_1)\,\mathbf{a} + (\lambda_3 - \lambda_1 - \lambda_2)\,\mathbf{b} + (\lambda_1 - \lambda_2 - \lambda_3)\,\mathbf{c}; \\ \mathbf{G} = \lambda_1 \mathbf{b} \times \mathbf{c} + \lambda_2 \mathbf{c} \times \mathbf{a} + \lambda_3 \mathbf{a} \times \mathbf{b}. \end{cases}$$

$$\mathbf{G}\,.\,\mathbf{R} = [\lambda_1(\lambda_2 - \lambda_3 - \lambda_1) + \lambda_3(\lambda_3 - \lambda_1 - \lambda_2) + \lambda_3(\lambda_1 - \lambda_2 - \lambda_3)]\,(\mathbf{abc})$$

$$= -(\lambda_1^2 + \lambda_2^2 + \lambda_3^2)\,[\mathbf{a\,b\,c}] = -(\lambda_1^2 + \lambda_2^2 + \lambda_3^2)k^2/\sqrt{2}.$$

Also

$$\mathbf{R}\,.\,\mathbf{R} = [(\lambda_2 - \lambda_3 - \lambda_1)^2 + (\lambda_3 - \lambda_1 - \lambda_2)^2 + (\lambda_1 - \lambda_2 - \lambda_3)^2]\,k^2$$

$$+2\,(\lambda_2 - \lambda_3 - \lambda_1)(\lambda_3 - \lambda_1 - \lambda_2) + (\lambda_3 - \lambda_1 - \lambda_2)(\lambda_1 - \lambda_2 - \lambda_3)$$

$$+ (\lambda_1 - \lambda_2 - \lambda_3)(\lambda_2 - \lambda_3 - \lambda_1)]\frac{1}{2}k^2$$

$$= 2\,(\lambda_1^2 + \lambda_2^2 + \lambda_3^2)\,k^2.$$

Thus, the required pitch $= \dfrac{\mathbf{G}\,.\,\mathbf{R}}{\mathbf{R}\,.\,\mathbf{R}} = \dfrac{k}{2\sqrt{2}}.$

Example 5. *Show that any system of forces can be reduced to a set of forces along the edge of a given tetrahedron OABC.*

If these forces are

$$\lambda\,\overrightarrow{BC},\ \mu\,\overrightarrow{CA},\ \nu\,\overrightarrow{AB}\ \ \textit{and}\ \ \lambda'\,\overrightarrow{OA},\ \mu'\,\overrightarrow{OB},\ \nu'\,\overrightarrow{OC},$$

prove that the forces reduce to a single force or a couple, if

$$\lambda\lambda' + \mu\mu' + \nu\nu' = 0,$$

Show that the forces reduce to a couple if

$$\lambda' + \mu - \nu = 0,\quad \mu' + \nu - \lambda = 0,\quad \nu' + \lambda - \mu = 0.$$

and prove that the axis of this couple makes angles α, β, γ with OA, OB, OC respectively where

$$\cos\alpha : \cos\beta : \cos\gamma = \frac{\lambda}{OA} : \frac{\mu}{OB} : \frac{\nu}{OC}.$$

Solution. We know that any system of forces can be replaced by three forces whose lines of action pass through three given points.

In the present case, we replace the given system by three forces whose lines of action pass through *A, B, C* and regard *A, B, C* as their points of application. Now the force at *A* can be replaced by three forces acting along the three edges *AB, AC, AO* through *A* and so also in relation to the other two forces. Hence, any system of forces can be reduced to a set of forces acting along the edges of a given tetrahedron.

Let

$$\vec{OA} = \mathbf{a},\ \vec{OB} = \mathbf{b},\ \vec{OC} = \mathbf{c}.$$

We have forces

$$\lambda\ (\mathbf{c} - \mathbf{b}),\ \mu\ (\mathbf{a} - \mathbf{c}),\ \ \nu\ (\mathbf{b} - \mathbf{c}),\ \ \lambda'\mathbf{a},\ \mu'\mathbf{b},\ \nu'\mathbf{c}$$

acting the points with position vectors

$$\mathbf{b},\ \mathbf{c},\ \mathbf{a},\qquad \mathbf{0},\ \mathbf{0},\ \mathbf{0}.$$

We have

$$\mathbf{R} = \lambda\ (\mathbf{c} - \mathbf{b}) + \mu\ (\mathbf{a} - \mathbf{c}) + \nu\ (\mathbf{b} - \mathbf{a}) + \lambda'\mathbf{a} + \mu'\mathbf{b} + \nu'\mathbf{c}$$
$$= (\lambda' + \mu - \nu)\ \mathbf{a} + (\mu' + \nu - \lambda)\ \mathbf{b} + (\nu' + \lambda - \mu)\ \mathbf{c},$$

and

$$\mathbf{G} = \mathbf{b} \times \lambda\ (\mathbf{c} - \mathbf{b}) + \mathbf{c} \times \mu\ (\mathbf{a} - \mathbf{c}) + \mathbf{a} \times \nu\ (\mathbf{b} - \mathbf{a})$$
$$= \lambda\ \mathbf{b} \times \mathbf{c} + \mu\ \mathbf{c} \times \mathbf{a} + \nu\ \mathbf{a} \times \mathbf{b}.$$

Thus,

$$\mathbf{R} \,.\, \mathbf{G} = [\ \lambda\ (\lambda' + \mu - \nu) + \mu\ (\mu' + \nu - \lambda) + \nu\ (\nu' + \lambda - \mu)]\ [\mathbf{a}\ \mathbf{b}\ \mathbf{c}]$$
$$= (\lambda\lambda' + \mu\mu' + \nu\nu')\ [\mathbf{a}\ \mathbf{b}\ \mathbf{c}].$$

Thus, the system will reduce to a single force or a couple if

$$\lambda\lambda' + \mu\mu' + \nu\nu' = 0. \qquad ...(i)$$

Also the system will reduce to a couple if $\mathbf{R} = 0$ which will be so if, and only if,

$$\lambda' + \mu - \nu = 0,\ \ \mu' + \nu - \lambda = 0,\ \ \nu' + \lambda - \mu = 0. \qquad ...(ii)$$

Axis of the couple is parallel to **G**. Now,

$$\mathbf{G} \,.\, \vec{OA} = |\mathbf{G}|\ OA \cos \alpha. \qquad \text{(By def.)}$$

Also

$$\mathbf{G} \,.\, \vec{OA} = \lambda\ [\mathbf{a}\ \mathbf{b}\ \mathbf{c}].$$

$$\Rightarrow \qquad \lambda [\mathbf{a}\ \mathbf{b}\ \mathbf{c}] = |\mathbf{G}|\ OA \cos \alpha$$

$$\Rightarrow \qquad \frac{\lambda}{OA \cos \alpha} = \frac{|\mathbf{G}|}{[\mathbf{a}\ \mathbf{b}\ \mathbf{c}]}.$$

Thus,

$$\frac{\lambda}{OA \cos \alpha} = \frac{\mu}{OB \cos \beta} = \frac{\nu}{OC \cos \gamma}.$$

Hence the result.

It may be easily seen that the condition (*i*) is a consequence of the conditions (*ii*).

Example 6. *Prove that the axis of the wrench equivalent to two forces* $\mathbf{R}_1$, $\mathbf{R}_2$ *acting along two non-coplanar lines intersects the line of shortest* *distance between the lines perpendicularly and divides the same in the ratio*

$$\mathbf{R}_1 \cdot (\mathbf{R}_1 + \mathbf{R}_2) \,/\, \mathbf{R}_2 \cdot (\mathbf{R}_1 + \mathbf{R}_2).$$

Solution. Take the mid-point O of the line of shortest distance AC between the lines of action of the forces as the origin of reference.

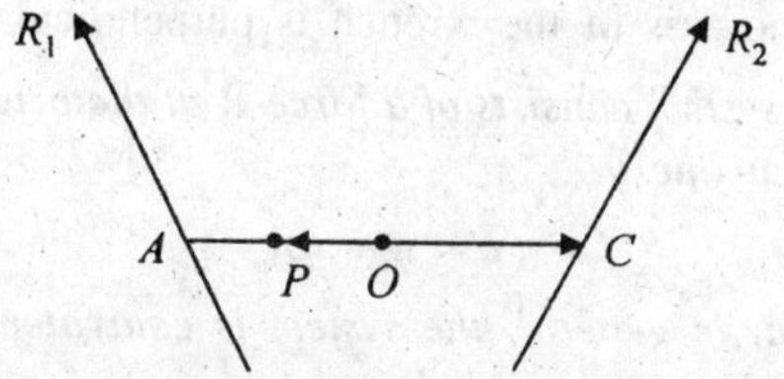

Fig. 8.11.

Let $\overrightarrow{OA} = \mathbf{a}$, so that $\overrightarrow{OC} = -\mathbf{a}$. We have

$$\mathbf{a} \cdot \mathbf{R}_1 = 0, \qquad \mathbf{a} \cdot \mathbf{R}_2 = 0.$$

The vector sum $\mathbf{R}$ and the moment sum $\mathbf{G}$ about the origin O are given by

$$\mathbf{R} = \mathbf{R}_1 + \mathbf{R}_2$$

$$\mathbf{G} = \mathbf{a} \times \mathbf{R}_1 - \mathbf{a} \times \mathbf{R}_2 = \mathbf{a} \times (\mathbf{R}_1 - \mathbf{R}_2).$$

Now $\quad \mathbf{R} \times \mathbf{G} = (\mathbf{R}_1 + \mathbf{R}_2) \times [\mathbf{a} \times (\mathbf{R}_1 - \mathbf{R}_2)]$

$$= [(\mathbf{R}_1 + \mathbf{R}_2) \cdot (\mathbf{R}_1 - \mathbf{R}_2)]\, \mathbf{a} - [(\mathbf{R}_1 + \mathbf{R}_2) \cdot \mathbf{a}]\, (\mathbf{R}_1 - \mathbf{R}_2)$$

$$= (\mathbf{R}_1^2 - \mathbf{R}_2^2)\, \mathbf{a}.$$

Thus, the central axis is

$$\mathbf{r} = \frac{\mathbf{R} \times \mathbf{G}}{\mathbf{R} \cdot \mathbf{R}} + t\mathbf{R} = \frac{(\mathbf{R}_1^2 - \mathbf{R}_2^2)\, \mathbf{a}}{(\mathbf{R}_1 + \mathbf{R}_2)^2} + t(\mathbf{R}_1 + \mathbf{R}_2)$$

so that it passes through a point, P, whose position vector is

$$\frac{(\mathbf{R}_1^2 - \mathbf{R}_2^2)\, \mathbf{a}}{(\mathbf{R}_1 + \mathbf{R}_2)^2}$$

which clearly lies on the line of shortest distance and is such that

$$\overrightarrow{OP} = \frac{\mathbf{R}_1^2 - \mathbf{R}_2^2}{(\mathbf{R}_1 + \mathbf{R}_2)^2}\, \overrightarrow{OA}$$

$$PA = OA - OP$$

$$= OA\left[1 - \frac{\mathbf{R}_1^2 - \mathbf{R}_2^2}{(\mathbf{R}_1 + \mathbf{R}_2)^2}\right] = \frac{2\mathbf{R}_2\,(\mathbf{R}_1 + \mathbf{R}_2)}{(\mathbf{R}_1 + \mathbf{R}_2)^2}\, OA.$$

$$CP = CO + OP$$

$$= OA + OP = OA\left[1 + \frac{\mathbf{R}_1^2 - \mathbf{R}_2^2}{(\mathbf{R}_1 + \mathbf{R}_2)^2}\right]$$

$$= \frac{2\mathbf{R}_1 \cdot (\mathbf{R}_1 + \mathbf{R}_2)}{(\mathbf{R}_1 + \mathbf{R}_2)^2} \; OA.$$

$$\therefore \quad \frac{\mathbf{CP}}{\mathbf{PA}} = \frac{\mathbf{R}_1 \cdot (\mathbf{R}_1 + \mathbf{R}_2)}{\mathbf{R}_2 \cdot (\mathbf{R}_1 + \mathbf{R}_2)^2}.$$

Also clearly the axis of the wrench is perpendicular to AC.

Example 7. *A system consists of a force* **R** *at the origin O and a couple of moment* **G** *and a line l*

$$\mathbf{r} = \mathbf{a} + t\mathbf{b}$$

is given. Show that, in general, the system is equivalent to a force along the line l and another force and that this second force is given by

$$\frac{\mathbf{G} \times (\mathbf{R} \times \mathbf{b}) - [\mathbf{a}\ \mathbf{R}\ \mathbf{b}]\,\mathbf{R}}{(\mathbf{G} - \mathbf{a} \times \mathbf{R}) \cdot \mathbf{b}}$$

where **b** *is a unit vector.*

Solution. If $p\mathbf{b}$ be the force along the given line, then the vector of the second force is $\mathbf{R} - p\mathbf{b}$. Suppose that **c** is the position vector of any point on the line of action of second force. We have, taking moments about the origin O.

$$\mathbf{r} \times p\mathbf{b} + \mathbf{c} \times (\mathbf{R} - p\mathbf{b}) = \mathbf{G}.$$

Multiplying scalarly with $\mathbf{R} - p\mathbf{b}$, we obtain

$$(\mathbf{a} \times p\mathbf{b}) \cdot (\mathbf{R} - p\mathbf{b}) = \mathbf{G} \cdot (\mathbf{R} - p\mathbf{b})$$

$$\Rightarrow \quad p\,[\mathbf{a}\ \mathbf{b}\ \mathbf{R}] = \mathbf{G} \cdot \mathbf{R} - p\,\mathbf{G} \cdot \mathbf{b}$$

$$\Rightarrow \quad p = \frac{\mathbf{G} \cdot \mathbf{R}}{[\mathbf{a}\ \mathbf{b}\ \mathbf{R}] + \mathbf{G} \cdot \mathbf{b}}$$

Thus, the second force is

$$= \mathbf{R} - \frac{\mathbf{G} \cdot \mathbf{R}}{[\mathbf{a}\ \mathbf{b}\ \mathbf{R}] + \mathbf{G} \cdot \mathbf{b}}\,\mathbf{b}$$

$$= \frac{\{[\mathbf{a}\ \mathbf{b}\ \mathbf{R}] + \mathbf{G} \cdot \mathbf{b}\}\,\mathbf{R} - (\mathbf{G} \cdot \mathbf{R})\,\mathbf{b}}{[\mathbf{a}\ \mathbf{b}\ \mathbf{R}] + \mathbf{G} \cdot \mathbf{b}} = \frac{[\mathbf{a}\ \mathbf{b}\ \mathbf{R}]\,\mathbf{R} + \mathbf{G} \times (\mathbf{R} \times \mathbf{b})}{(\mathbf{G} - \mathbf{a} \times \mathbf{R}) \cdot \mathbf{b}}$$

Example 8. *Show that the null point of the plane* $\mathbf{r} \cdot \mathbf{n} = p$, *is* $(\mathbf{n} \times \mathbf{G} + p\mathbf{R}) / (\mathbf{G} \cdot \mathbf{R})$, *for the system of forces equivalent to a couple* **G** *and force* **R**, *whose line of action passes through the origin of reference O.*

Solution. Given plane is $\mathbf{r} \cdot \mathbf{n} = p$ ...(1)

Let the null point be **a**. Then, the equation of the null plane of the point **a** is

$$\mathbf{r} \cdot (\mathbf{G} - \mathbf{a} \times \mathbf{R}) = \mathbf{a} \cdot \mathbf{G} \qquad ...(2)$$

Comparing (1) and (2), we get

$$\mathbf{G} - \mathbf{a} \times \mathbf{R} = \mathbf{n} \quad ...(3)$$

and

$$\mathbf{a} \cdot \mathbf{G} = p \quad ...(4)$$

$$(3) \Rightarrow \quad \mathbf{G} \times \mathbf{G} - (\mathbf{a} \times \mathbf{R}) \times \mathbf{G} = \mathbf{n} \times \mathbf{G}$$

$$\Rightarrow \quad -[(\mathbf{a} \cdot \mathbf{G})\, \mathbf{R} - (\mathbf{R} \cdot \mathbf{G})\, \mathbf{a}] = \mathbf{n} \times \mathbf{G}$$

$$\Rightarrow \quad -p\mathbf{R} + (\mathbf{R} \cdot \mathbf{G})\, \mathbf{a} = \mathbf{n} \times \mathbf{G}$$

$$\Rightarrow \quad \mathbf{a} = [\mathbf{n} \times \mathbf{G} + p\mathbf{R}] \,/\, (\mathbf{G} \cdot \mathbf{R}).$$

EXERCISES

1. Force proportional to the lengths of the edges of a tetrahedron act along the edges *OA, OB, OC, BC, CA, AB* of a tetrahedron *OABC;* show that the sum of the moment vectors about *O* is perpendicular to the plane *ABC.* Also show that the central axis is parallel to the line *OG; G* being the centroid of the tetrahedron.

2. Forces act at the vertices of a tetrahedron along the outward drawn normals to the opposite faces and proportional to their areas; show that the forces are in equilibrium.

3. Forces are acting at the vertices of a tetrahedron along the lines joining them to the centroids of the opposite faces and proportional to the lengths of the same; show that the system is in equilibrium.

4. The lines of action of two non-intersecting forces are at right angles to one another; show that their distances from the central axis are in the ratio of the squares of the magnitudes of the forces.

5. Every line whatsoever of the null plane of a point for a given system is a null line; show that the parameter of the system is zero.

6. Show that the parameter of any system of forces is the sum of the parameters of the pairs of forces of the system.

7. *P* is the null point of a plane and *A, B* are two points where any two conjugate lines for a system of forces meet the plane; show that the points *A, B, P* are collinear.

8. Show that for a system of forces, the conjugate line of the line of intersection of two planes is the line joining their null points.

9. If the null point of a plane π_1 is a point of a plane π_2, then show that the null point of π_2 is a point of π_1.

10. Show that the conjugate lines of a system of coplanar lines is a system of concurrent lines.

11. Prove that the line of shortest distance between two conjugate lines of a system of forces intersects perpendicularly the central axis of the system.

12. Prove that a system of forces may, in general, be reduced to two forces one of which passes through a given point and the other lies in a given plane.

    (If *Q* be the null point of a given plane π, then the conjugate of the line joining the given point to *Q* lies in the plane π).

13. A system of forces is equivalent to two forces one of which acts at a fixed point and lies in a fixed plane through the point; prove that the other passes through another fixed point and lies in a fixed plane.

14. A system of four forces is in equilibrium; show that the parameter of any two of them equals the parameter of the other two and the parameter of any three is zero.

15. The sum of the moments of a system of forces about each of three non-coplanar lines is zero; show that the system cannot be equivalent to a couple.

16. Equal forces are acting along three non-intersecting edges of a rectangular parallelopiped; find the condition for the system of forces to be equivalent to a single force.

17. Forces act along the edges BA, BC, OA, OC of a tetrahedron $OABC$ in the directions indicated by the order of the letters and each force is inversely proportional to the length of the edge in which it acts. Prove that, if the forces have a single resultant,

$$AO \,.\, BC = AB \,.\, CO.$$

18. Wrenches of the same pitch p act along the edges of a regular tetrahedron $ABCD$ of side a. If the intensities of the wrenches along AB, DC are the same, and also those along BC, DA and DB, CA, are the same show that the pitch of the equivalent wrench is $p + a/2\sqrt{2}$.

19. Wrenches of equal intensities act along the edges BC, CA, AB, AD, BD, CD of a regular tetrahedron $ABCD$. Prove that the pitch of the resultant wrench is $\frac{1}{2}(p+q+r)+\frac{1}{2}k$, where p, q, r are the pitches of the wrenches which act along AD, BC; BD, CA; AB, CD; and k is the length of the shortest distance between any pair of opposite edges of the tetrahedron.

20. OX, OY, OZ is a system of right-handed rectangular coordinate axes and the axes of two wrenches are along OX and OY. I_1, I_2 are the intensities and p_1, p_2 are the pitches of these wrenches. Show that the intensity and the pitch of the resultant wrench are

$$\sqrt{(I_1^2 + I_2^2)} \text{ and } (p_1 I_1^2 + p_2 I_2^2)/(I_1^2 + I_2^2),$$

and show that the axis of the resultant wrench meets OZ at a point distant

$$(p_2 - p_1)\,[I_1 I_2 / \sqrt{(I_1^2 + I_2^2)}]$$

from O and makes an angle

$$\tan^{-1}(I_2 / I_1)$$

with OX.

21. A system of forces is equivalent to a couple $\mathbf{G}$ and a force $\mathbf{R}$ through the origin O. Show that the line which is conjugate to the line

$$\mathbf{r} = \mathbf{a} + t\mathbf{b};$$

$\mathbf{b}$ being a unit vector, is given by

$$\mathbf{G} - \mathbf{r} \times \mathbf{R} = \lambda\,(\mathbf{a} - \mathbf{r}) \times \mathbf{b},$$

where $\lambda = \dfrac{\mathbf{G} \cdot \mathbf{R}}{\mathbf{G} \cdot \mathbf{b} + [\mathbf{a}\ \mathbf{b}\ \mathbf{R}]}$.

22. The line of shortest distance between the axes of the two wrenches ($\mathbf{R}_1$, $p_1\ \mathbf{R}_1$) and ($\mathbf{R}_2$, $p_2\ \mathbf{R}_2$) meets the axes in the points A, C; show that the axis of the resultant equivalent wrench divides AC in the ratio

$$\frac{d^2\mathbf{R}_2 \cdot (\mathbf{R}_1 + \mathbf{R}_2) - (p_2 - p_1)\, V}{d^2\mathbf{R}_2 \cdot (\mathbf{R}_1 + \mathbf{R}_2) + (p_2 - p_1)\, V}$$

and its pitch is

$$\frac{V + p_1\mathbf{R}_1 \cdot (\mathbf{R}_1 + \mathbf{R}_2) + p_2\mathbf{R}_2 \cdot (\mathbf{R}_1 + \mathbf{R}_2)}{(\mathbf{R}_1 + \mathbf{R}_2)^2}$$

where

$$V = [\mathbf{R}_1 \mathbf{R}_2\ \overrightarrow{AC}] \text{ and } d = AC.$$

23. Coplanar forces all drawn inwards (or outwards) are acting at the middle points of the sides of a plane convex polygon perpendicular and proportional to the corresponding sides; show that the system is in equilibrium.

24. Prove that three forces represented in magnitude, direction and line of action by three sides of a triangle, taken in order, are equivalent to a couple, whose magnitude is twice the area of the triangle. Hence, or otherwise, show that the system of forces represented in magnitude, direction and line of action by the four sides taken in order, of a plane quadrilateral of area A is equivalent to a couple of moment $2A$.

25. Forces represented by $\overrightarrow{AC}$ and $\overrightarrow{DB}$ act along the diagonals AC and DB of a parallelogram $ABCD$. Show that their resultant is represented in magnitude, and direction by $2\overrightarrow{AB}$. Determine the line of action of the resultant of the forces represented in magnitude, direction and line of action by

$$\overrightarrow{AB},\ \overrightarrow{BC},\ \overrightarrow{CD},\ \overrightarrow{DA},\ \overrightarrow{AC} \text{ and } \overrightarrow{DB}.$$

26. Calculate the moment about the point C (1, 1, 1) of a force of 5 lb wt. acting along the line $\overrightarrow{AB}$, where A, B are the points (2, 3, 4), (3, 5, 6) respectively, the distances being measured in feet.

27. A force of magnitude 6 in the direction of the vector $\mathbf{i} - 2\mathbf{j} + 2\mathbf{k}$ acts at the point whose position vector is $\mathbf{i} - \mathbf{j}$. A second force acting at the point with position vector $\mathbf{j} - \mathbf{k}$ forms a couple with the first force. Find the vector moment of the couple.

28. A force of unit magnitude acts along a line which passes through the origin O of a rectangular system of co-ordinate axes Ox, Oy and Oz and has equal moments about each of the points (1, 1, 0) and (2, 0, 1). Find the possible values of the components of the force in the directions Ox, Oy and Oz.

SUMMARY

1. The moment **M** of a force **F** about a point O is, by def., $\mathbf{r} \times \mathbf{F}$, where $\mathbf{r} = \overrightarrow{OP}$ and P is a point on the line of action of the force **F**.

2. The equation of the line of action of the force **F** whose moment about a point O is **M** is

$$\mathbf{r} = \frac{-1}{\mathbf{F}\cdot\mathbf{F}}(\mathbf{F} \times \mathbf{M}) + t\mathbf{F};$$

O being the origin of reference.

3. The force vector sum **R** and the moment vector sum **G** about point O of a system are given as follows :

$$\mathbf{R} = \mathbf{F}_1 + \mathbf{F}_2 + \mathbf{F}_3 + \ldots$$

and

$$\mathbf{G} = \mathbf{r}_1 \times \mathbf{F}_1 + \mathbf{r}_2 \times \mathbf{F}_2 + \mathbf{r}_3 \times \mathbf{F}_3 + \ldots$$

$\mathbf{r}_1, \mathbf{r}_2, \mathbf{r}_3, \ldots$ being the position vectors of the points $P_1, P_2, P_3 + \ldots$ on the lines of action of the forces $\mathbf{F}_1, \mathbf{F}_2, \mathbf{F}_3, \ldots$.

4. If **G** denotes the moment about a point O, then the moment about O' is

$$\mathbf{G} - \mathbf{s} \times \mathbf{R} \text{ where } \mathbf{s} = \overrightarrow{OO'}.$$

5. The moment of a force about a line. The moment of a force **F** about a directed line $\overrightarrow{OI}$ is

$$\mathbf{r} \times \mathbf{F} \cdot \overrightarrow{OI}$$

where **r** is the position vector $\overrightarrow{OP}$ of a point P on the line of action of the force **F** and $\overrightarrow{OI}$ is a unit vector.

The moment of a force about a line is a scalar.

6. The moment of a system of forces about a line is the sum of the moments of the forces of the system about the line.

7. Null lines. Any line about which the moment of a system is zero, is called a null line of the system.

8. Null planes. Null lines through a point lie on a plane through the point called a null plane.

9. A pair of equal, unlike parallel forces is called a *couple.* The moment sum of the forces of a couple about every point is the same and is perpendicular to the plane of the couple. The vector so determined is called the *Axis of the couple.*

10. Poinsot's Central Axis. Every system of forces is equivalent to a force and a couple, such that the line of action of the force is the axis of the couple. This axis is called the *Poinsot's Central Axis* and its equation is

$$\mathbf{r} = \frac{\mathbf{R} \times \mathbf{G}}{\mathbf{R} \cdot \mathbf{R}} + t\mathbf{R}.$$

11. Wrench. A system consisting of a force and a couple such that the line of the force is the axis of the couple is called a *wrench.*

12. Conditions of Equilibrium. A system is in equilibrium when any one of the following conditions hold :

(*i*) $\mathbf{R} = \mathbf{O}$ and $\mathbf{G} = \mathbf{O}$.

(*ii*) The moment sum of the system about each of three non-collinear points is zero.

(*iii*) Each of the six edges of a tetrahedron is a null line for the system of forces.

OBJECTIVE QUESTIONS

For each of the following questions, four alternatives are given for the answer. Only one of them is correct. Choose the correct alternative.

1. If the resultant of two forces P and Q be equal in magnitude to one of the components P and perpendicular to it in direction, then the value of Q is

   (*a*) P (*b*) $\frac{1}{\sqrt{2}} P$ (*c*) $\sqrt{2}\, P$ (*d*) $\sqrt{3}\, P$

2. A, B, C are fixed points and P a variable point such that the resultant of forces at P represented by $\overrightarrow{PA}$ and $\overrightarrow{PB}$ always passes through C and D is the mid-point of AB, then locus of P is

   (*a*) circle passing through A, B, C and D
   (*b*) straight line passing through C and D
   (*c*) straight line passing through A and B
   (*d*) None of these

3. Moment about the point $\mathbf{i} + 2\mathbf{j} - \mathbf{k}$ of a force represented by $\mathbf{i} + 2\mathbf{j} + \mathbf{k}$ acting through the point $2\mathbf{i} + 3\mathbf{j} + \mathbf{k}$ is

   (*a*) $3\mathbf{i} + \mathbf{j} - \mathbf{k}$ (*b*) $3\mathbf{i} - \mathbf{j} + \mathbf{k}$
   (*c*) $-3\mathbf{i} + \mathbf{j} + \mathbf{k}$ (*d*) $3\mathbf{i} + \mathbf{j} + \mathbf{k}$

4. Sum of moments of two intersecting forces about a point in their plane is equal to

(*a*) Moment of their resultant about that point
(*b*) Negative of moment of their resultant about that point
(*c*) zero
(*d*) None of these

5. A single force and coplanar couple together acting on a rigid body are equivalent to
(*a*) a couple
(*b*) a single force equal to the given force and in any direction
(*c*) a single force equal and parallel to the given force
(*d*) None of these

6. Any number of coplanar forces acting on a rigid body can be reduced to
(*a*) a single force
(*b*) a single couple
(*c*) either to a single force or a single couple
(*d*) a single force together with a couple

7. If in a wrench either the force or the couple of the wrench is zero, then wrench is called
(*a*) null wrench (*b*) degenerate wrench
(*c*) non-degenerate wrench (*d*) central wrench

8. The quantities which are invariants for any given system of forces acting on a rigid body are
(*a*) **R** and **G** (*b*) **K** and **G**
(*c*) **R** and $\mathbf{R} \times \mathbf{G}$ (*d*) $\mathbf{R} \cdot \mathbf{R}$ and $\mathbf{G} \cdot \mathbf{R}$

9. If in a system of forces acting on a rigid body $\mathbf{R} \cdot \mathbf{G} \neq 0$, then the given system of forces is equivalent to
(*a*) degenerate wrench (*b*) non-degenerate wrench
(*c*) a couple (*d*) a single force

10. A line about which the moment of a given system of forces is zero is known as
(*a*) central axis (*b*) null line
(*c*) line of force (*d*) arm of couple

ANSWERS

1. (*c*)	**2.** (*b*)	**3.** (*c*)	**4.** (*a*)	**5.** (*c*)
6. (*d*)	**7.** (*b*)	**8.** (*d*)	**9.** (*b*)	**10.** (*b*)

Answers

Page 14-15

1. (*i*) $\vec{BC} = \vec{AC} - \vec{AB}$. $\vec{AD} = \frac{1}{2}\vec{AB} + \frac{1}{2}\vec{AC}$.

$\vec{BE} = \frac{1}{2}\vec{AC} - \vec{AB}$. $\vec{CF} = \frac{1}{2}\vec{AB} - \vec{AC}$.

(*ii*) $\vec{AB} = -\frac{4}{3}\vec{BE} + \frac{4}{3}\vec{CF}$. $\vec{BC} = \frac{2}{3}\vec{BE} - \frac{2}{3}\vec{CF}$.

$\vec{CA} = \frac{2}{3}\vec{BE} + \frac{4}{3}\vec{CF}$. $\vec{AD} = -\vec{BE} - \vec{CF}$.

(*iii*) $\vec{AC} = 2\vec{AB} + 2\vec{BE}$. $\vec{BC} = \vec{AB} + 2\vec{BE}$.

$\vec{AD} = \frac{3}{2}\vec{AB} + \vec{BE}$. $\vec{CF} = -\frac{3}{2}\vec{AB} - 2\vec{BE}$.

2. (*i*) $\vec{AC} = \vec{AB} + \vec{AD}$. $\vec{BD} = -\vec{AB} + \vec{AD}$.

(*ii*) $\vec{AB} = \frac{1}{2}\vec{AC} - \frac{1}{2}\vec{BD}$. $\vec{AD} = \frac{1}{2}\vec{AC} + \frac{1}{2}\vec{BD}$.

(*iii*) $\vec{AB} = \vec{AD} - \vec{BD}$. $\vec{AC} = 2\vec{AD} - \vec{BD}$.

3. (*i*) $\vec{BC} = \vec{OC} - \vec{OB}$, $\vec{CA} = \vec{OA} - \vec{OC}$. $\vec{AB} = \vec{OB} - \vec{OA}$.

(*ii*) $\vec{OA} = \vec{OC} - \vec{AB} - \vec{AC}$. $\vec{OB} = \vec{OC} - \vec{BC}$. $\vec{CA} = -\vec{AB} - \vec{BC}$

4. $\mathbf{a} = 5\mathbf{c} + 3\mathbf{d}$, $\mathbf{b} = 3\mathbf{c} + d$

5. (*i*) $\mathbf{a} = \frac{2\mathbf{c} + 3\mathbf{d}}{4}$, $\mathbf{b} = \frac{10\mathbf{c} + 3\mathbf{d}}{8}$. (*ii*) $\mathbf{a} = \mathbf{d}$, $\mathbf{b} = \mathbf{d}$.

7. (*c*) is the only correct statement.

8. (*b*), (*d*) are correct statements.

Page 21-22

1. (*i*) (4, – 2, 2), (*ii*) (10, 0, 7), (*iii*) (3, – 4, 8).

3. (*i*) $-\mathbf{b} + 2\mathbf{c} = 2(\mathbf{a} - 2\mathbf{b} + 3\mathbf{c}) + (-2\mathbf{a} + 3\mathbf{b} - 4\mathbf{c})$.

(*ii*) $3\mathbf{a} + 20\mathbf{b} + 5\mathbf{c} = 2(5\mathbf{a} + 6\mathbf{b} + 7\mathbf{c}) - (7\mathbf{a} - 8\mathbf{b} + 9\mathbf{c})$.

5. Linearly dependent.

6. (0, – 7/5, 1/5).

7. $x' = \frac{1}{2}x - y + \frac{1}{2}z$, $y' = -\frac{1}{2}x + \frac{1}{2}z$, $z' = y$.

8. $\frac{1}{2}(-x + y + z)$, $\frac{1}{2}(x - y + z)$, $\frac{1}{2}(x + y - z)$.

9. (*i*) $(7/2)(\mathbf{a} - \mathbf{b} + \mathbf{c}) + (\mathbf{b} + \mathbf{c} - \mathbf{a}) - \frac{1}{2}(\mathbf{c} + \mathbf{a} + \mathbf{b}) = 2\mathbf{a} - 3\mathbf{b} + 4\mathbf{c}$.

(*ii*) $-8(\mathbf{a}-2\mathbf{b}+\mathbf{c})+6(-\mathbf{b}+2\mathbf{c}+\mathbf{a})-3(\mathbf{c}-\mathbf{a}+3\mathbf{b})$
$=\mathbf{a}+\mathbf{b}+\mathbf{c}.$

10. (*i*) $h=-4/3,\ k=-5/3.$ (*ii*) $h=-\dfrac{1}{2},\ k=1.$

11. (*i*) $h=-13/5,\ k=-2/5,\ l=-4.$ (*ii*) $h=4/3,\ k=-4/3,\ l=1.$

Page 41-42

1. (*i*) $2\mathbf{b}-2\mathbf{a}$. (*ii*) Origin. (*iii*) $5\mathbf{a}+3\mathbf{b}$. (*iv*) Parallel. (*v*) $\mathbf{a}+\mathbf{b}$.

2. (*i*) Collinear, (*ii*) Non-collinear, (*iii*) Collinear, (*iv*) Non-collinear.

3. $\mathbf{a}+\mathbf{b}+\mathbf{c}$. **4.** $OC/CE=4/1,\ BC/CF=3/2.$

5. $5\mathbf{a}-\mathbf{b},\ \mathbf{r}=\mathbf{b}-\mathbf{a}+\lambda(\mathbf{a}+\mathbf{b}),$
$\mathbf{r}=2\mathbf{a}-4\mathbf{b}+\mu(\mathbf{b}-3\mathbf{a}),\ -4\mathbf{a}-2\mathbf{b}.$

6. $2:1,\ 2:1,\ 2:1.$ **7.** $3\mathbf{a}-\mathbf{b},\ 3\mathbf{b}-\mathbf{a},\ \dfrac{3\mathbf{a}-\mathbf{b}}{4}.$

8. $-1:6.$ **13.** $1:2.$

Page 48-50

1. $\mathbf{r}=t\mathbf{a}+p\mathbf{b},\ \mathbf{r}=t\mathbf{b}+p\mathbf{c}.$
$\mathbf{r}=t\mathbf{a}+p\mathbf{c},\ \mathbf{r}=\mathbf{a}+t(\mathbf{b}-\mathbf{a})+p(\mathbf{c}-\mathbf{a}).$

2. $\mathbf{r}=t\mathbf{a}+p\mathbf{b},$ $\mathbf{r}=\mathbf{c}+t\mathbf{a}+p\mathbf{b}.$
$\mathbf{r}=t\mathbf{b}+p\mathbf{c},$ $\mathbf{r}=\mathbf{a}+t\mathbf{b}+p\mathbf{c}.$
$\mathbf{r}=t\mathbf{a}+p\mathbf{c},$ $\mathbf{r}=\mathbf{b}+t\mathbf{a}+p\mathbf{c}.$

3. (*i*) $\mathbf{r}=(2+t+5p)\,\mathbf{a}+(2+2t-2p)\,\mathbf{b}+(-2+4t+8p)\mathbf{c}.$
(*ii*) $\mathbf{r}=(1+4p)\,\mathbf{a}+(1+t+2p)\,\mathbf{b}+(1+3p)\,\mathbf{c}.$

4. (*i*) Coplanar (*ii*) Coplanar (*iii*) Non-Coplanar.

7. Line lies in each of the planes.

8. $2\mathbf{a}+3\mathbf{b}-4\mathbf{c}.$

9. (*i*) The lines intersect at the point $\mathbf{a}+2\mathbf{b}-\mathbf{c}$.
(*ii*) The lines intersect at the point $3\mathbf{a}+5\mathbf{b}-2\mathbf{c}$.
(*iii*) The lines intersect at the point $\mathbf{b}-2\mathbf{c}$.

10. (*i*) $\mathbf{r}=t(2\mathbf{a}+3\mathbf{b}-\mathbf{c})+(\mathbf{b}-\mathbf{c}).$
(*ii*) Planes are parallel. (*iii*) Planes are parallel.
(*iv*) $\mathbf{r}=-2\mathbf{a}+t(\mathbf{b}+\mathbf{c}).$

11. $5\mathbf{a}-7\mathbf{b}+6\mathbf{c}.$

13. $\dfrac{1}{2}(\mathbf{a}-\mathbf{b}-3\mathbf{c})$ is the point of intersection.

Miscellaneous Exercises 1

Page 63-69

1. (*d*) is correct. **4.** Parallelogram

5. $\overrightarrow{AB}=\mathbf{b}-\mathbf{a},\ \overrightarrow{PA}=\dfrac{1}{2}\mathbf{a},\ \overrightarrow{PQ}=\dfrac{1}{2}\mathbf{b},\ \overrightarrow{AR}\dfrac{1}{2}\mathbf{b}-\mathbf{a},$

where $\overrightarrow{OA}=\mathbf{a}$ and $\overrightarrow{OB}=\mathbf{b}$.

6. $\overrightarrow{PQ}=\overrightarrow{OQ}-\overrightarrow{OP},\ \overrightarrow{SR}=\dfrac{1}{2}\left(\overrightarrow{OQ}=\overrightarrow{OP}\right),$

$\vec{SQ} = \vec{OQ} - \frac{1}{2}\vec{OP}$, $\vec{RP} = \vec{OP} - \frac{1}{2}\vec{OQ}$.

7. $\vec{BE} = (1/3)(-2\mathbf{a} + \mathbf{b})$, $\vec{DC} = 2\mathbf{a}$.

9. (*i*) (1/4) **a** + (1/3) **b**, (*ii*) (1/3) **b** – (3/4) **a.**
$\lambda = 4/3$, $\mu = 4/9$, $AP/PQ = 3/1$, $OQ/OB = 4/5$.

11. B is **a** + **p**, C is **a** + **p** + **q**, D is **a** + **q**, E is $\frac{1}{2}(2\mathbf{a} + \mathbf{p} + \mathbf{q})$.

12. R is (1/11) (**a** + 6**b** + 4**c**), 6 : 1.

13. $\vec{AQ} = 3\mathbf{b} - \mathbf{a}$, $\vec{BP} = 2\mathbf{a} - \mathbf{b}$.
S divides AB in the ratio 3 : 4.

14. $AF : FB = 3 : 4$, $AL \div LC = -2$.

15. $\vec{OS} = \frac{1}{2}(\mathbf{a} + h\mathbf{b})$, $\vec{AM} = h\mathbf{b} - \mathbf{a}$.

(*i*) $\vec{OS} = (3/4)\left[\mathbf{b}k + (1-k)\,\mathbf{a}\right]$, $\vec{OL} = \mathbf{b}k + (1-k)\,\mathbf{a}$

$h = \frac{1}{2}$, $k = \frac{1}{3}$,

(*ii*) $AL / LB = \frac{1}{2}$, $OM / OB = 1$, $ON / NA = 2$.

16. (*i*) $\vec{BC} = 2(\mathbf{b} - \mathbf{a})$, (*ii*) $\vec{PQ} = \mathbf{b} - \mathbf{a}$, (*iii*) $\vec{PC} = 2\mathbf{b} - \mathbf{a}$.

(*iv*) $\vec{BQ} = \mathbf{b} - 2\mathbf{a}$, $BC \parallel PQ$ and $BC - 2PQ$.

17. $\vec{OY} = \mathbf{a} + 2\mathbf{b}$, $OX / XY = \frac{1}{2}$.

18. $\vec{PQ} = (1/3)(2\mathbf{c} - \mathbf{b} - \mathbf{a})$, $\vec{SR} = (1/3)(\mathbf{d} + \mathbf{c} - 2\mathbf{a})$

19. $\vec{AD} = \mathbf{d} - \mathbf{a}$, $\vec{OC} = \frac{1}{2}\mathbf{a} + 3\mathbf{d}$, $\vec{DC} = \frac{1}{2}\mathbf{a} + 2\mathbf{d}$.
$OX/XC = 2/5$, $AX/XD = 6/1$.

20. P is the mid-point of AB, $XP/PY = 3/1$.

21. $2LM = NL$.

23. (*i*) $\vec{OP} = \frac{1}{2}\mathbf{a}$, $\vec{OQ} = \mathbf{b}$, $\vec{OR} = \frac{1}{2}(\mathbf{b} + \mathbf{c})$, $\vec{OS} = \frac{1}{2}(\mathbf{a} + \mathbf{c})$.

(*iv*) $\vec{OG} = \frac{1}{4}(\mathbf{a} + \mathbf{b} + \mathbf{c}) = \vec{OH}$.
PR and QS bisect each other.

24. $\vec{AE} = \frac{1}{2}(\mathbf{a} + \mathbf{b})$, $\vec{BD} = \mathbf{b} - \mathbf{a}$, $\vec{MB} = \frac{1}{2}(\mathbf{a} - 2\mathbf{b})$.

26. (*ii*) $\overrightarrow{PR} = \mathbf{a} + \mathbf{b}$, $\overrightarrow{SM} = \frac{1}{2}(4\mathbf{a} - \mathbf{b})$, $\overrightarrow{SX} = (2/5)(4\mathbf{a} - \mathbf{b})$,

$\overrightarrow{PX} = (3/5)(\mathbf{a} + \mathbf{b})$.

(*i*) $PX/PR = 3/5$. (*iii*) P, X, R are collinear.

28. $\overrightarrow{AB} = \mathbf{b} - \mathbf{a}$, $\overrightarrow{AC} = (k-1)\,\mathbf{a} + l\mathbf{b}$, $\overrightarrow{BC} = k\mathbf{a} + (1-1)\,\mathbf{b}$,

A, B, C are collinear, $AC/CB = l/k$, $\overrightarrow{XY} = (2/3)(3\mathbf{b} - \mathbf{a})$.

29. $\overrightarrow{AB} = \mathbf{b} - \mathbf{a}$, $\overrightarrow{BC} = -(3\mathbf{a} + \mathbf{b})$,

$\overrightarrow{AD} = -(3\mathbf{b} + \mathbf{a})$, $\overrightarrow{DC} = 3(\mathbf{b} - \mathbf{a})$, $\overrightarrow{AB} \,||\, \overrightarrow{DC}$.

30. $(1/4)\ \mathbf{a} + (1/3)\ \mathbf{b}$, $(1/3)\ \mathbf{b} - (3/4)\ \mathbf{a}$

$h = 4/3$, $k = 4/9$,

$\frac{AP}{PQ} = 3/1$, $\frac{OQ}{OB} = 4/5$.

31. $\overrightarrow{OD} = \frac{1}{2}(\mathbf{a} + \mathbf{b})$, $\overrightarrow{OE} = \frac{1}{3}(\mathbf{b} + 2\mathbf{c})$, $\overrightarrow{OF} = 4(\mathbf{a} + \mathbf{b} + 2\mathbf{c})$

$AF : FE = 3 :1$.

32. $OY/YA = 2/7$, $XC/CY = 9/2$.

34. (*i*) $\mathbf{m} = \frac{1}{2}(\mathbf{p} + \mathbf{q})$. (*ii*) $\mathbf{t} = \frac{1}{3}(\mathbf{p} + \mathbf{q} + \mathbf{s})$, (*iii*) $\mathbf{r} = \mathbf{q} - \mathbf{p} - \mathbf{s}$.

Page 88–89

1. $\sqrt{14}, \sqrt{5}, \sqrt{45}$.

2. (*i*) $\frac{\pi}{2}$, (*ii*) $\cos^{-1} \frac{8}{5\sqrt{29}}$, (*iii*) $\cos^{-1} \frac{1}{\sqrt{39}}$.

3. (*i*) $a_1\mathbf{i} + a_2\mathbf{j} + a_3\mathbf{k}$, where $a_1 = -1/\sqrt{14}$; $a_2 = -2/\sqrt{14}$; $a_3 = -3/\sqrt{14}$.

(*ii*) $a_1\mathbf{i} + a_2\mathbf{j} + a_3\mathbf{k}$, where $d_1 = -1/\sqrt{6}$; $a_2 = -1/\sqrt{6}$,

$a_3 = 2/\sqrt{6}$. (*iii*) $\frac{1}{\sqrt{6}}(\mathbf{i} - \mathbf{j} - 2\mathbf{k})$, (*iv*) $\frac{1}{\sqrt{35}}(3\mathbf{i} - \mathbf{j} - 5\mathbf{k})$.

4. $(-\mathbf{j} + \mathbf{k})/\sqrt{2}$ and $\frac{\sqrt{2}\,(4\mathbf{i} - \mathbf{j} + \mathbf{k})}{6}$.

6. $\frac{1}{\sqrt{2}}(\mathbf{i} + \mathbf{j})$; $\mathbf{k}$; Yes.

7. (*i*) $c = \frac{1}{\sqrt{6}}(\mathbf{i} + \mathbf{j} + 2\mathbf{k})$. $\mathbf{d} = \frac{1}{\sqrt{2}}(\mathbf{i} + \mathbf{j} - \mathbf{k})$.

(*ii*) $\mathbf{c} = l(\mathbf{i} + \mathbf{j} + \mathbf{k}) + m\left(\mathbf{i} + \sqrt{2}\mathbf{j} - \sqrt{6}\mathbf{k}\right)$.

where l and m satisfy the relations

$$3l = \left(\sqrt{6} - \sqrt{2} - 1\right) m \text{ and } 9m^2 - 3l^2 = 1,$$

$$\mathbf{d} = \frac{1}{k}\left[\left(-\sqrt{6} - \sqrt{2}\right)\mathbf{i} + \left(1 + \sqrt{6}\right)\mathbf{j} + \left(\sqrt{2} - 1\right)\mathbf{k}\right]$$

where $k^2 = 18 + 2\left(\sqrt{12} + \sqrt{6} - \sqrt{2}\right)$.

8. (*i*) 1. (*ii*) – 2k (*iii*) 7.

9. (*i*) $\dfrac{19 - 9\sqrt{2} + 8\sqrt{6}}{54}, \dfrac{12 - 9\sqrt{2} + 8\sqrt{6}}{51}$. (*ii*) $\dfrac{\sqrt{3}+1}{4}, \dfrac{3+\sqrt{3}}{3}$.

Page 94-95

1. (*i*) $\sqrt{65}$, (*ii*) $3\sqrt{14}$. **2.** (*i*) $\sqrt{26}$, (*ii*) $3\sqrt{2}$.

3. (*i*) $\cos^{-1}\left(\dfrac{91}{3\sqrt{6}}\right)$, (*ii*) $\cos^{-1}\left(\sqrt{\dfrac{7}{12}}\right)$.

4. (*i*) 1/3, 2/3, – 2/3, (*ii*) 3/7, – 6/7, 2/7.

6. The lines are parallel. **12.** All statements are false.

13. $\cos^{-1}\left(\dfrac{1}{3}\right)$. **15.** (*ii*) (1, 3, 5).

Page 105

2. $(\mathbf{r} \cdot \mathbf{n}_1 - 1)(\mathbf{n}_2 \cdot \mathbf{n}_3) = (\mathbf{r} \cdot \mathbf{n}_2 - 1)(\mathbf{n}_1 \cdot \mathbf{n}_3)$.

3. $\mathbf{a} + 2\dfrac{q - \mathbf{a} \cdot \mathbf{n}}{\mathbf{n}^2}\mathbf{n}$. $\mathbf{r} = \mathbf{a} + 2\dfrac{q - \mathbf{a} \cdot \mathbf{n}}{\mathbf{n}^2}\mathbf{n} + t\left[\mathbf{b} - 2\dfrac{\mathbf{b} \cdot \mathbf{n}}{\mathbf{n}^2}\mathbf{n}\right]$.

4. $2\mathbf{b} - \mathbf{a} + 2\dfrac{(\mathbf{a} - \mathbf{b}) \cdot \mathbf{c}}{\mathbf{c}^2}\mathbf{c}$.

7. $\dfrac{|q - \mathbf{a} \cdot \mathbf{n}|}{|\mathbf{c} \cdot \mathbf{n}|}|\mathbf{c}|$.

Page 132-133

1. (*a*) $-\mathbf{i} + \mathbf{j} + 3\mathbf{k}, \sqrt{11}$. (*b*) $3\mathbf{j} + 2\mathbf{k}, \sqrt{13}$. (*c*) $4\mathbf{i} - 3\mathbf{j} - 7\mathbf{k}, \sqrt{74}$.

2. (*a*) 12 (*b*) – 14.

3. (*a*) – 15**i** – 25**j** + 15**k**. (*b*) **i** + 16**j** – 9**k**. **6.** 7

Page 138-139

7. (*a*) $\dfrac{\sqrt{150}}{2}$, (*b*) $\dfrac{\sqrt{3}}{2}$,

(*c*) $\dfrac{\sqrt{89}}{2}$, (*d*) $\dfrac{\sqrt{\left(a^2b^2 + b^2c^2 + c^2a^2\right)}}{2}$.

8. (*a*) 2/3, (*b*) 6.

10. [**b c d**] + [**c a d**] + [**a b d**] = [**a b c**]

Page 152-154

1. $\mathbf{r} = \mathbf{a} + t\mathbf{n}_1 \times \mathbf{n}_2$.

2. $\mathbf{r} = \mathbf{a} + t\mathbf{n}_1 \times \mathbf{n}_2$. 3. $\mathbf{r} = \mathbf{c} + t\mathbf{b} \times \mathbf{n}$.

4. $\mathbf{r} = \mathbf{a} + t\left[(\mathbf{a}-\mathbf{b}) - \dfrac{(\mathbf{a}-\mathbf{b})\cdot\mathbf{n}}{\mathbf{c}\cdot\mathbf{n}}\mathbf{c}\right]$

5. $[\mathbf{r}\ \mathbf{n}_1\ \mathbf{n}_2] = [\mathbf{a}\ \mathbf{n}_1\ \mathbf{n}_2]$.

6. $[\mathbf{r}\ \mathbf{n}\ \mathbf{c}] = [\mathbf{a}\ \mathbf{n}\ \mathbf{c}]$.

7. $\mathbf{r}\cdot[\mathbf{b}-\mathbf{a}]\times\mathbf{n} = [\mathbf{a}\ \mathbf{b}\ \mathbf{n}]$

8. $[\mathbf{r}\ \mathbf{b}\ \mathbf{n}] = [\mathbf{a}\ \mathbf{b}\ \mathbf{n}]$.

9. $[\mathbf{n}_2\ \mathbf{n}_3\ \mathbf{n}_4]\ (\mathbf{r}\ .\ \mathbf{n}_1 - \mathbf{q}_1) = [\mathbf{n}_1\ \mathbf{n}_3\ \mathbf{n}_4]\ (\mathbf{r}\ .\ \mathbf{n}_2 - \mathbf{q}_2)$.

10. $\mathbf{r}\cdot(\mathbf{b}-\mathbf{a})\times\mathbf{c} = [\mathbf{a}\ \mathbf{b}\ \mathbf{c}]$.

11. $a.c = b.c;\ \dfrac{1}{[a\,b\,c]}\{(b.a)(b\times c)+(a.b)(c\times b)+(a.c)(a\times b)\}$.

12. $\mathbf{a} + \mathbf{b}$.

13. $\mathbf{r} = \mathbf{a} + t\ \{(\mathbf{b}-\mathbf{a})\times\mathbf{c}\}\times\{(\mathbf{d}-\mathbf{a})\times\mathbf{c}\}$.

14. $[\mathbf{r}\ \mathbf{b}\ \mathbf{d}] = [\mathbf{a}\ \mathbf{b}\ \mathbf{d}]$. The required shortest distance is the distance of any point, say **c**, on the line $\mathbf{r} = \mathbf{c} + p\mathbf{d}$ from this plane so that it is

$$\frac{|[\mathbf{c}\ \mathbf{b}\ \mathbf{d}] - [\mathbf{a}\ \mathbf{b}\ \mathbf{d}]|}{|\mathbf{b}\times\mathbf{d}|}.$$

16. (*i*) The line *AB* is normal to the plane *OCD*.
(*ii*) The line *AB* is parallel to the plane *OCD*.

23. $\cos^{-1}\dfrac{x^2}{\sqrt{(x^2+z^2)}\sqrt{(x^2+y^2)}};\ \cos^{-1}\dfrac{xz}{\sqrt{(x^2y^2+y^2z^2+z^2x^2)}}$.

27. The given ratio should be 1 : 1.

Page 166-168

1. (*i*) (15, – 13, – 2); (*ii*) (– 9, 9, 12); (*iii*) (7, – 10, – 3).
2. (*i*) (6, 4, 1); (12, 8, 2); (*ii*) (1, 1, 3), (2, 2, 6); (*iii*) (2, 3, 4) (4, 6, 8).
3. (9, 5, – 2).
5. (*a*) Co-planar, (*b*) Co-planar, (*c*) Non-coplanar.
6. (*i*) $3x - 2y + 3z = 4$, (*ii*) $x - y - 7z + 23 = 0$.
9. (*i*) $\sqrt{90}$, (*ii*) (6/7) $\sqrt{14}$, (*iii*) 6, (*iv*) $3/5\sqrt{10}$, (*v*) $\sqrt{6}$.
10. $2p^3/3lmn$. 11. $x - y - z = 2$.
13. $\left(\frac{1}{2}, -\frac{1}{2}, -3/2\right);\ x - 2y + z = 0$.
14. $(a + a', b + b', c + c');\ \Sigma x\ (bc' - b'c) = 0$.
15. (5, – 7, 6); $11x = 6y + 5z + 67$.
16. (*i*) $x = y = z;\ 4\sqrt{3}$, (*ii*) $(x-4) = (y-2)/3 = -z\,(z+3)/5;\ \sqrt{35}$.
17. (11, 11, 31) and (3, 5, 7). 18. 0.

Miscellaneous Exercises II

Page 169-174

7. 27 : 1.

12. $\frac{1}{2}\mathbf{a}, \frac{1}{2}\mathbf{b}, \frac{1}{2}\mathbf{c}, \frac{1}{2}(\mathbf{b}+\mathbf{c}), \frac{1}{2}(\mathbf{c}+\mathbf{a}), \frac{1}{2}(\mathbf{a}+\mathbf{b});$

$$r = \frac{1}{2}\mathbf{a} + t(\mathbf{b}+\mathbf{c}-\mathbf{a}), \quad \mathbf{r} = \frac{1}{2}\mathbf{b} + t(\mathbf{c}+\mathbf{a}-\mathbf{b}),$$

$$\left(\frac{1}{4}\right)(\mathbf{a}+\mathbf{b}+\mathbf{c}).$$

14. (3, 7, 11), (0, 5, 4).

15. $p = 1$.

16. P is $\mathbf{i} + 2\mathbf{j} + \mathbf{k}$.

17. $p = \sqrt{3/2},\ q = \sqrt{3}/\sqrt{2}, r = 0.$

18. $\pm(1/13)(-12\mathbf{i}+3\mathbf{j}+4\mathbf{k}),\ \cos^{-1}(1/\sqrt{3}),\ \pi - \cos^{-1}(1/\sqrt{3}).$

20. $\mathbf{a} = (3/\sqrt{2})(\mathbf{i}+\mathbf{j}),\ \mathbf{b} = (1/\sqrt{2})(\mathbf{i}-\mathbf{j}),\ \mathbf{c} = l\mathbf{i} - l\mathbf{j} - \sqrt{6}k$

where $l = 5\sqrt{\frac{3}{2}}$.

21. $\mathbf{x} = \mathbf{j} + t(\mathbf{i} + 2\mathbf{j} + \mathbf{k})$ where t is any scalar.

23. $\cos^{-1}\sqrt{\frac{1}{30}},\ \cos^{-1}\left(\frac{1}{3\sqrt{30}}\right).$

25. B is (4, 1), C is (4, 6), C, E, F are collinear and E is the mid-point of CF.

27. $\mathbf{r} \cdot (\mathbf{i} + \mathbf{j}) = 2, \qquad (\mathbf{r} - \mathbf{j})(\mathbf{i} + 2\mathbf{j} - \mathbf{k}) = 0.$

29. $l_1 : x = 1 - y = z,\ l_2 : x - 1 = y + 1 = z.$

$\pi_1 : x + 2y + z = 2,\ \pi_2 : x - 2y + z = 1.\ AB : 1/\sqrt{2}.$

The equation of AB is

$\mathbf{r} = (1/4)(5\mathbf{i} - \mathbf{j} + 5\mathbf{k}) + t(\mathbf{i} - \mathbf{k}).$

30. $\mathbf{r} \cdot (3\mathbf{i} + \mathbf{j}) = 6,\ \mathbf{r} \cdot (\mathbf{i} + 3\mathbf{j}) = 6.$

31. $\pi_1 : \mathbf{r}\cdot(\mathbf{i}-\mathbf{j}+\mathbf{k}) = 4,\ \pi_2 : \mathbf{r}\cdot(\mathbf{i}+2\mathbf{j}+\mathbf{k}) = 6,$

$\pi_2 : \mathbf{r}\cdot(\mathbf{i}-4\mathbf{j}+\mathbf{k}) = 2,\ \sqrt{2}/3.$

32. $\cos^{-1}\left(-\frac{1}{4}\right),\ \cos^{-1}\left(\frac{1}{4}\right).$

35. The vectors **p, q** are parallel.

36. $x = \mathbf{q} - \frac{1}{2\mathbf{p}^2}(\mathbf{p}\cdot\mathbf{q})\mathbf{p}$ **39.** $\lambda = d/|\mathbf{a}|^2,\ \mu = 0,\ \gamma = -1/|\mathbf{a}|.$

40. (*i*) $\vec{OB} = \mathbf{c} - (2/3)\mathbf{a},\ \vec{AB} = \mathbf{c} - \frac{1}{3}\mathbf{a},\ \vec{AM} = \frac{1}{2}\mathbf{c} - \mathbf{a},$ (*iv*) 4.6 cm.

41. $\left(\frac{1}{3}\right)\mathbf{b}-\mathbf{a},\ (3/4)\,\mathbf{a}-\mathbf{b},\ 0.922.$

44. (*i*) The locus is the circle on the line joining the points **p** and **q** as diameter.

(*ii*) The locus is the line joining the points **p** and **q**.

45. (*i*) $\sqrt{6}a/3$ (*ii*) $a/\sqrt{2}$.

46. *N* is (1/4) $\mathbf{b}+\frac{1}{2}\mathbf{c}$, 3:2, 3:1, 1:4

52. The general value of **v** = **b** + *t***a**,

$$\mathbf{v}=(3/2)\,\mathbf{i}+\left(\frac{1}{2}\right)\mathbf{k} \text{ if } \mathbf{a}\,.\,\mathbf{v}=0.$$

Page 185

1. The general solution is

$$\mathbf{r}=\frac{\mathbf{n}_2^{\,2}-\mathbf{n}_2\,.\,\mathbf{n}_1}{(\mathbf{n}_2\times\mathbf{n}_1)^2}\,\mathbf{n}_1+\frac{\mathbf{n}_1^{\,2}-\mathbf{n}_1\,.\,\mathbf{n}_2}{(\mathbf{n}_1\times\mathbf{n}_2)^2}+t\mathbf{n}_1\times\mathbf{n}_2;$$

being any scalar.

(Start with assuming $\mathbf{r}=x\mathbf{n}_1+y\mathbf{n}_2+t\mathbf{n}_1\times\mathbf{n}_2$.

2. $\mathbf{b}\,.\,\mathbf{c}+\mathbf{a}\,.\,\mathbf{d}=0,\ \frac{1}{[\mathbf{a}\,\mathbf{b}\,\mathbf{c}]}\left[-(\mathbf{d}\cdot\mathbf{b})\,\mathbf{a}+(\mathbf{d}\cdot\mathbf{a})\,\mathbf{b}+(\mathbf{b}\cdot\mathbf{b})\,\mathbf{c}\right],$

If $[\mathbf{a}\,\mathbf{b}\,\mathbf{c}]\neq 0$. [Assume $\mathbf{r}=x\mathbf{a}+y\mathbf{b}+z\mathbf{c}$].

Page 197-198

2. $\sqrt{(3/8)k},\ \frac{\sqrt{6}}{12}k$; *k* being length of each edge.

6. $\frac{1}{2\sqrt{2}}\sqrt{(\mathbf{a}^2+\mathbf{b}^2+\mathbf{c}^2)},$

$$\frac{\sqrt{\left[(\mathbf{b}^2+\mathbf{c}^2-\mathbf{a}^2)(\mathbf{c}^2+\mathbf{a}^2-\mathbf{b}^2)(\mathbf{a}^2+\mathbf{b}^2-\mathbf{c}^2)\right]}}{2\sqrt{2}\sqrt{(2\mathbf{a}^2\mathbf{b}^2+2\mathbf{b}^2\mathbf{c}^2+2\mathbf{c}^2\mathbf{a}^2-\mathbf{a}^4-\mathbf{b}^4-\mathbf{c}^4)}}.$$

7. Take *O*, as the origin of reference. Then the axis is given by

$$\mathbf{r}=t\,\frac{|\mathbf{b}\times\mathbf{c}|\,\mathbf{a}+|\mathbf{c}\times\mathbf{a}|\,\mathbf{b}+|\mathbf{a}\times\mathbf{b}|\,\mathbf{c}}{|\mathbf{b}\times\mathbf{c}|+|\mathbf{c}\times\mathbf{a}|+|\mathbf{a}\times\mathbf{b}|}.$$

where $\overrightarrow{OA}=\mathbf{a},\ \overrightarrow{OB}=\mathbf{b},\ \overrightarrow{OC}=\mathbf{c}.$

Page 235-237

26. $\frac{5}{3}(-2\mathbf{i}+\mathbf{j}).$ 27. $-2\,(2\mathbf{i}+\mathbf{j}).$

28. $\frac{1}{\sqrt{3}},\frac{-1}{\sqrt{3}},\frac{1}{\sqrt{3}},\frac{-1}{\sqrt{3}},\frac{1}{\sqrt{3}},\frac{-1}{\sqrt{3}}.$

VECTOR ALGEBRA

Contents : Vector Addition • Centroid • Vector equations of Straight Line and Plane • Scalar Product of Two Vectors • Vectors Product of two Vectors • Multiple Products • Vector Equations of Straight Line and Plane • Solution of Vector Euqations • Volume of Tetratedron • Vector Equation of Sphere • Applicatioon of Vectors tp Mechanics.

COORDINATE GEOMETRY

Contents : Coordinates • The Straight Line • Transformation of Co-ordinates • Pair of Straight Lines • Circles • System of Circles • Parabola • Ellopse • Hyperbola • The Polar Equation of a Conic • General Equation of the Second Degree • General Conic • System or Conic • Conlocal Conics • Circle of Curvature.

DIFFERENTIAL CALCULUS

Contents: Real Numbers • Limit and Continuity • Differentiation or Derivatives • Successive Differentiation • Indetrminate Forms • Tangents amd Normals • Some General Theorems and Exapansions of Fucntions • Maxima and Minima • Curvature • Asymptotes • Partial Differentiation • Points of Inflexion and Singular Points • Curve Tracing

IMPORTANT BOOKS ON MATHEMATICS

PROBABILITY THEORY AND RANDOM PROCESSES

S.P. Eugene Xavier

The purpose of this book is to present an introductory, yet comprehensive treatment of Probability Theory and Random Processes, with a strong emphasis on numerical examples. The author has attempted to explain these concepts and indicate their usefulness through discussion, examples and exercises.

CONTENTS: Probability Theory • Random Variables • Statistical Averages • Random Processes • Linear Systems and Random Noise Processes • Appendices • Bibliography.

14 369 **ISBN: 81-219-1395-0** **pp. 160**

A COURSE OF MATHEMATICAL ANALYSIS

Shanti Narayan

Designed for students who desire to proceed to a course preparatory to a serious study of the purely arithmetical theory of functions of real variables in its general aspects.

CONTENTS: Real Numbers • Bounded Sets, Open and Closed Sets • Real Sequences • Real Valued Functions of a Single Real Variable, Limit and Continuity • Real Valued Functions of a Single Real Variable • Derivability • Riemann Integrability • Sequences of Functions Point-wise and Uniform Convergence • Elementary Functions • Improper Integrals • Fourier Series • Euclidean Spaces • Open and Closed Sets • Compact Sets •

Real Valued Functions of Several Real Variables. • Limit • Continuity • Partial Derivatives • Invertible Functions Implicit Functions • Integrals as Functions of a Parameter • Integration in R^2 Line Integrals. Double Integrals • Curve Lengths. Surface Areas • Integration in R^3 Gauss's and Stoke's Theorems Mix-Exercises • Answers • Appendix.

14 045 **ISBN: 81-219-0472-2** **pp. 440**

MATHEMATICAL STATISTICS

J.N. Kapur & H.C. Saxena

This book is on the same model as books on calculus, differential equation, statics, dynamics, etc. A special feature is the large number of exercises especially constructed to illustrate the theory.

CONTENTS: Introduction • Frequency Distribution and Measures of Location • Measures of Dispersions, Skewness & Kurtosis, Moments of Frequency Distributions • Theory of Probability • Discrete Probability Distributions • Special Discrete Probability Distributions • Univariate Continuous Probability Distributions • Special Continuous Probability Distributions • Principle of Least Squares, Fittings of Curves & Orthagonal Polynomials • Correlation and Regression • Multiple and Partial Correlation • Theory of Sampling • Exact Sampling Distributions • Tests of Significance Based on the *T, F* and *Z* Distributions • Tests of Significance Based on the Chi-Square Distribution • Statistical Theory of Point Estimation • Testing of EHypotheses, Sequential Analysis Distribution-Free Methods, Statistical Decision Theory • Elements of Stochastic Processes • Appendices • Index • Log Tables.

14 046 **ISBN: 81-219-1246-6** **pp. 784**

ANALYTICAL SOLID GEOMETRY

Shanti Narayan

This book is intended as an introduction to Analytical Solid Geometry and is suitable for students of B.A., B.Sc., Pass and Honours courses. The book contains numerous exercises of varied types in a graded form.

CONTENTS: Co-ordinates • The Plane • Right Line • Interpretation of Equations-Loci • Transformation of Co-ordinates • The Sphere • Cones, and Cylinders • Homogeneous Cartesian Co-ordinates-Elements at infinity • The Conicoid • Plane Sections of Conicoids • Generating Lines of Conicoids • General Equation of the Second Degree: Reduction to Canonical forms • Appendix • Index.

14 0471 **ISBN:81-219-0473-0** **pp. 320**

A TEXTBOOK ON DYNAMICS

M. Ray & G.C. Sharma

A thoroughly revised edition where the chapters on Central Forces, Moment of Inertia and D'Alembert's Principle have been substantially retouched for the B.A. and B.Sc. students of Dynamics.

CONTENTS: Kinematics and Kinetics • Rectilinear Motion • Uniplanar Motion • Work Energy and Impulse • Impact • Circular and Harmonic Motions • Hodograph • Central Forces • Resisting Medium • Constrained Motion • Moment of Inertia • D'Almbert's Principle and the Equations of Motion.

14 048 **ISBN: 81-219-0342-4** **pp. 328**

DIFFERENTIAL CALCULUS

Shanti Narayan

The text is marked with unified and graphical presentation. The complicated issues are well illustrated.

CONTENTS: Real Numbers • Functions and Graphs-Elementary Functions • Continuity and Limit • Differentiation • Successive Differentiation • Tangents and Normals • Mean Value Theorem • Maxima and Minima • Indeterminate Forms • Partial Differentiation • Equality of Repeated Derivatives • Concavity and Points of Inflexion • Curvature & Evolutes Asymptotes • Singular Points • Curve Tracing • Envelopes • Partial Differentiation • Equality of Repeated Derivatives.

14 049 **ISBN: 81-219-0471-4** **pp. 400**

INTEGRAL CALCULUS

Shanti Narayan

Has a new chapter on `Centre of Gravity and Moment of Inertia'. Many illustrations are given so as to enlighten the subject-matter.

CONTENTS: Table of Standard Results • Methods of Integration • Integration of Algebraic Rational Functions • Integration of Trigonometric Functions • Integration of Irrational Functions • Definite Integral as the Limit of a Sum • Areas of Plane Regions • Rectification • Lengths of Plane Curves • Volumes and Surfaces of Revolution • Centre of Gravity: Moment of Inertia • Differential Equations of First Order and First Degree • Equations of First Order but not of the First Degree • Trajectories of a Family of Curves • Linear Equations • Appendices Mise. Exercises.

14 050 **ISBN: 81-219-0681-4** **pp. 360**

A TEXTBOOK OF VECTOR CALCULUS

Shanti Narayan & J.N. Kapur

Gives a systematic and self- contained treatment to the subject- matter. Good many illustrations are there to make understanding comprehensive.

CONTENTS: Vector Valued Functions of Scalar Variables • Differential Geometry I—Curves • Differential Geometry II—Surfaces • Mechanics I—Particles • Mechanics II—Rigid Body • Differential Operators—Point Functions • Integral Transfor-mations—Line, Surface, Volume Integrals • Hydrodynamics • Electromagnetic Theory.

14 051 **ISBN:81-219-0161-8** **pp. 328**

ORDINARY AND PARTIAL DIFFERENTIAL EQUATIONS

M.D. Raisinghania

Caters to the needs of the honours and postgraduate students. Part I deals with Elementary Differential Equations. Part II deals with ordinary differential equations, and Part III deals with partial differential equations.

CONTENTS: Part I: Elementary Differential Equations— Introduction • Equations of first order and first degree • Trajectories • Linear Equations with constant coefficients • Homogeneous linear equations or Cauchy-Euler equations • Equations of the first order but not of the first degree and singular solutions- Extraneous loci • Ordinary Simultaneous differential equations • **Part II: Ordinary Differential Equations**—Numerical Integration • Existence and uniqueness theorems • Independence of solutions of linear differential equations-Wronskian • Exact differential equations and equations of special forms • Linear Equations of second order • Simultaneous equations of the form dx/P = dy/Q = dz/R • Total differential equations • Riccati's equations • Series solutions of linear differential equations • Legendre Polynomials and Functions • Legendre functions of second kind • Bessel functions • Hermite Polynomials • Laguerre Polynomials • Hyper-geometric Function • **Part III:** Partial Differential Equations— Linear partial differential equations of order one • Non-linear partial-differential equations of Order one • Partial differential equation of the second order • Linear partial differential equations with constant coefficients • Monge's methods • Appendices.

14 282 **ISBN:81-219-0892-2** **pp. 576**

EXAMPLES IN FINITE DIFFERENCES AND NUMERICAL ANALYSIS

H.C. Saxena

This book comprises of twelve chapters with an introduction to important concepts and the list of important formulae needed in the solution of the examples.

CONTENTS: Differences, Operators, Interpolation with Equal Intervals • Interpolation for Unequal Intervals of the Argument • Central Difference Interpolation Formulae • Numerical Differentiation • Inverse Interpolation • Numerical Quadrature • Summation of Series • Difference Equations • Roots of Algebraic and Transcendental Equations • Simultaneous Linear Algebraic Equations • Numerical solution of Ordinary Differential Equations • Linear Programming • Log Tables.

14 175 **ISBN:81-219-0753-5** **pp. 368**

ADVANCED DIFFERENTIAL EQUATIONS
(Including Boundary Value Problems and Difference Equations)

M. D. Raisinghania

This book has been written for Honours, M.A., M.Sc. (Mathematics and Physics) and B.E. students of various universities. The students preparing for P.C.S., I.A.S., A.M.I.E. will also find it useful.

CONTENTS: Part I: Ordinary Differential Equations • Numerical Integration • Existence and Uniqueness Theorem • Independence of solutions of Linear Differential Equations – Wronskian • Exact Differential Equations and Equations of Special Forms • Linear Equations of Second Order • Simultaneous Equations of the form dx/P = dy/Q = dz/R • Total Differential Equations • Riccati's Equation • Series Solutions of Linear Differential Equations • Legendre Polynomials and Functions • Legendre Functions of Second Kind • Bessel Functions • Hermite Polynomials • Languerre Polynomials • Hypergeometric Function • • **Part II: Partial Differential Equations** • Linear Partial Differential Equations of Order One • Non-linear Partial Differential Equations of the Second Order • Linear Partial Differential Equations with Constant Coefficients • Monge's Methods • More an Partial Differential Equations • **Part III: Boundary Value Problems • Part IV: Difference Equations.**

14 271 **ISBN:81-219-0893-0** **pp. 648**

A TEXTBOOK OF VECTOR ALGEBRA

Shanti Narayan

An attractively well-written book which deals rigorously and systematically with three dimensional Vector algebra and its application to Euclidean Geometry and Statics.

CONTENTS: Multiplication of Vectors by Scalars and Addition of Vectors • Geometry with Vectors Affine Geometry • Scalar Product • Applications to Metric Geometry • Vector Product and Scalar Triple Product • Application of Scalar Triple Products to Geometry • Geometry with Cartesian Co-ordinates • Some Miscellaneous Topics • Some Properties of Tetrahedra • Statics with Vectors • Answers.

14 064 **ISBN: 81-219-0952-X** **pp. 240**

FLUID DYNAMICS

M.D. Raisinghania

The present text has been well-organised and made up-to-date in the light of the latest syllabi. A detailed index has been provided at the end of the book and more solved examples have been added so that the readers may gain confidence in the technique of problem solving. Almost all the chapters have been rewritten so that in the present form, readers will easily understand the subject matter.

CONTENTS: Introduction • Kinematics of Fluids in Motion • Equations of Motion of Inviscid Fluid • One-Dimensional Inviscid Incompressible Flow (Bernoulli's equation and its applications) • Motion in Two-Dimensions and Sources and Sinks • General Theory of Irrotational Motion • Motion of Cylinders • Irrotational Motion in three Dimensions (Motion of a sphere) Stoke's Stream Function • Vortex Motion (Rectilinear vortices) • Waves • General Theory of Stress and Rate of Strain • The Navier-Stokes Equations and the Energy Equation • Dimensional Analysis • Dimensionless Constants (numbers) • Similarity of Flows • Laminar Flow of Viscious Incompressible Fluids • Boundary Layer Theory • Discontinuous Motion • Index.

14 349 **2nd Edn.** **Rep.1998**

ISBN: 81-219-0869-8 **pp. 872**

INTEGRAL TRANSFORMS

M.D. Raisinghania

This book is useful for Honours, M.A., M.Sc. (Mathematics) and Physics and B.E. students. It is also useful for students preparing for A.M.I.E., I.A.S. and other competitive examinations. Various types of problems have been fully illustrated by solved examples.

CONTENTS: Laplace Transform • Inverse Laplace Transform • Applications of Laplace Transforms • Fourier Transforms • Finite Fourier Transform • Hankel Transforms • The Finite Hankel Transforms • The Mellin Transform • Index.

14 283 **2nd Edn. 1995**

ISBN:81-219-0886-8 **pp. 296**

RAMANUJAN NUMBERS (Mathematical Thoughts and Ideas)

M. Meyyappan

This book is intended to provide a modest knowledge of mathematics with an excursion in the number world. The book has a lot of illustrations and through activities the students are helped to acquire basic concepts and facts.

The standard of students in the subject mathematics is gradually deteriorating in our country. Popular Science books may interest and help divert students' attention on to developing skills but, this book in its humble effort tries to rejuvenate the mathematical inclination of Indian students.

To remove the fear of mathematics at an early stage and to create strong interest in this vital subject, the mother of all Sciences, mathematics, may be introduced as fun and entertainment. In this book such an attempt has been made.

The book also presents algebra and problem solving in an exciting and amusing manner. On the whole the book endeavours to arouse the curiosity in the students and thereby develop interest.

CONTENTS: Ramanujan— Mathematician Extraordinary • Second Order Ramanujan Numbers • Third Order Ramanujan Numbers • Fourth Order Ramanujan Numbers • Appendix.

14 368 1st Edn. Rep. 1998 ISBN: 81-219-1272-5 pp. 128

ELEMENTS OF REAL ANALYSIS

Shanti Narayan

This book is an attempt to make presentation of Elements of Real Analysis more lucid. The book contains examples and exercises meant to help a proper understanding of the text.

CONTENTS : Sets and Statements • The Real Numbers • Limit Points of a Set • Sequences • Infinite Series with positive terms • Infinite Series with positive and negative terms • Real Functions—Limit and Continuity • Real Functions— The Derivative • Riemann Integrability.

14 052 6th Edn. Rep. 1998 ISBN: 81-219-0306-8 pp. 216

VECTOR ALGEBRA

M.D. Raisinghania

The present edition of the book has been thoroughly revised, and updated in the light of latest syllabi of various universities and institutions of India. More solved examples have been added and a detailed index has been provided at the end of the book inorder to make it more useful to students and teachers.

CONTENTS: Vector Addition • Centroid • Vector equations of straight line and plane (Parametric form) • Scalar products of two vectors • Vector product of two vectors • Product of three and four vectors (Multiple products) • Vector equations of straight line and plane (non-parametric forms) • Volume of tetrahedron • Solution of vector equation • Vector equation of sphere • Applications of vectors to mechanics (Dynamics and Statics) • Dynamics with vectors • statics with vectors **Section I: Concurrent Forces • Section II : Vector moment (or torque) and work • Section III : Parallel Forces • Section IV Couple • Section V : Coplaner non-concurrent forces acting on a rigid body • Section VI : Forces in three dimensions** • Index.

14 363 3rd Edn. 1997 ISBN : 81-219-4447-7 pp. 312

FINITE DIFFERENCES AND NUMERICAL ANALYSIS

H.C. Saxena

This thoroughly revised edition of the book completely covers the syllabi in the calculus of Finite Differences of various Indian universities. Examples given at the end of each chapter have been specially constructed, taken from university papers, and standard books.

CONTENTS : Part I—Calculus of Finite Differences — Finite Differences, Interpolation with Equal Intervals • Interpolation of Unequal Intervals of the Argument • Central Difference Interpolation Formulae • Inverse Interpolation • Numerical Differentiation • Numerical Quadrature • Summation of Series • Difference Equations • Generating Functions and Theorems • Bernoulli and Euler Polynomials • Remainder Terms in (or Errors associated with) • Interpolation Formulas **Part II —Numerical Analysis—** Eigen Value and Eigen Vector • Roots of Polynomial and Transcendental equation in one variable • Simultaneous Linear Algebraic Equations • Numerical Solution of Ordinary Differential Equations • Errors • Partial Differential Equations • Appendices • Answers • Index • Log Tables.

14 056 14th Rev. Edn. 1998 ISBN: 81-219-0339-4 pp. 420